AF581647

Novel Approaches for Nondestructive Testing and Evaluation

Novel Approaches for Nondestructive Testing and Evaluation

Editors

Jinyi Lee
Hoyong Lee
Azouaou Berkache

MDPI • Basel • Beijing • Wuhan • Barcelona • Belgrade • Manchester • Tokyo • Cluj • Tianjin

Editors
Jinyi Lee
Chosun University
Korea

Hoyong Lee
Gwangju University
Korea

Azouaou Berkache
Chosun University
Korea

Editorial Office
MDPI
St. Alban-Anlage 66
4052 Basel, Switzerland

This is a reprint of articles from the Special Issue published online in the open access journal *Applied Sciences* (ISSN 2076-3417) (available at: https://www.mdpi.com/journal/applsci/special_issues/nondestructive_testing_evaluation).

For citation purposes, cite each article independently as indicated on the article page online and as indicated below:

LastName, A.A.; LastName, B.B.; LastName, C.C. Article Title. *Journal Name* **Year**, *Volume Number*, Page Range.

ISBN 978-3-0365-3599-9 (Hbk)
ISBN 978-3-0365-3600-2 (PDF)

Contents

About the Editors

Jinyi Lee studied mechanical design at Chonbuk University, Jeonju, Korea, in 1992. He received his M.S. and Ph.D degrees in mechanical design, and aeronautics, and space engineering from Tohoku University, Sendai, Japan, from 1993 to 1998. Afterwards, he also worked as a scientist at Tohoku University, Iwate University, and Saitama University from 1997 to 2000. He also worked as a research director for Gloria Tech Co.Ltd and Lacomm Co. Ltd, Korea from 2000 to 2003. Since 2003, he has been the director of the IT-based Real Time NDT Center and a professor at Chosun University (Gwangju, Korea). Additionally, he has been a Specially Appointed Visiting Professor, NICHe, Tohoku University (Sendai, Japan). In 2010, 2011, and 2012, he received the Innovation Leader in Korea news maker award; a silver prize from the Military Science and Technologies Conference; and an award for an excellent paper from the KSME, respectively. He is a member of the World Conference on Non-Destructive Testing (WCNDT) and a member of the Korean Society for Nondestructive Testing (KSNT). His research interests include the application of magneto-optical and laser sensors and the development of magnetic cameras so that they can cover a wider range of applications in the field of non-destructive testing. He is the author and co-author of about 200 patents and scientific journals.

Hoyong Lee, Ph.D., is a Senior Researcher at Korea Aerospace Industries, LTD (Sachon, South Korea). He previously worked at Gwangju University and the Air Force Research Institute in South Korea. He received his M.S. degree (2005) in structural analysis from Seoul National University and received his Ph.D. (2012) in the mechanical reliability of avionics system from the State University of New York at Binghamton. His research interests are the application of nondestructive inspection methods in the aerospace industry and the development of aircraft structural repair methods. Currently, He is developing nondestructive inspection and structural repair methods at Korea Aerospace Industries, Ltd.

Azouaou Berkache, Ph.D., is a postdoctoral researcher at the IT-based Real-Time NDT Center of Chosun University (Gwangju, South Korea). He previously worked at the University Mouloud Mammeri (Tizi-Ouzou, Algeria), where he received his B.S. and M.S. degrees in electrical drive and his PhD degrees in the modeling and in the design of electromagnetic systems in 2013, 2015, and 2019, respectively. His scientific interests are centered on electromagnetic systems, and non-destructive testing as well as the development of evaluation methods for a wide spectrum of applications. To this end, together with Professor Jinyi Lee, he has developed an eddy current test setup and three-dimensional metal grain imaging technique for industry and laboratory applications.

MDPI

Editorial

Special Issue: Novel Approaches for Nondestructive Testing and Evaluation

Jinyi Lee [1,2,3,*], Hoyong Lee [4] and Azouaou Berkache [1]

1 IT-Based Real-Time NDT Center, Chosun University, Gwangju 61452, Korea; azouaoubrk@yahoo.fr
2 Department of Electronics Engineering, Chosun University, Gwangju 61452, Korea
3 Interdisciplinary Program in IT-Bio Convergence System, Chosun University, Gwangju 61452, Korea
4 KF-X Support System Development Team, Korea Aerospace Industries Ltd., Sacheon 52529, Korea; hoyong.lee@koreaaero.com
* Correspondence: jinyilee@chosun.ac.kr

Citation: Lee, J.; Lee, H.; Berkache, A. Special Issue: Novel Approaches for Nondestructive Testing and Evaluation. *Appl. Sci.* **2022**, *12*, 565. https://doi.org/10.3390/app12020565

Received: 29 December 2021
Accepted: 5 January 2022
Published: 7 January 2022

1. Introduction

Nondestructive testing and evaluation (NDT&E) is one of the most important techniques for determining the quality and safety of materials, components, devices, and structures. NDT&E technologies include ultrasonic testing (UT), magnetic particle testing (MT), magnetic flux leakage testing (MFLT), eddy current testing (ECT), radiation testing (RT), penetrant testing (PT), and visual testing (VT), and these are widely used throughout modern industries. However, some NDT processes, such as cleaning specimens and removing paint, cause environmental pollution and must be inspected in limited environments (time, space, and sensor selection). Thus, NDT&E is classified as a typical 3D (dirty, dangerous, and difficult) job. In addition, the NDT operator judges the presence of damage by experience and subjective judgment, so in some cases, a flaw that exists may not be detected during the test. Therefore, to obtain clearer test results, a means for the operator to determine the flaw more easily should be provided. In addition, the test results should be organized systemically, in order to identify the cause of the abnormality in the test specimen and to identify the progress of the damage quantitatively.

Thus far, from a total of 18 submitted papers to this Special Issue, 13 have been published. The next sections provide a brief summary of each of the papers published.

2. Ultrahigh Resolution Pulsed Laser-Induced Photoacoustic Detection of Multi-Scale Damage in CFRP Composites by Wang et al.

This paper [1] presented a photoacoustic nondestructive evaluation (pNDE) system with an ultrahigh resolution for the detection of multi-scale damage in carbon-fiber-reinforced plastic (CFRP) composites. The pNDE system consisted of three main components: a picosecond pulsed laser-based ultrasonic actuator, an ultrasound receiver, and a data acquisition/computing subsystem. During the operation, high-frequency ultrasound was generated by a pulsed laser and recorded by an ultrasound receiver. By implementing a two-dimensional back-projection algorithm, pNDE images could be reconstructed from the recorded ultrasound signals, to represent the embedded damage. Both potential macroscopic and microscopic damages, such as surface notches and delamination in CFRP, could be identified by examining the reconstructed pNDE images. Three ultrasonic presentation modes, i.e., A scan, B scan, and C scan, were employed to analyze the recorded signals for the representation of the detected micro-scale damage in two-dimensional and three-dimensional images, with a high spatial resolution of up to 60 μm. Macro-scale delamination and transverse ply cracks were clearly visualized, identifying the edges of the damaged area. The results of the study demonstrate that the developed pNDE system provides a nondestructive and robust approach for multi-scale damage detection in composite materials.

3. Fast Terahertz Coded-Aperture Imaging Based on Convolutional Neural Network by Gan et al.

Terahertz coded-aperture imaging (TCAI) has many advantages such as forward-looking imaging, staring imaging, low cost, etc. However, it is difficult to resolve the target under a low signal-to-noise ratio (SNR), and the imaging process is time consuming. In this study [2], the authors provided an efficient solution to tackle this problem. A convolution neural network (CNN) was leveraged to develop an off-line, end-to-end imaging network whose structure is highly parallel and free of iterations. Additionally, it can simply have a general and powerful mapping function. Once the network is well trained and adopted for TCAI signal processing, the target of interest can be recovered immediately from the echo signal. Additionally, the method to generate training data was shown, and the authors found that the imaging network trained with simulation data was of good robustness against noise and model errors. The feasibility of the proposed approach was verified by simulation experiments, and the results show that it has a competitive performance with state-of-the-art algorithms.

4. Indirect Method for Measuring Absolute Acoustic Nonlinearity Parameter Using Surface Acoustic Waves with a Fully Non-Contact Laser-Ultrasonic Technique by Jun et al.

This paper [3] proposed an indirect method to measure absolute acoustic nonlinearity parameters using surface acoustic waves by employing a fully non-contact laser-ultrasonic technique. For this purpose, the relationship between the ratio of relative acoustic nonlinearity parameters measured using the proposed method in two different materials (a test material and a reference material) and the ratio of absolute acoustic nonlinearity parameters in these two materials was theoretically derived. Using this relationship, when the absolute nonlinearity parameter of the reference material was known, the absolute nonlinearity parameter of the test material could be obtained using the ratio of the measured relative parameters of the two materials. For experimental verification, aluminum and copper specimens were used as reference and test materials, respectively. The relative acoustic nonlinearity parameters of the two materials were measured from surface waves generated and received using lasers. Additionally, the absolute parameters of aluminum and copper were measured using a conventional direct measurement method, with the former being used as a reference value and the latter being used for comparison with the estimation result. The absolute parameter of copper estimated by the proposed method showed good agreement with the directly measured result.

5. Proposal of UWB-PPM with Additional Time Shift for Positioning Technique in Nondestructive Environments by Huyen et al.

The ultra-wideband (UWB) technology has many advantages in positioning and measuring systems; however, the powers of UWB signals rapidly reduce while traveling in propagation environments; hence, detecting UWB signals are difficult. Various modulation techniques are applied for UWB signals to increase the ability for detecting the reflected signal from transmission mediums, such as pulse amplitude modulation (PAM), pulse position modulation (PPM), etc. In this paper [4], the authors proposed an ultra-wideband pulse position modulation technique with an optimized additional time shift (UWB-PPM-ATS), to enhance the accuracy in locating buried objects in nondestructive environments. Moreover, the Levenberg–Marquardt–Fletcher algorithm (LMFA) was applied to determine the medium parameters and buried object location simultaneously. The influences of the proposed modulation technique on determining the system's parameters, such as propagation time, distance, and properties of the medium were analyzed. Calculation results indicate that the proposed UWB-PPM-ATS provided higher accuracy than conventional methods such as UWB-OOK and UWB-PPM, in both homogeneous and heterogeneous environments. Furthermore, the LMFA approach with the proposed UWB-PPM-ATS outperformed the LMFA with the traditional modulation method, especially for unknown propagation environments.

6. An Attention-Based Network for Textured Surface Anomaly Detection by Liu et al.

Textured surface anomaly detection is a significant task in industrial scenarios. In order to further improve the detection performance, the authors of this study proposed a novel two-stage approach with an attention mechanism [5]. Firstly, in the segmentation network, the feature extraction and anomaly attention modules were designed to capture detailed information as much as possible and focused on the anomalies, respectively. To strike dynamic balances between these two parts, an adaptive scheme in which learnable parameters are gradually optimized was introduced. Subsequently, the weights of the segmentation network were frozen, and the outputs were fed into the classification network, which was trained independently in this stage. Finally, the proposed approach was evaluated on the DAGM 2007 dataset, which consists of diverse textured surfaces with weakly labeled anomalies; the experiments revealed that this method can achieve 100% detection rates in terms of true-positive rate (TPR) and true-negative rate (TNR).

7. A Comparison of Power Quality Disturbance Detection and Classification Methods Using CNN, LSTM, and CNN-LSTM by Garcia et al.

The use of electronic loads has improved many aspects of everyday life, permitting more efficient, precise, and automated processes. As a drawback, the nonlinear behavior of these systems entails the injection of electrical disturbances on the power grid that can cause distortion of voltage and current. In order to adopt countermeasures, it is important to detect and classify these disturbances. To this end, several machine learning algorithms are currently being exploited. Among them, for the present work [6], the long short-term memory (LSTM), convolutional neural networks (CNNs), convolutional neural network–long short-term memory (CNN-LSTM), and the CNN-LSTM with adjusted hyperparameters were compared. As a preliminary stage of the research, the voltage and current time signals were simulated using MATLAB Simulink. From the simulation results, it is possible to acquire a current and voltage dataset with which the identification algorithms are trained, validated, and tested. These datasets include simulations of several disturbances such as Sag, Swell, Harmonics, Transient, Notch, and Interruption. Data augmentation techniques were used in order to increase the variability of the training and validation dataset, to obtain a generalized result. Afterward, the networks were fed with an experimental dataset of voltage and current field measurements containing the disturbances mentioned above. The networks were compared, resulting in a 79.14% correct classification rate with the LSTM network versus 84.58% for the CNN, 84.76% for the CNN-LSTM. and 83.66% for the CNN-LSTM with adjusted hyperparameters. All of these networks were tested using real measurements.

8. Leaky Lamb Wave Radiation from a Waveguide Plate with Finite Width by Park et al.

In this paper [7], leaky Lamb wave radiation from a waveguide plate with finite width was investigated to gain a basic understanding of the radiation characteristics of the plate-type waveguide sensor. Although the leaky Lamb wave behavior has already been theoretically revealed, most studies have only dealt with two-dimensional radiations of a single leaky Lamb wave mode in an infinitely wide plate, and the effect of the width modes (that are additionally formed by the lateral sides of the plate) on leaky Lamb wave radiation has not been fully addressed. This work aimed to explain the propagation behavior and characteristics of the Lamb waves induced by the existence of the width modes and to reveal their effects on leaky Lamb wave radiation for the performance improvement of the waveguide sensor. To investigate the effect of the width modes in a waveguide plate with finite width, propagation characteristics of the Lamb waves were analyzed by the semi-analytical finite element (SAFE) method. Then, the Lamb wave radiation was computationally modeled on the basis of the analyzed propagation characteristics and was also experimentally measured for comparison. From the modeled and measured results of the leaky radiation beam, it was found that the width modes could affect leaky Lamb wave radiation with the mode superposition, and radiation characteristics were significantly changed depending on the wave phase of the superposed modes on the radiation surface.

9. Evaluation of Cracks on the Welding of Austenitic Stainless Steel Using Experimental and Numerical Techniques by Berkache et al.

This paper [8] dealt with the investigation and characterization of weld circumferential thin cracks in austenitic stainless steel (AISI 304) pipe with eddy current nondestructive testing technique (EC-NDT). During the welding process, the heat source applied to the AISI 304 was not uniform, accompanied by a change in physical property. To take this change into consideration, the relative magnetic permeability was considered as a gradiently changed variable in the weld and heat-affected zone (HAZ), which was generated by the Monte Carlo method based on pseudo-random number generation (PRNG). Numerical simulations were performed by means of MATLAB software, using the 2D finite element method to solve the problem. To verify, results from the modeling works were conducted and contrasted with findings from experimental ones. Indeed, the results of the comparison agreed well. In addition, they showed that the consideration of this change in magnetic property allows distinguishing the thin cracks in the weld area.

10. Measurement of Thinned Water-Cooled Wall in a Circulating Fluidized Bed Boiler Using Ultrasonic and Magnetic Methods by Lee et al.

In this paper [9], a nondestructive inspection system was proposed to detect and quantitatively evaluate the size of the near- and far-side damages on the tube, membrane, and weld of the water-cooled wall in a fluidized bed boiler. The shape and size of the surface damages were evaluated from the magnetic flux density distribution measured by the magnetic sensor array on one side from the center of the magnetizer. The magnetic sensors were arrayed on a curved shape probe according to the tube's cross-sectional shape, membrane, and weld. On the other hand, the couplant was doped to the water-cooled wall, and a thin film was formed thereon by polyethylene terephthalate. Then, the measured signal of the flexible ultrasonic probe was used to detect and evaluate the depth of the damages. The combination of magnetic and ultrasonic methods helped to detect and evaluate both near and far-side damages. Near-side damages with a minimum depth of 0.3 mm were detected, and the depth from the surface of the far-side damage was evaluated, with a standard deviation of 0.089 mm.

11. Micromagnetic Characterization of Operation-Induced Damage in Charpy Specimens of RPV Steels by Rabung et al.

The embrittlement of two types of nuclear pressure vessel steel, 15Kh2NMFA and A508 Cl.2, was studied using two different methods of magnetic nondestructive testing: micromagnetic multiparameter microstructure and stress analysis (3MA-X8) and magnetic adaptive testing (MAT) [10]. The microstructure and mechanical properties of reactor pressure vessel (RPV) materials are modified due to neutron irradiation; this material degradation can be characterized using magnetic methods. For the first time, the progressive change in material properties due to neutron irradiation was investigated on the same specimens, before and after neutron irradiation. A correlation was found between magnetic characteristics and neutron-irradiation-induced damage, regardless of the type of material or the applied measurement technique. The results of the individual micromagnetic measurements proved their suitability for characterizing the degradation of RPV steel caused by simulated operating conditions. A calibration/training procedure was applied on the merged outcome of both testing methods, producing excellent results in predicting transition temperature, yield strength, and mechanical hardness for both materials.

12. Three-Dimensional Imaging of Metallic Grain by Stacking the Microscopic Images by Lee et al.

Three-dimensional observation of metal grains (MGs) has a wide potential application serving the interdisciplinary community. It can be used for industrial applications and basic research to overcome the limitations of nondestructive testing methods, such as ultrasonic testing, magnetic particle testing, and eddy current testing. This study [11] proposed a method and its implementation algorithm to observe metal grains (MGs) in three dimensions, in a general laboratory environment equipped with a polishing machine and a metal microscope. An image was taken by a metal microscope while polishing the mounted object to be measured. Then, the metal grains (MGs) were reconstructed into three dimensions through local positioning, binarization, boundary extraction, MG selection, and stacking. The goal was to reconstruct the 3D MG in a virtual form that reflects the real shape of the MG. The usefulness of the proposed method was verified using the carbon steel (SA106) specimen.

13. THz-TDS Techniques of Thickness Measurements in Thin Shim Stock Films and Composite Materials by Im et al.

Terahertz wave (T-ray) scanning applications are one of the most promising tools for nondestructive evaluation. T-ray scanning applications use a T-ray technique to measure the thickness of both thin Shim stock films and glass-fiber-reinforced plastic (GFRP) composites, of which the samples were selected because the T-ray method could penetrate the non-conducting samples. Notably, this method is nondestructive, making it useful for analyzing the characteristics of the materials. Thus, the T-ray thickness measurement can be found for both non-conducting Shim stock films and GFRP composites. In this work [12], a characterization procedure was conducted to analyze electromagnetic properties, such as the refractive index. The obtained estimates of the properties are in good agreement with the known data for polymethyl methacrylate (PMMA) for acquiring the refractive index. The T-ray technique was developed to measure the thickness of the thin Shim stock films and the GFRP composites. The study tests obtained good results on the thickness of the standard film samples, with the different thicknesses ranging from around 120 μm to 500 μm. In this study, the T-ray method was based on the reflection mode measurement, and the time of flight (TOF) and resonance frequencies were utilized to acquire the thickness measurements of the films and GFRP composites. The results showed that the thickness of the frequency samples matched those obtained directly by time-of-flight (TOF) methods.

14. Time-Resolved Neutron Bragg-Edge Imaging: A Case Study by Observing Martensitic Phase Formation in Low-Temperature Transformation (LTT) Steel during GTAW by Griesche et al.

Griesche et al. [13] used neutron imaging to visualize the sample remelting during the welding process. Polychromatic and wavelength-selective neutron transmission radiography were applied during bead-on-plate welding on 5 mm thick sheets on the face side of martensitic low-temperature transformation (LTT) steel plates using gas tungsten arc welding (GTAW). The in situ visualization of austenitization upon welding and subsequent α'-martensite formation during cooling could be achieved with a temporal resolution of 2 s for monochromatic imaging using a single neutron wavelength and of 0.5 s for polychromatic imaging using the full spectrum of the beam (whitebeam). The spatial resolution achieved in the experiments was approximately 200 μm. The transmitted monochromatic neutron beam intensity at a wavelength of $\lambda = 0.395$ nm was significantly reduced during cooling below the martensitic start temperature Ms since the emerging martensitic phase had a ~10% higher attenuation coefficient than the austenitic phase. Neutron imaging was significantly influenced by coherent neutron scattering caused by the thermal motion of the crystal lattice (Debye–Waller factor), resulting in a reduction in the neutron transmission by approx. 15% for monochromatic and approx. 4% for polychromatic imaging

Author Contributions: Conceptualization, J.L. and A.B; methodology, A.B. and H.L.; validation, J.L., H.L. and A.B. All authors have read and agreed to the published version of the manuscript.

Funding: This research received no external funding.

Acknowledgments: We would like to thank the authors and reviewers for their valuable contributions to this Special Issue. This issue would not be possible without their valuable and professional work. In the same way, we would also like to take this opportunity to express our gratitude to the *Applied Sciences* editorial team for their work, thanks to whom this Special Issue has been a success.

Conflicts of Interest: The authors declare no conflict of interest.

References

1. Wang, S.; Echeverry, J.; Trevisi, L.; Prather, K.; Xiang, L.; Liu, Y. Ultrahigh Resolution Pulsed Laser-Induced Photoacoustic Detection of Multi-Scale Damage in CFRP Composites. *Appl. Sci.* **2020**, *10*, 2106. [CrossRef]
2. Gan, F.; Luo, C.; Liu, X.; Wang, H.; Peng, L. Fast Terahertz Coded-Aperture Imaging Based on Convolutional Neural Network. *Appl. Sci.* **2020**, *10*, 2661. [CrossRef]
3. Jun, J.; Jhang, K.-Y. Indirect Method for Measuring Absolute Acoustic Nonlinearity Parameter Using Surface Acoustic Waves with a Fully Non-Contact Laser-Ultrasonic Technique. *Appl. Sci.* **2020**, *10*, 5911. [CrossRef]
4. Huyen, N.T.; Le Cuong, N.; Thanh Hiep, P. Proposal of UWB-PPM with Additional Time Shift for Positioning Technique in Nondestructive Environments. *Appl. Sci.* **2020**, *10*, 6011. [CrossRef]
5. Liu, G.; Yang, N.; Guo, L. An Attention-Based Network for Textured Surface Anomaly Detection. *Appl. Sci.* **2020**, *10*, 6215. [CrossRef]
6. Garcia, C.I.; Grasso, F.; Luchetta, A.; Piccirilli, M.C.; Paolucci, L.; Talluri, G. A Comparison of Power Quality Disturbance Detection and Classification Methods Using CNN, LSTM and CNN-LSTM. *Appl. Sci.* **2020**, *10*, 6755. [CrossRef]
7. Park, S.-J.; Kim, H.-W.; Joo, Y.-S. Leaky Lamb Wave Radiation from a Waveguide Plate with Finite Width. *Appl. Sci.* **2020**, *10*, 8104. [CrossRef]
8. Berkache, A.; Lee, J.; Choe, E. Evaluation of Cracks on the Welding of Austenitic Stainless Steel Using Experimental and Numerical Techniques. *Appl. Sci.* **2021**, *11*, 2182. [CrossRef]
9. Lee, J.; Choe, E.; Pham, C.-T.; Le, M. Measurement of Thinned Water-Cooled Wall in a Circulating Fluidized Bed Boiler Using Ultrasonic and Magnetic Methods. *Appl. Sci.* **2021**, *11*, 2498. [CrossRef]
10. Rabung, M.; Kopp, M.; Gasparics, A.; Vértesy, G.; Szenthe, I.; Uytdenhouwen, I.; Szielasko, K. Micromagnetic Characterization of Operation-Induced Damage in Charpy Specimens of RPV Steels. *Appl. Sci.* **2021**, *11*, 2917. [CrossRef]
11. Lee, J.; Berkache, A.; Wang, D.; Hwang, Y.-H. Three-Dimensional Imaging of Metallic Grain by Stacking the Microscopic Images. *Appl. Sci.* **2021**, *11*, 7787. [CrossRef]
12. Im, K.-H.; Kim, S.-K.; Cho, Y.-T.; Woo, Y.-D.; Chiou, C.-P. THz-TDS Techniques of Thickness Measurements in Thin Shim Stock Films and Composite Materials. *Appl. Sci.* **2021**, *11*, 8889. [CrossRef]
13. Griesche, A.; Pfretzschner, B.; Taparli, U.A.; Kardjilov, N. Time-Resolved Neutron Bragg-Edge Imaging: A Case Study by Observing Martensitic Phase Formation in Low Temperature Transformation (LTT) Steel during GTAW. *Appl. Sci.* **2021**, *11*, 10886. [CrossRef]

Article

Ultrahigh Resolution Pulsed Laser-Induced Photoacoustic Detection of Multi-Scale Damage in CFRP Composites

Siqi Wang [1], Jesse Echeverry [1], Luis Trevisi [1], Kiana Prather [1], Liangzhong Xiang [1,*] and Yingtao Liu [2,*]

1 School of Electrical and Computer Engineering, University of Oklahoma, Norman, OK 73019, USA; sqwang@ou.edu (S.W.); jesse.echeverry@ou.edu (J.E.); lmtrevisi@ou.edu (L.T.); kiana_prather@ou.edu (K.P.)
2 School of Aerospace and Mechanical Engineering, University of Oklahoma, Norman, OK 73019, USA
* Correspondence: xianglzh@ou.edu (L.X.); yingtao@ou.edu (Y.L.)

Received: 15 February 2020; Accepted: 11 March 2020; Published: 20 March 2020

Featured Application: The developed photoacoustic nondestructive detection system is a novel approach for precise identification of embedded structural damage in composite laminates.

Abstract: This paper presents a photoacoustic non-destructive evaluation (pNDE) system with an ultrahigh resolution for the detection of multi-scale damage in carbon fiber-reinforced plastic (CFRP) composites. The pNDE system consists of three main components: a picosecond pulsed laser-based ultrasonic actuator, an ultrasound receiver, and a data acquisition/computing subsystem. During the operation, high-frequency ultrasound is generated by pulsed laser and recorded by an ultrasound receiver. By implementing a two-dimensional back projection algorithm, pNDE images can be reconstructed from the recorded ultrasound signals to represent the embedded damage. Both potential macroscopic and microscopic damages, such as surface notches and delamination in CFRP, can be identified by examining the reconstructed pNDE images. Three ultrasonic presentation modes including A-scan, B-scan, and C-scan are employed to analyze the recorded signals for the representation of the detected micro-scale damage in two-dimensional and three-dimensional images with a high spatial resolution of up to 60 μm. Macro-scale delamination and transverse ply cracks are clearly visualized, identifying the edges of the damaged area. The results of the study demonstrate that the developed pNDE system provides a non-destructive and robust approach for multi-scale damage detection in composite materials.

Keywords: composites; multi-scale; embedded damage; non-destructive testing; photoacoustic; ultrasonic representation

1. Introduction

High-performance carbon fiber-reinforced plastic (CFRP) composite materials are well known for their high strength to weight ratio, being light in weight, and resistance to corrosion [1,2]. However, aging-related damage and low-velocity impact damage in composites, such as fatigue cracks and delamination, can significantly reduce their structural integrity and durability. In addition, manufacturing imperfections can result in embedded defects, including voids, cracks, and inclusions [3,4]. Since the size, location, and properties of embedded defects in composites are generally unknown and difficult to detect, there is an urgent need to develop new non-destructive evaluation (NDE) and structural health monitoring (SHM) technologies to help assess the quality of composite products and to help provide accurate inspections throughout a composite's service life.

Early detection of embedded and barely visible damage in composites is imperative for long-term operation, risk management, and prognostics of complex composite structures.

Recent advance of NDE technologies has led to efficient damage detection in composites. Currently well-accepted NDE technologies include acoustic emission, infrared thermography, and ultrasonic testing. Acoustic emission identifies damage initiation and tracks damage growth by continuously analyzing elastic waves generated via energy release from localized sources within the tested structures [5]. This method can be potentially used for in situ damage characterization, since the failure events are detected as they occur. Infrared thermography has been used to detect subsurface cracking and embedded delamination in composites [6,7]. The obtained thermal patterns induced, either by directly heating the sample or applying a mechanical oscillatory load, have been analyzed to study embedded damage in composites. However, it is difficult to measure through-thickness locations of the damage using infrared thermography. Ultrasonic techniques are some of the most popular NDE methods for damage detection in composite structures and have been extensively reported in the literature [8–10]. For example, Kessler et al. used Lamb wave methods to detect delamination, transverse ply cracks and through-holes in quasi-isotropic graphite/epoxy composites [11]. Chang et al. developed a tomographic damage imaging approach by combining inverse acoustic wave propagation by combining the k-space method with the adjoint method [12]. Although NDE technologies have been used for a broad range of engineering applications, most of the NDE equipment cannot detect micro-scale damage initiation in composites in depth. High-frequency ultrasound can detect micro-scale damage. However, the detectable depth is limited.

Real-time NDE has been under investigation as a method to monitor the integrity of materials and structures over the past two decades. Progress has been made in developing and improving real-time NDE, which allows early detection of material defects, providing timely warning to those at stake [13–17]. Real-time NDE technologies utilizing advanced sensors (i.e., piezoelectric ceramic sensors [18,19], impedance-based sensors [20,21], piezoresistive sensors [22,23], and fiber Brag grating sensors [24,25]) have been referred to as structural health monitoring (SMH) and prognostics. In order to detect structural defects under regular load conditions, innovative signal processing and pattern recognition algorithms have been developed [26–28]. While progress has been made in the development of SHM and prognostics, this technology has not been implemented in industries that require large-scale applications, especially in the aerospace industry, due to limitations pertaining to the sensors, the power supplies, and real-time data processing.

Laser-induced ultrasound has been recognized as a promising technical solution for NDE and SHM of CFRP composites. Current laser-induced ultrasonic NDE systems use Q-switched lasers with nanosecond pulses and pulse energy levels of several millijoules (mJ), generating ultrasonic signal frequencies ranging from tenths of kilohertz (kHz) up to tens of megahertz (MHz) [29,30]. Both through-transmission and pulse-echo ultrasonic spectroscopy methods are able to detect CFRP composites up to several centimeters [31–33]. Although remote ultrasonic energy generation and data collection are the ideal approaches, complex and relatively expensive instruments are required for ultrasonic interferometric detection [34,35]. To the best of the authors' knowledge, currently, sophisticated laser-based ultrasound imaging systems for remote evaluation of CFRP composites are expensive and technologically immature.

In this paper, we developed a picosecond pulsed laser-induced photoacoustic non-destructive evaluation (pNDE) system for the detection of multi-scale damage in CFRP composites using a picosecond pulsed laser and high-frequency ultrasound transducer. At the micro scale, the damage precursors of surface notches and matrix cracks were successfully detected and represented in 3D images. The size and position of the micro-scale defects in composites were evaluated with a high spatial resolution of 60 μm. Scanning electron microscopy (SEM) images were obtained to validate all micro-scale surface notches on composites. At the macro scale, both delamination and transverse ply cracks were successfully detected and represented using the developed pNDE system. The size of delamination at different depths and the locations of transverse ply cracks were accurately measured.

2. Materials and Methods

2.1. Materials and Experimental Procedure

CFRP composite samples were fabricated using commercial prepreg carbon fiber fabrics (#2511 semi-toughened epoxy resin-coated T700G-12K-31E carbon fiber fabrics) manufactured by Toray Industries, Inc. Each prepreg lamina had a standard resin content of 35.3 wt.% and fiber areal weight of 150.6 g/m^2. Four plies of composite lamina were stacked in a [+45°/−45°]$_s$ sequence and manufactured using a hot press, following the vendor's curing instructions. The scanned composite thickness is approximately 0.8 mm based on the measurement of a caliper. To generate micro-scale damage precursors in composites, shallow X-shaped notches were cut on the composite sample's surface using a sharp razor blade. The dimensions of the notches were measured using SEM images. Macro-scale damage in composites was generated under the velocity impact load using an in-house developed drop-weight impactor. The impact energy absorbed by the composite sample was 4 J. Both embedded delamination and transverse ply cracks were generated in the composite sample due to the applied impact load.

Throughout the pNDE damage detection process, the scanned composite target was secured on a scanning platform (LMS203 Fast XY Scanning stage, Thorlabs, Newton, NJ, USA), which has a maximum linear translation speed of 100 mm/s and a peak acceleration of 10 m/s^2 in both lateral directions. In addition, a step length of 0.01 mm was used during the pNDE detection. The position of the scanning stage is synchronized with the laser source excitation sequence.

2.2. Theory of pNDE Method

In this paper, the pNDE mechanism is based on the photoacoustic effect. Acoustic and ultrasonic waves are generated following the local temporal thermal elastic deformation and pressure caused by the optical absorption of pulsed laser in materials. The relationship between the generated photoacoustic pressure p(r, t) (at location r and time t) and the deposited pulsed laser heat H(r, t) is described using the following equation [36]:

$$(\nabla^2 - \frac{1}{v_s^2}\frac{\partial^2}{\partial t^2})p(\vec{r},t) = -\frac{\beta}{C_p}\frac{\partial H(\vec{r},t)}{\partial t} \tag{1}$$

where v_s is the compressional wave speed in CFRP composites, β is the thermal coefficient of volume, and C_p represents the specific heat capacity at constant pressure. We designated the position of the transducer as the origin of the coordinate system for convenience. The acoustic pressure p (r, t) at transducer position r and time t is, therefore, expressed as:

$$p(r,t) = \frac{\beta}{4\pi C_p}\iiint \frac{dr}{r}\frac{\partial H(r,t')}{\partial t'}\bigg|_{t'=t-\frac{r}{v_s}} \tag{2}$$

Since optical absorption is proportional to acoustic signal strength in photoacoustic imaging, the absorption difference between an undamaged solid material and a damaged area with material vacancy provides the imaging contrast. In a photoacoustic microscopy image, the optical absorption at each illuminated point can be derived from the time of flight of the acoustic wave detected at each ultrasound transducer location. Therefore, the micro-structure within the composite can be mapped with photoacoustic microscopy in 3D to reveal any underlying defects.

2.3. pNDE Imaging System

The developed pNDE system consisted of three major components: (i) a picosecond pulsed laser for the generation of ultrasound signals in composite samples, (ii) a PZT ultrasonic receiver and signal amplifiers, and (iii) a data acquisition, processing, and imaging subsystem. A picosecond laser

(COMPILER 532/266, Passat, Ltd., Vaughan, ON, Canada) was used to provide ultrafast laser pulses with a pulse duration of less than 7 ps. The laser pulse repetition rate (PRR) was adjusted between 1 and 400 Hz. The generated photoacoustic signals were captured by two PZT ultrasound transducers in the experiments. In the multi-scale damage detection test, a transducer (V2062, Olympus NDT, Waltham, MA, USA) with a center frequency of 125 MHz and a bandwidth greater than 87% −6 dB was used for high spatial resolution. In the macro-scale damage detection test, a transducer (U8517149, Olympus NDT) with a center frequency of 20 MHz and a bandwidth greater than 50% −6 dB was used. A low-noise preamplifier (ZFL-1000LN+, Mini-Circuits, Brooklyn, NY, USA) with a bandwidth of 0.1–1000 MHz at −3 dB and a typical gain of 20 dB was used to prepare weak electrical signals from the transducer and deliver the noise-tolerant output signals to the second-stage amplifier (ZFL-500+, Mini-Circuits) with a bandwidth of 0.05–500 MHz at −3 dB, and a gain of 25 dB to further improve the signal-to-noise ratio (SNR). Finally, the pre-processed ultrasonic signals were recorded by the data acquisition card (NI PCI-5153EX, National Instruments, Austin, TX, USA). The schematic diagram of the developed pNDE system is shown in Figure 1a. In addition, Figure 1b illustrates the developed pNDE hardware system.

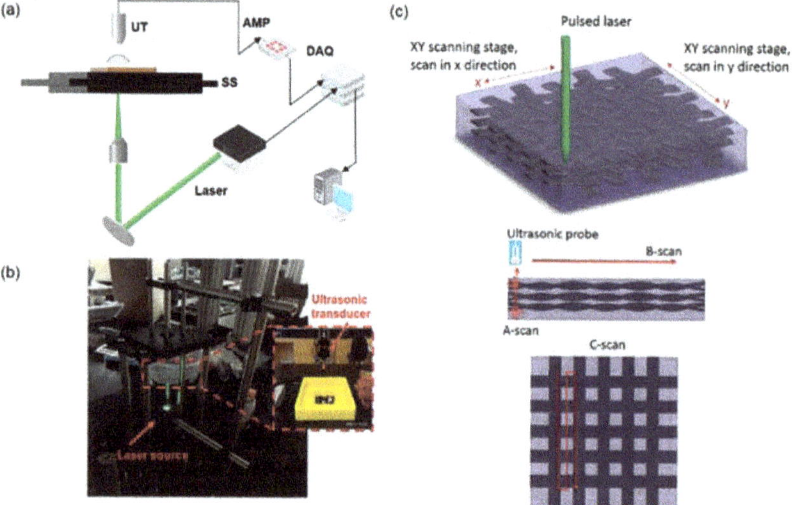

Figure 1. Developed photoacoustic non-destructive evaluation (pNDE) system. (**a**) A schematic of the pNDE system, showing the following key components: UT, ultrasonic transducer; SS, scanning stage; OL, objective lens; AMP, amplifiers; DAQ, data acquisition card; Laser, green laser (532 nm). (**b**) Experimental setup of the pNDE system, the transducer and the 3D printed water container are further magnified to demonstrate the sample set up. (**c**) A schematic of A-scan, B-scan, and C-scan for ultrasonic signal presentation.

For micro-scale damage detection, the CFRP composite samples were horizontally placed on the bottom glass window of the 3D-printed water container (Figure 1b). The ultrasound transducers were submerged in the water to keep good coupling of the ultrasound propagation between the sample and the transducer. The PPR of the laser was set at 30 Hz so that an adequate signal-to-noise ratio was obtained. A sampling rate of 500 MHz was used to record the ultrasonic signals during data collection at each scanning location. One set of pNDE damage detection data included 500 × 500 positions with a step length of 10 μm per step and detected micro-scale damage in an area of 5 mm × 5 mm. All the ultrasonic data was collected when the scanning stage (LMS203 Fast XY Scanning stage, Thorlabs) traveled in the X-Y plane and was controlled using an in-house LABVIEW software. Back projection-based photoacoustic reconstruction was performed in MATLAB [37–40]. It is noted that both the pulsed laser beam and ultrasound transducer were kept stationary, while the composite sample was shifted on a 2D translation stage at an average translation speed of 0.05 mm/s. The position-synchronized output of the translator triggered the pulsed laser and turned the laser on and off during the scanning. Similar experimental procedures were adopted for the pNDE detection of macro-scale impact-induced damage in composites. The PRR of the laser was set at 30 Hz, a sampling rate of 500 MHz was used to record the ultrasonic signals, and the translation speed of the stage was 0.05 mm/s. Laser-induced ultrasonic signals were generated and then recorded using an ultrasonic probe. The pNDE damage detection area was 15 mm × 15 mm, and the recorded data included 500 × 500 positions with a step length of 30 μm per step.

Accurate 3D imaging of the measured microstructures and potential damage in composites is critical in order to demonstrate the developed system for NDE applications. In our study, all the collected pNDE data was studied in the ultrasonic A-scan, B-scan, and C-scan presentation modes. A schematic of the three ultrasonic presentation modes is shown in Figure 1c. Each presentation mode provided a different way to evaluate the inspected region. A-scan displayed the ultrasonic signal energy as a function of time in the ultrasonic propagation direction. B-scan provided the display of ultrasonic signal energy regarding the linear position of the transducer, resulting in the plot to show the transverse cross-section of the detected composite plate. C-scan allowed a plan-type view of the location and size of damage in the detected sample. The combination of the three ultrasonic presentation modes allowed a comprehensive demonstration of the detected multi-scale damage in the composite structures.

3. Multi-Scale Damage Detection Results Using pNDE

3.1. Typical A-Scan and Correction for Micro-Scale Damage Detection

A typical photoacoustic A-scan signal obtained during experiment is shown in Figure 2a. The beginning and the end of the 'A' lines were cut such that images only show the reconstruction of the CFRP. For acoustic attenuation compensation in the scanned material, a time gain correction (TGC) function was applied to all recorded photoacoustic signals as shown in Figure 2b. The ultrasound attenuation compensation was derived the exponential law [41]:

$$A_{TGC}(z_k) = A_0(z_k)\exp(2\alpha(z_k - z_0)) \tag{3}$$

where the first corrected sample along the z-direction was denoted as z_0, corresponding to the frame of the CFRP board reconstruction. The last corrected sample was denoted as z_k, corresponding to the last frame of the CFRP board reconstruction. The acoustic attenuation coefficient α was measured to be 5.64 cm-1 for all A-scans [41].

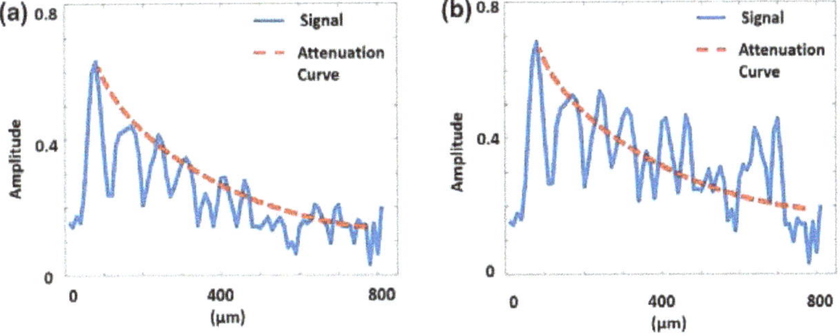

Figure 2. Photoacoustic A-scan signal. (**a**) Typical full bandwidth signal with an assumed signal attenuation function (red dashed curve) and (**b**) corrected A-scan signal by normalizing using the exponent of Equation (3).

3.2. Ultrasonic C-Scan for Micro-Scale Damage Detection

Figure 3 shows pNDE C-scan images in the X-Y plane parallel to the top surface of the composite sample at various depths. As illustrated in Figure 3a, the pNDE C-scan image clearly represented the X-shaped notches on the tested CFRP sample. The width of the X-shape notch shown on the C-scan matched well with that measured in the SEM image, indicating the maximum width of the notch was approximately 270 µm, as shown in Figure 3b. In addition, the tow orientation and the woven structures of the carbon fiber fabrics were visible in the reconstructed images. The photoacoustic signal profile across two grid lines on the surface of the material was extracted along the solid blue line shown in Figure 3a. Based on the extracted grid line profiles, the lateral resolution of the system is estimated to be approximately 60 µm, corresponding to the smaller full width at half maximum (FWHM) of the line spread function in Figure 3c.

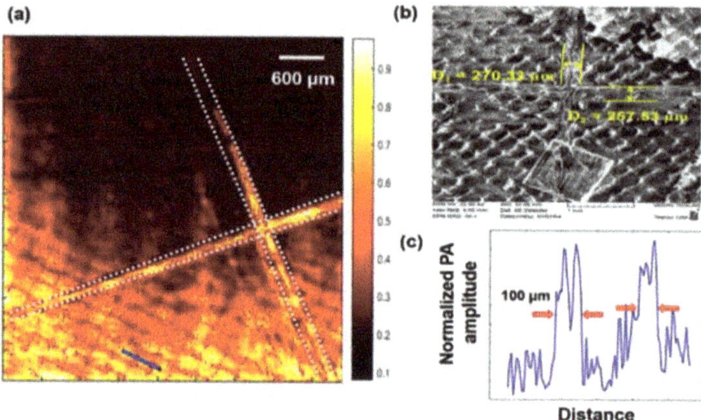

Figure 3. X-Y plane image. (**a**) pNDE X-Y plane image; the 'X' marking is highlighted. (**b**) SEM image; the width of the 'X' marking is calculated. (**c**) Full width at half maximum (FWHM) measurements along the blue line on the pNDE image, which are taken as the lateral resolution of the image.

Figure 4a–e shows the typical pNDE C-scans in the horizontal cross-sections (X-Y plane) that were parallel to the top surface of the composite structure at different depths. The C-scan presentation provided the top view of the locations and sizes of the defects featured in the tested CFRP composites. In the C-scan images of Figure 4a,b, the X-shape notch was clearly visible. Only partial micro-scale matrix cracks can be visualized in Figure 4a,b because of the scanning plane (X-Y plane) is not strictly parallelized to the surface of the composite plate. In Figure 4e, the dense and highly distributed micro-scale matrix cracks, which lead to potential delamination, were detected. Due to the distribution of microscale matrix cracks throughout the entire layer in the composites, the ultrasonic signals were dispersed. Therefore, carbon fiber fabric yarns were not observed in this layer. In Figure 4f, the photoacoustic maximum-amplitude image projected from the top view of the CFRP composite indicated the complex microstructure inside the sample. Both the surface notches and embedded matrix cracks were shown in this image. The 3D image of the detected composite structure with micro-scale notch damage is shown in Figure 4g.

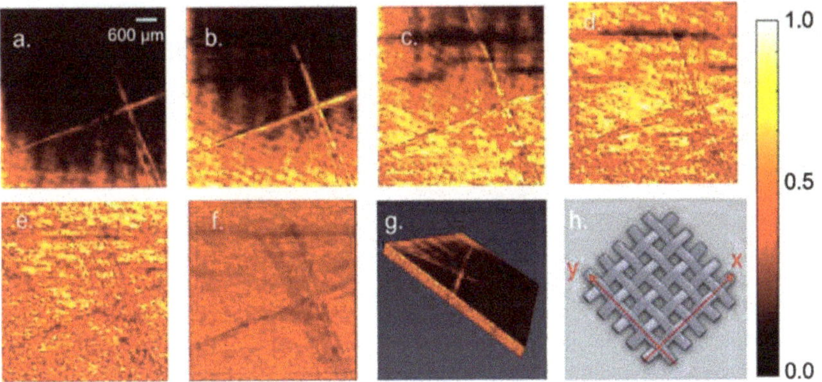

Figure 4. The 3D C-scan images of micro-scale damage in composites. (**a–e**) X-Y plane slices at a depth of 0.001, 0.025, 0.05, 0.065, and 0.15 mm, respectively. (**f**) The photoacoustic maximum-amplitude image projection (average) from the top of the carbon fiber-reinforced plastic (CFRP) composites. (**g**) A 3D C-scan image of a rotating 3D model. (**h**) Grid pattern explanation image.

3.3. *Ultrasonic B-Scan for Micro-Scale Damage Detection*

A typical B-scan image in the vertical cross-section (Y-Z plane, perpendicular to the top surface) of the tested composite structure is shown in Figure 5. Each B-scan image was generated by analyzing the ultrasonic A-scan data in the same cross-section in the vertical cross-section. Since high accuracy A-scan steps were enabled by a LABVIEW programmed stepper motor, the pulsed laser was precisely triggered on the predefined detection position, and accurate B-scan images were created to represent the cross-section conditions in composites. Figure 5a,b shows the typical B-scan images from two x positions in the Y-Z plane. The surface notches were observed from multiple B-scan images. One or two notches were visualized in the selected B-scan images. The B-scan images matched with the C-scan results, indicating accurate pNDE detection of micro-scale damage in the composite sample with notches on the surface.

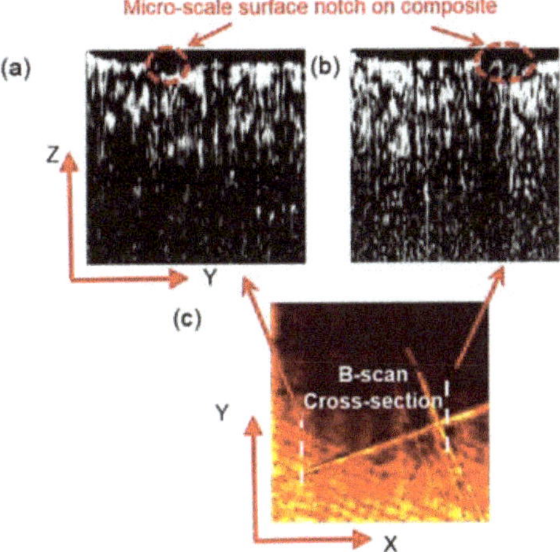

Figure 5. (**a**,**b**) Typical B-scan images showing the vertical cross-section of the detected notch in the composite plate, at positions: x = 1.6 and 4.1 mm, respectively (from left to right). (**c**) C-scan image showing the position of B-scan images in the X-Y plane.

3.4. Macro-Scale Damage Detection in Composites

Low-velocity impact load can cause severe structural damage, such as delamination and transverse ply crack, in composite laminates. Demonstration of the reliable detection of macro-scale damage in composites using the developed pNDE method is critical for the engineering application. As shown in Figure 6, the macro-scale delamination and transverse ply cracks were successfully detected and represented using C-scan images at 5 different depths in the range 0–1.6 mm. The edge of transverse ply cracks was clearly drawn using the dashed line. In addition, the area of delamination at different depths was also highlighted in each subfigure. In Figure 6a–e, the size of the delamination area gradually reduced, indicating that the size of delamination was relatively large near the composite surface. This observation was reasonable since the excitation energy was reduced as it penetrated deep into composites, resulting in reduced delamination area as the depth increased. The detected macro-scale delamination and transverse ply cracks matched with the optical image of the damaged composite sample shown in Figure 6f. A typical photoacoustic signal received by the 20 MHz center frequency transducer is shown in Figure 6g. The corresponding single-sided photoacoustic signal frequency amplitude spectrum is also shown in Figure 6h. Due to the increased scan steps, the spatial resolution of macro-scale pNDE scanning was lower than that of the micro-scale pNDE detection. This is reasonable, since the extremely high spatial resolution was not necessary to identify macro-scale damage in composites. In addition, the optimization of pNDE parameters for macro-scale damage detection was able to significantly increase the scanning speed by choosing the relatively large scanning step length during the detection. Therefore, it is critical to adjust the pNDE parameters following the potential damage size and required spatial resolution, allowing the developed system to be suitable for both micro- and macro-scale damage detection in composites.

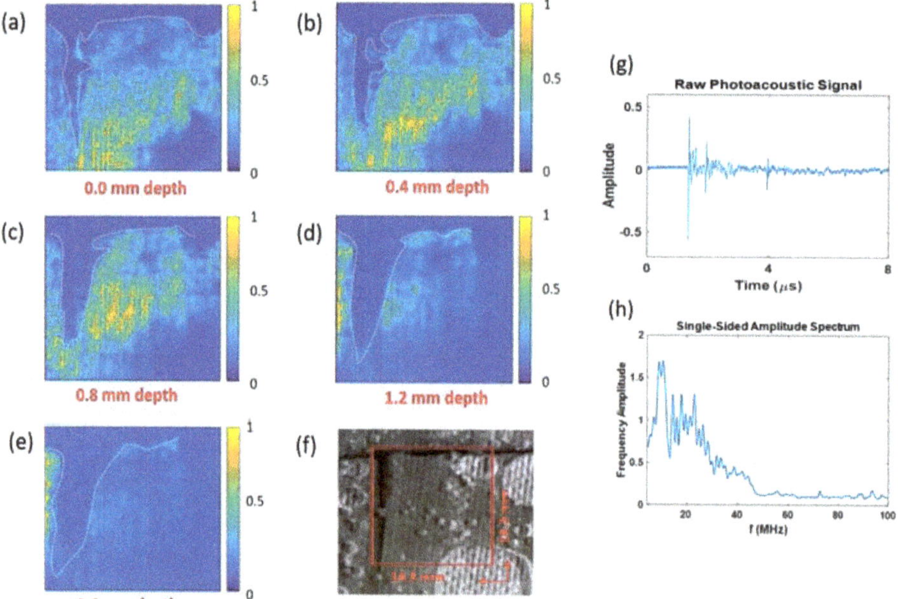

Figure 6. A pulsed laser-induced pNDE system to detect macro-scale delamination and transverse ply cracks in composites. (**a–e**) The 2D C-scan images showing the size of delamination and transverse ply cracking. (**f**) Optical images using the detected impact damage of composites. (**g**) A plot of the raw photoacoustic signal. (**h**) Single-sided photoacoustic signal frequency amplitude spectrum plot.

4. Discussion

Detection sensitivity is a major parameter in pNDE. Our theoretical model and calculations showed that the pulse width of the excitation laser beam was a crucial factor for the effective generation of photoacoustic waves. Mathematical analysis revealed that the resultant photoacoustic pressure was proportional to the time derivative of the excitation pulse [42,43]. Compared to the typical laser-induced ultrasound system (nanoseconds pump laser) [44,45], the implementation of the 7 picosecond pulsed laser can produce an increase in photoacoustic signal conversion efficiency, which will lead to improved detection sensitivity. Further, a higher frequency photoacoustic signal will lead to better depth resolution when paired with a high-frequency ultrasound transducer.

The lateral resolution of pNDE was estimated by the diffraction-limited spot size of the optical focus [42]. For a wavelength of 532 nm and a numerical aperture (NA) of 0.10, the lateral resolution of the pNDE micro-scale scan was theoretically determined to be 2.71 μm. This resolution was sufficient to resolve individual carbon fibers (diameter: 7–10 μm) inside the CFRP composite in a non-invasive manner (Figure 3a). The axial resolution in the depth direction was jointly determined by the laser pulse width and detection bandwidth of the ultrasonic transducer. In the macro-scale damage detection experiment, a 20 MHz ultrasonic transducer was used, which resulted in an axial imaging resolution of approximately 0.30 mm in the depth direction. In future studies, lateral resolution of the pNDE scan can be further improved by (i) increasing the NA of the objective lens, and (ii) using a shorter excitation source wavelength with the maximum imaging depth scaled accordingly. If a higher frequency transducer is employed, submicrometer spatial resolution can be achievable. This analysis indicates a new approach for microstructural determination inside CFRP composites with photoacoustic imaging for future research.

The current pNDE detection speed was limited by the single ultrasonic probe for signal collection during the scanning. However, implementing a pair of galvanometer mirrors with optical scanning can dramatically improve the imaging speed (up to 30 kHz, the limitation of a galvanometer) and should be much faster than any mechanical scanning imaging system. Further increase in laser PRR has been limited mostly by the time required to store raw data and the ultrasound propagation time inside the CFRP composites. Additionally, the system can be further optimized for real-time imaging in field applications by increasing the laser PRR of the developed pNDE system, and the goal is to reach the multi-kHz laser PRR range.

Conventional ultrasound methods and thermography are only capable of providing a contour of the damaged area rather than the detailed layer-by-layer distribution information demonstrated in the resulted images of the pDNE test [46]. Although current ultrasound-based non-destructive evaluation technologies can detect barely visible and embedded geometries, they do not have the adequate lateral resolution to identify micro-scale damage initiation in composites, especially for complex layered materials like CFRP composites. Our proposed technology has the advantage of being able to focus the excitation source to achieve a much higher lateral resolution than the ultrasound-based imaging system. Thus, the pNDE scanning method described in this paper shows great potential compared to current methods for the characterization of impact damage at multiple length scales via in situ imaging within the upper part of an in-depth damage distribution. With future hardware modifications, pNDE scans can theoretically be completed within minutes.

5. Conclusions

In this paper, we developed a pNDE system for the detection of multi-scale damage with extremely high resolution in CFRP composites. Micro-scale damage precursors in composite laminate samples were identified and represented using 2D and 3D images with a high spatial resolution of 60 μm. SEM images taken from the same location were used to validate the length and width of the detected notches on the composite surface. By adjusting the pNDE scanning parameters, the macro-scale impact damage, including delamination and transverse ply cracks, was quickly detected and represented using the ultrasonic C-scan mode. Experimental results and high-resolution 3D images generated by the described technology can be further used in complex mechanical models for prediction of the deep-layer damage propagation of composites and to evaluate the remaining useful life of composites subjected to impact and fatigue loads. Therefore, the developed pNDE system shows great potential for damage detection and quality assessment in a broad range of engineering applications, including aerospace, automobile, and civil industries.

Author Contributions: Conceptualization, L.X. and Y.L.; methodology, S.W.; software, J.E.; validation, L.T.; formal analysis, S.W. and K.P.; writing—original draft preparation, S.W. and K.P.; writing—review and editing, L.X. and Y.L. All authors have read and agreed to the published version of the manuscript.

Funding: This research received no external funding.

Conflicts of Interest: The authors declare no conflict of interest.

References

1. Mouritz, A.P.; Bannister, M.K.; Falzon, P.; Leong, K. Review of applications for advanced three-dimensional fibre textile composites. *Compos. Part A Appl. Sci. Manuf.* **1999**, *30*, 1445–1461. [CrossRef]
2. Agarwal, B.D.; Broutman, L.J.; Chandrashekhara, K. *Analysis and Performance of Fiber Composites*; John Wiley & Sons: Hoboken, NJ, USA, 2017.
3. Croft, K.; Lessard, L.; Pasini, D.; Hojjati, M.; Chen, J.; Yousefpour, A. Experimental study of the effect of automated fiber placement induced defects on performance of composite laminates. *Compos. Part A Appl. Sci. Manuf.* **2011**, *42*, 484–491. [CrossRef]
4. Heslehurst, R.B. *Defects and Damage in Composite Materials and Structures*; CRC Press: Boca Raton, FL, USA, 2014.
5. Dahmene, F.; Yaacoubi, S.; Mountassir, M.E. Acoustic emission of composites structures: Story, success, and challenges. *Phys. Procedia* **2015**, *70*, 599–603. [CrossRef]
6. Kordatos, E.; Aggelis, D.; Matikas, T. Monitoring mechanical damage in structural materials using complimentary NDE techniques based on thermography and acoustic emission. *Compos. Part B Eng.* **2012**, *43*, 2676–2686. [CrossRef]
7. Steinberger, R.; Leitão, T.V.; Ladstätter, E.; Pinter, G.; Billinger, W.; Lang, R. Infrared thermographic techniques for non-destructive damage characterization of carbon fibre reinforced polymers during tensile fatigue testing. *Int. J. Fatigue* **2006**, *28*, 1340–1347. [CrossRef]
8. Scott, I.; Scala, C. A review of non-destructive testing of composite materials. *NDT Int.* **1982**, *15*, 75–86. [CrossRef]
9. Su, Z.; Ye, L.; Lu, Y. Guided Lamb waves for identification of damage in composite structures: A review. *J. Sound Vib.* **2006**, *295*, 753–780. [CrossRef]
10. Birt, E.; Smith, R. A review of NDE methods for porosity measurement in fibre-reinforced polymer composites. *Insight Non Destr. Test. Cond. Monit.* **2004**, *46*, 681–686. [CrossRef]
11. Kessler, S.S.; Spearing, S.M.; Soutis, C. Damage detection in composite materials using Lamb wave methods. *Smart Mater. Struct.* **2002**, *11*, 269. [CrossRef]
12. Chang, Q.; Peng, T.; Liu, Y. Tomographic damage imaging based on inverse acoustic wave propagation using k-space method with adjoint method. *Mech. Syst. Signal Process.* **2018**, *109*, 379–398. [CrossRef]
13. Sohn, H.; Farrar, C.R.; Hemez, F.M.; Shunk, D.D.; Stinemates, D.W.; Nadler, B.R.; Czarnecki, J.J. A review of structural health monitoring literature: 1996–2001. *Los Alamos Natl. Lab. USA* **2003**, 1–7, Technical Repts LA-13976-MS.
14. Ou, J.; Li, H. Structural health monitoring in mainland China: Review and future trends. *Struct. Health Monit.* **2010**, *9*, 219–231.
15. Farrar, C.R.; Worden, K. *Structural Health Monitoring: A Machine Learning Perspective*; John Wiley & Sons: Hoboken, NJ, USA, 2012.
16. Peng, Y.; Dong, M.; Zuo, M.J. Current status of machine prognostics in condition-based maintenance: A review. *Int. J. Adv. Manuf. Technol.* **2010**, *50*, 297–313. [CrossRef]
17. Liu, Y.; Nayak, S. Structural health monitoring: State of the art and perspectives. *JOM* **2012**, *64*, 789–792. [CrossRef]

18. Duan, W.H.; Wang, Q.; Quek, S.T. Applications of piezoelectric materials in structural health monitoring and repair: Selected research examples. *Materials* **2010**, *3*, 5169–5194. [CrossRef] [PubMed]
19. Liu, Y.; Chattopadhyay, A. Low-velocity impact damage monitoring of a sandwich composite wing. *J. Intell. Mater. Syst. Struct.* **2013**, *24*, 2074–2083. [CrossRef]
20. Baptista, F.G.; Budoya, D.E.; de Almeida, V.A.; Ulson, J.A.C. An experimental study on the effect of temperature on piezoelectric sensors for impedance-based structural health monitoring. *Sensors* **2014**, *14*, 1208–1227. [CrossRef]
21. Min, J.; Park, S.; Yun, C.-B.; Lee, C.-G.; Lee, C. Impedance-based structural health monitoring incorporating neural network technique for identification of damage type and severity. *Eng. Struct.* **2012**, *39*, 210–220. [CrossRef]
22. Chiacchiarelli, L.M.; Rallini, M.; Monti, M.; Puglia, D.; Kenny, J.M.; Torre, L. The role of irreversible and reversible phenomena in the piezoresistive behavior of graphene epoxy nanocomposites applied to structural health monitoring. *Compos. Sci. Technol.* **2013**, *80*, 73–79. [CrossRef]
23. Vertuccio, L.; Guadagno, L.; Spinelli, G.; Lamberti, P.; Tucci, V.; Russo, S. Piezoresistive properties of resin reinforced with carbon nanotubes for health-monitoring of aircraft primary structures. *Compos. Part B Eng.* **2016**, *107*, 192–202. [CrossRef]
24. López-Higuera, J.M.; Cobo, L.R.; Incera, A.Q.; Cobo, A. Fiber optic sensors in structural health monitoring. *J. Lightwave Technol.* **2011**, *29*, 587–608. [CrossRef]
25. Kinet, D.; Mégret, P.; Goossen, K.; Qiu, L.; Heider, D.; Caucheteur, C. Fiber Bragg grating sensors toward structural health monitoring in composite materials: Challenges and solutions. *Sensors* **2014**, *14*, 7394–7419. [CrossRef] [PubMed]
26. Liu, Y.; Kim, S.B.; Chattopadhyay, A.; Doyle, D.T. Application of system-identification techniquest to health monitoring of on-orbit satellite boom structures. *J. Spacecr. Rocket.* **2011**, *48*, 589–598. [CrossRef]
27. Liu, Y.; Mohanty, S.; Chattopadhyay, A. Condition based structural health monitoring and prognosis of composite structures under uniaxial and biaxial loading. *J. Nondestruct. Eval.* **2010**, *29*, 181–188. [CrossRef]
28. Liu, Y.; Fard, M.Y.; Chattopadhyay, A.; Doyle, D. Damage assessment of CFRP composites using a time–frequency approach. *J. Intell. Mater. Syst. Struct.* **2012**, *23*, 397–413. [CrossRef]
29. Scruby, C.B.; Drain, L.E. *Laser Ultrasonics Techniques and Applications*; CRC Press: Boca Raton, FL, USA, 1990.
30. Gusev, V.E.; Karabutov, A.A. Laser optoacoustics. *NASA STI/Recon Tech. Rep. A* **1991**, *93*.
31. Karabutov, A.; Podymova, N. Nondestructive evaluation of fatigue-induced changes in the structure of composites by an ultrasonic method using a laser. *Mech. Compos. Mater.* **1995**, *31*, 301–304. [CrossRef]
32. Karabutov, A.; Podymova, N. Quantitative analysis of the influence of voids and delaminations on acoustic attenuation in CFRP composites by the laser-ultrasonic spectroscopy method. *Compos. Part B Eng.* **2014**, *56*, 238–244. [CrossRef]
33. Karabutov, A.; Podymova, N. Nondestructive porosity assessment of CFRP composites with spectral analysis of backscattered laser-induced ultrasonic pulses. *J. Nondestruct. Eval.* **2013**, *32*, 315–324. [CrossRef]
34. Monchalin, J.P. Laser-Ultrasonics: From the Laboratory to Industry. In *Proceedings of the AIP Conference Proceedings*; American institue of physics: College Park, MD, USA, 2004; pp. 3–31.
35. Kim, J.; Jhang, K.-Y. Non-contact measurement of elastic modulus by using laser ultrasound. *Int. J. Precis. Eng. Manuf.* **2015**, *16*, 905–909. [CrossRef]
36. Wang, L.V. Tutorial on photoacoustic microscopy and computed tomography. *IEEE J. Sel. Top. Quantum Electron.* **2008**, *14*, 171–179. [CrossRef]
37. Lao, Y.; Xing, D.; Yang, S.; Xiang, L. Noninvasive photoacoustic imaging of the developing vasculature during early tumor growth. *Phys. Med. Biol.* **2008**, *53*, 4203. [CrossRef] [PubMed]
38. Tang, S.; Chen, J.; Samant, P.; Stratton, K.; Xiang, L. Transurethral photoacoustic endoscopy for prostate cancer: A simulation study. *IEEE Trans. Med Imaging* **2016**, *35*, 1780–1787. [CrossRef] [PubMed]
39. Yang, D.; Xing, D.; Yang, S.; Xiang, L. Fast full-view photoacoustic imaging by combined scanning with a linear transducer array. *Opt. Express* **2007**, *15*, 15566–15575. [CrossRef] [PubMed]
40. Treeby, B.E.; Cox, B.T. k-Wave: MATLAB toolbox for the simulation and reconstruction of photoacoustic wave fields. *J. Biomed. Opt.* **2010**, *15*, 021314. [CrossRef]
41. Pelivanov, I.; Ambroziński, Ł.; Khomenko, A.; Koricho, E.G.; Cloud, G.L.; Haq, M.; O'Donnell, M. High resolution imaging of impacted CFRP composites with a fiber-optic laser-ultrasound scanner. *Photoacoustics* **2016**, *4*, 55–64. [CrossRef]

42. Wang, L.V.; Hu, S. Photoacoustic tomography: In Vivo imaging from organelles to organs. *Science* **2012**, *335*, 1458–1462. [CrossRef]
43. Winkler, A.M.; Maslov, K.I.; Wang, L.V. Noise-equivalent sensitivity of photoacoustics. *J. Biomed. Opt.* **2013**, *18*, 097003. [CrossRef]
44. Zhang, J.; Li, W.; Cui, H.-L.; Shi, C.; Han, X.; Ma, Y.; Chen, J.; Chang, T.; Wei, D.; Zhang, Y. Nondestructive evaluation of carbon fiber reinforced polymer composites using reflective terahertz imaging. *Sensors* **2016**, *16*, 875. [CrossRef]
45. Dunkers, J.P.; Sanders, D.P.; Hunston, D.L.; Everett, M.J.; Green, W.H. Comparison of optical coherence tomography, X-ray computed tomography, and confocal microscopy results from an impact damaged epoxy/E-glass composite. *J. Adhes.* **2002**, *78*, 129–154. [CrossRef]
46. Burch, S.; Bealing, N. A physical approach to the automated ultrasonic characterization of buried weld defects in ferritic steel. *NDT Int.* **1986**, *19*, 145–153. [CrossRef]

Article

Fast Terahertz Coded-Aperture Imaging Based on Convolutional Neural Network

Fengjiao Gan, Chenggao Luo *, Xingyue Liu, Hongqiang Wang and Long Peng

College of Electronic Science and Technology, National University of Defense Technology, Changsha 410073, China

* Correspondence: luochenggao@nudt.edu.cn; Tel.: +86-731-8457-5714

Received: 14 February 2020; Accepted: 7 April 2020 ; Published: 12 April 2020

Abstract: Terahertz coded-aperture imaging (TCAI) has many advantages such as forward-looking imaging, staring imaging and low cost and so forth. However, it is difficult to resolve the target under low signal-to-noise ratio (SNR) and the imaging process is time-consuming. Here, we provide an efficient solution to tackle this problem. A convolution neural network (CNN) is leveraged to develop an off-line end to end imaging network whose structure is highly parallel and free of iterations. And it can just act as a general and powerful mapping function. Once the network is well trained and adopted for TCAI signal processing, the target of interest can be recovered immediately from echo signal. Also, the method to generate training data is shown, and we find that the imaging network trained with simulation data is of good robustness against noise and model errors. The feasibility of the proposed approach is verified by simulation experiments and the results show that it has a competitive performance with the state-of-the-art algorithms.

Keywords: terahertz; coded-aperture imaging; convolution neural network (CNN); fast image reconstruction

---

1. Introduction

Since the terahertz wave (0.1–10 THz) lies between the visible and microwave frequencies, it has stronger penetration capability than light and higher resolution than microwave, allowing for visualization of hidden objects at the millimeter level. Moreover, it does little harm to the human body compared to X-rays. Therefore, THz technology has attracted increasing attention, and the generation and detection of THz have also been extensively researched utilizing various approaches. Generation by nonlinear optical effects such as optical parametric oscillation [1] and detection by GaSe electro-optic sensors [2] are one of the typical approaches. Solid-state electronic devices [3] and a low-temperature-grown GaAs photoconductive antenna gated [4] are also used as emitters and detectors. In order to overcome the limitations of THz band, some practical methods have been proposed [5,6]. With the development of THz technology and radar imaging technology, great progress has been made in various industries and research fields. A graphene-based THz ring resonator is considered a potential application for label-free sensing [7]. The application of THz wave in modulation technology was also reported [8]. For nondestructive detection, a millimeter wave radar imaging method based on synthetic aperture radar was presented [9]. In addition, THz radar imaging technology is attractive for security screening [10,11].

As a promising THz radar imaging technology, Terahertz coded-aperture imaging (TCAI), is derived from optical coded-aperture imaging [12] and radar coincidence imaging [13], it utilizes an aperture coded antenna [14] to generate a spatiotemporal independent wave distribution. By modeling the imaging system, the matrix-imaging equation can be established. And the target of interest is reconstructed through computational imaging [15] method. TCAI has many advantages such as high resolution, all-time functionality, low complexity, and low cost and so forth. Besides, the forward-looking and staring imaging capability can be obtained without relying on any relative motion between imaging platform and target, which is different from synthetic aperture radar [16]. As a result, the TCAI technique has a potential application in security screening, battlefield reconnaissance, nondestructive detection and so on.

In recent years, many methods and systems have been proposed that promote the development of TCAI. The authors of Reference [17] designed a single-pixel pulsed terahertz camera, which utilizes random patterns for imaging and the theory of compression sensing (CS) [18] to solve the imaging equation. In 2013, metamaterial apertures that support custom-designed complex measurement modes were introduced into microwave imaging, which avoids mechanical scanning [15]. Xu et al. utilized randomly programmable metasurface to solve the inverse-scattering problem in the single-sensor imaging [19]. Furthermore, the matrix-imaging equation based on the theory of physical optics was derived and high-resolution TCAI was achieved [20]. Besides, the Bayesian estimation [21] and sparsity-driven methods [22] have greatly inspired TCAI research. Although these methods and systems are effective, there are still some great challenges in TCAI. First, the accuracy of system modeling cannot easily be guaranteed whether it is based on the time-delay signal model [23] or Fresnel diffraction theory [20], and some errors are introduced into the reference signal matrix, which leads to the ability to solve the target scattering coefficient is poor at low SNR. Second, the iterative algorithms are used to reconstruct the target, which are too time-consuming to high frame rate imaging, and they are not quite stable and robust to modeling errors and noise. Considering the importance of these problems in practical applications, a fast TCAI method needs to be devised.

Deep learning (DL) has greatly inspired the research in object detection, image classification, signal processing, and among many others. For the inverse problem community, learning-based methods have been successfully employed in multiple scattering media imaging [24], holographic image reconstruction [25], lensless computational imaging [26], computational ghost imaging [27,28] and so forth, but they usually take a lot of effort to collect data set, which is not easily affordable. To reduce the cost of training, some researchers have proposed training imaging network with simulation data set. In References [29] and [30], the practically usable networks that were trained using simulation data set show competitive imaging performance in real-world scenarios, and the simulation results of Reference [31] also demonstrate the effectiveness using simulation data set to train the network. Thus, we investigated the TCAI based on DL to tackle a series of problems mentioned above. In this paper, we design an end to end neural network, which is trained with simulation data. Once trained, the target of interest can be restored instantly by inputting echo signal into the imaging network, and the simulation experiment also proves that the imaging quality at low SNR superior to state-of-the-art iterative approaches for TCAI.

2. Method

2.1. Signal Model and Learning-Based Approach

For the convenience, we take TCAI model based on single input single output technology as an example. The schematic diagram is shown in Figure 1 and it mainly contains a controlling and processing terminal, transmitter module, transmitter, coded aperture, receiver module, receiver. The transmitting module includes mixer and frequency multiplier, and the receiving module includes low noise amplifier and mixer. The THz wave transmitted from the transmitter, and then it propagates to the coded aperture. The coded aperture, controlled by the controlling and processing terminal, randomly or pseudo-randomly modulates the amplitude or phase of incident THz wave to produce a spatiotemporal-independent radiation field in the imaging plane, which is divided into M grid-cells. After being reflected from the imaging area, the pseudo-random signal is collected by receiver and then sent to the controlling and processing terminal for reconstruction the target.

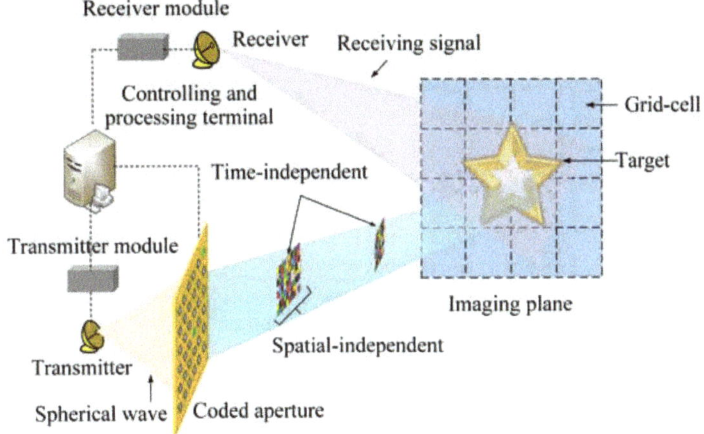

Figure 1. Schematic of Terahertz coded-aperture imaging (TCAI) system.

Suppose the transmitter emits a THz linear frequency modulation signal, the reference signal of the m-th grid cell at time t_n can be deduced as

$$S\left(t_n, m\right) = \sum_{q=1}^{Q} A \exp\left[j2\pi\left(f_c\left(t_n - \frac{r}{c}\right) + \frac{K}{2}\left(t_n - \frac{r}{c}\right)^2\right)\right] \exp(j\varphi_{t_n,q}), \tag{1}$$

where A is the amplitude, f_c is the center frequency, K is the chirp rate, c is the speed of light, r represents the signal propagation distance, Q stands for the number of transmitting element of the coded aperture, $\varphi_{t_n,q}$ is the random phase modulation factor for the q-th coding aperture element at time t_n . Then, the echo signal at time t_n is expressed as

$$S_b\left(t_n\right) = \sum_{m=1}^{M} \beta_m S(t_n, m), \tag{2}$$

where β_m stands for the scattering coefficient corresponding to the m-th grid cell. Based on the time discretion of Equation (2), the matrix-imaging formula can be written as

$$\begin{bmatrix} S_b(t_1) \\ S_b(t_2) \\ \vdots \\ S_b(t_N) \end{bmatrix} = \begin{bmatrix} S(t_1,1) & S(t_1,2) & \cdots & S(t_1,M) \\ S(t_2,1) & S(t_2,2) & \cdots & S(t_2,M) \\ \vdots & \vdots & \cdots & \vdots \\ S(t_N,1) & S(t_N,2) & \cdots & S(t_N,M) \end{bmatrix} \cdot \begin{bmatrix} \beta_1 \\ \beta_2 \\ \vdots \\ \beta_M \end{bmatrix} + \begin{bmatrix} \omega_1 \\ \omega_2 \\ \vdots \\ \omega_N \end{bmatrix}, \tag{3}$$

where $\omega = \{\omega_n\}_{n=1}^{n=N}$ is the additive measurement noise, $S_b = \{S_b(t_n)\}_{n=1}^{n=N}$ is echo signal vector. The reference signal matrix is

$$S = \begin{bmatrix} S(t_1,1) & S(t_1,2) & \cdots & S(t_1,M) \\ S(t_2,1) & S(t_2,2) & \cdots & S(t_2,M) \\ \vdots & \vdots & \cdots & \vdots \\ S(t_N,1) & S(t_N,2) & \cdots & S(t_N,M) \end{bmatrix}. \tag{4}$$

The row vector and column vector of S are the time-domain samples at $\{t_n\}_{n=1}^{n=N}$ and the spatial-domain samples at $\{m\}_{m=1}^{m=M}$, respectively.

For the previous TCAI [23,32], the compressed sensing [18] recovery algorithms are the standard method to solve $\boldsymbol{\beta}$ and the imaging problem is treated as an optimization problem after obtaining the evaluation of $\boldsymbol{S}$

$$\tilde{\boldsymbol{\beta}} = \arg\min_{\boldsymbol{\beta}} \psi(\boldsymbol{\beta})\ s.t.\ \|\boldsymbol{S}_b - \boldsymbol{S}\boldsymbol{\beta}\|_2^2 < \sigma^2, \tag{5}$$

where $\|\boldsymbol{S}_b - \boldsymbol{S}\boldsymbol{\beta}\|_2^2$ is the fitting error, σ^2 is usually the variance of ω and ψ represents the prior information on $\boldsymbol{\beta}$. Typically, $\psi(\boldsymbol{\beta}) = \|\boldsymbol{\beta}\|_1$, where $\|\cdot\|_1$ is L_1 Norm that is used to constrain the imaging domain. For the solving (5), a large number of iterations are usually required which essentially limit the imaging speed. However, the method we presented breaks the bottleneck, it uses neural network to approximate function g so that target scattering coefficient can be estimated directly from echo signal.

$$\hat{\boldsymbol{\beta}} = g_{\Theta}(\boldsymbol{S}_b), \tag{6}$$

where Θ is the set of all possible parameters. Here they can be learned from a training set each of which pairs up a known original target $\boldsymbol{\beta}^d$ and the corresponding echo signal $\boldsymbol{S}_b^d$, where $d = 1, 2, \ldots, D$, enumerates the total D different training data pairs. Thus, this parametric reconstruction process can be expressed as

$$g_{learn} = \arg\min_{g_\theta, \theta \in \Theta} \sum_{d=1}^{D} L\left(\boldsymbol{\beta}^d, g_\theta\left(\boldsymbol{S}_b^d\right)\right), \tag{7}$$

where $g_\theta(\cdot)$ stands for the target of recovery from the imaging network under parameter θ, $L(\cdot)$ is the loss function.

As can be seen from Equation (4), the large-scale reference-signal matrix creates a heavy computational burden. Due to the short wavelength and precise resolving ability of THz waves, the imaging area is divided into smaller grids, which means that the more the number of elements in M and the more complicated calculation of reference-signal matrix. As a result, it is time-consuming to solve the target scattering coefficient through the algorithms consist of iterative-based processes as shown in Equation (5). Therefore, it is very necessary to use the parameter reconstruction algorithm as shown in shown Equation (6) to achieve fast TCAI. In contrast, the proposed approach is in significant ways different. It does not need some time-consuming operations like previous imaging technique. Instead, it demands plenty of data set which includes the groundtruth image and corresponding echo signal and spends some time in training network, but these can be done in advance. Once the training procedure is completed, the designed imaging network can blindly reconstruct target from echo signal.

2.2. Network Structure and Data Generation

Recently, convolution neural networks (CNN) have been extended and applied to solve inverse problems. The theoretical motivations for using CNNs as the learning architecture and the design strategies of CNN-based imaging framework have been discussed [33]. Moreover, the relationship between CNN and iterative optimization algorithms has also been surveyed [34]. Motivated by this research, we propose a neural network for fast TCAI which includes the nonlinear part of the encoded information and the linear part of the decoded information. Figure 2 shows the schematic diagram of the network architecture. We suppose that the input of the network is the echo signal with the length of $(H \times W) \times 1$. It is down-sampled by $\times 1$, $\times 2$, creating two flow structure, with spatial dimensions of $H \times W$, $H/2 \times W/2$, respectively. And the output is the expected target with different scales. The number of feature maps in each layer is 16. After the down-sample, the two tensors flow to the residue blocks, which is constructed by two convolutional layers with batch normalization and two rectified linear units (ReLU), that is, $\mathrm{Re}LU(x) = \max(0, x)$. And a shortcut is utilized between the block's input and output, as indicated by the red arrows, which mitigates the divergent gradient problem and accelerates the convergence of the deep neural network. Following residue block, the spatial dimensions of this feature map from $H/2 \times W/2$ to $H \times W$ via up-sampling block, each block includes one convolutional layer with batch normalization, a ReLU operation and one up-sampling layer that facilitates super-resolution. Finally, the fusion tensor can be obtained by a connection operation in third-dimension of the output tensor of each flow structure. Here, all nonlinear operations are completed, and the low-level and high-level semantic features are learned from the echo signal. For the linear part, which is made up of convolutional layers, it can transform these extracted features into the output in imaging domain. It is important to note that some tricks are adopted to reduce network parameters and achieve cross-channel information interaction. In testing, to avoid over-fitting, the neurons are randomly ignored.

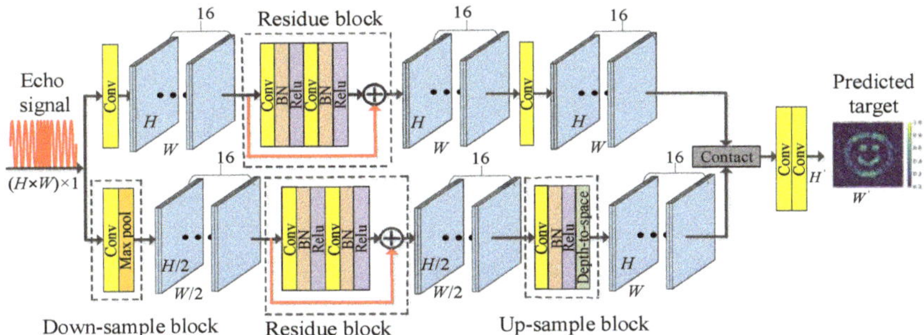

Figure 2. The developed neural network for TCAI.

In the developed network, the spatial correlations between convolution layers can be modeled as

$$v_{i,j}^{x,y} = \sum_{r} \sum_{p=0}^{P-1} \sum_{b=0}^{B-1} w_{i,j,r}^{p,b} v_{i-1,r}^{x+p,y+b}, \tag{8}$$

where r stands for the number of channels in $i-1$ layer, P and B are the size of the weight matrix, that is, convolutional kernels. In i-th convolutional layer, $v_{i,j}^{x,y}$ is the value of x, y-th pixel in the j-th feature map and $w_{i,j,r}^{p,b}$ is the weight of p, b-th position in the j-th convolutional kernels. These parameters are included in Θ, and they can be optimized by minimizing the loss function of the predicted target and the original target

$$L = \frac{\sum_{d=1}^{D^*} \left\| g_\theta(S_b^d) - \boldsymbol{\beta}^d \right\|_2^2}{numel(\boldsymbol{\beta}^d) D^*}, \tag{9}$$

where $\|\cdot\|_2$ refers to L_2 Norm, $numel(\cdot)$ indicates the quantity of pixels in the original target, and $D^* = 8$ is the mini-batch size in the stochastic gradient descent (SGD) method [35]. And the Adam optimizer [36] was adopted to optimize imaging parameters.

As mentioned above, the training of an imaging network usually requires a large amount of the original target and corresponding echo signal. For the generation of the original target data set, we first randomly generate the number of scattering points and their positions on the 60×60 imaging plane, which is designed to imitate real-target cases and guarantee the diversity and richness of the target. Then the scattering coefficients are generated randomly. Subsequently, the corresponding echo data set can be obtained by Equation (3), each of which is $N = 3600$ in length. In particular, the white Gaussian noise is added to the echo signal to acquire the data set with noise. Eventually, we generated two different data sets, one set of which corresponds to data without noise and another with SNR = 5 dB. Each data set containing $D = 50{,}000$ training pairs $(S_b^d, \boldsymbol{\beta}^d)_{d=1}^{D}$. One tenth of this is used as a verification set and the rest as a training set. In the quantitative analysis, the testing data are generated in the same way as the training data.

3. Results

In this section, the feasibility of the DL based approach is verified by numerical experiments. Parameters in the imaging experiment are given in Table 1. The frequency range of the THz signal is from 330 GHz to 350 GHz. The 1-bit transmission-type coded aperture has 60×60 elements that are employed to modulate the phase of the transmitted signal by 0^0 or 180^0. The imaging area is divided into $M = 60 \times 60$, and it is easy to calculate that the pixel interval is no larger than 0.005 m. The size of the target (T) is expressed as the number of non-zero scattering coefficients in imaging area [37], the ratio between T and M is considered the complexity of the target. Also, it is easy to know from the table that the spatial dimension of $\mathbf{S}$ is 3600×3600, which creates high imaging complexity. In practice, the scale of $\mathbf{S}$ may be larger and this calculation is more complicated. Therefore, it is necessary to use a neural network to model $\mathbf{S}$ implicitly.

The network is implemented in Python version 3.6 using DL frameworsorFlow version 1.8 and trained on a desktop computer with GPU NVIDIA 2080 and the CUDA edition is 9.0. The training took approximately 10 h, which is time-consuming but this procedure can be done in advance. Once trained, the target of interest can be restored instantly by inputting echo signal into the imaging network. The Adam optimization algorithm is employed to optimize weights, and the initial learning is 10^{-3} which decays with the factor of 0.98 after each epoch. The batch size is 8 and the training lasts for 500 epochs. Two typical reconstruction algorithms for TCAI, the Sparse Bayesian Learning (SBL) algorithm and TVAL3 algorithm, were chosen to compete with the presented approach. To analyze the imaging performance of various algorithms, the Mean Square Error (MSE) is used as the quantitative index. In this paper, these test targets are composed of scattering points, and the corresponding scattering coefficients are a random value between zero and unity. All reconstruction results and MSE calculations were done on the computer with Inter Xeon Silver 4116 CPU except as specifically stated, and each MSE represents the average results 100 Monte Carlo trials and the shape of the target changes randomly in each trial.

Table 1. Parameters Used in the TCAI Simulation Experiment.

Parameters	Values
Center frequency f_c	340 GHz
Bandwidth	20 GHz
Imaging distance	2 m
Size of coded aperture Δb	0.3 m $\times$ 0.3 m
Size of imaging plane Δs	0.3 m $\times$ 0.3 m
Number of time sampling N	3600
Number of coded aperture elements	60×60
Number of grid cells in the imaging plane	60×60
The distance between coded aperture and receiver	0.15 m
The distance between coded aperture and transmitter	1 m

To investigate the validity of the imaging network, we first carried out simulation experiments at various targets and the results with SNR = 20 dB are shown Figure 3, and the ratio between the target size and the number of grid cells is $T/M = 11/3600$, $T/M = 17/120$, $T/M = 431/3600$, $T/M = 731/3600$, $T/M = 71/200$, $T/M = 37/60$, respectively. It can be seen that the target can be successfully reconstructed whether it is made up of unit ideal point scatters or multi-value point scatters. For further investigating the imaging performance of the proposed method, the SBL algorithm [38] and TVAL3 algorithm [39] were implemented as comparisons. Figure 4 shows the reconstruction results of the "smile" whose scattering coefficients of all the point scatters are a random value between zero and unity. One can clearly see that the object can be recovered from the echo signal through the deep network-based algorithm as low as SNR = −5 dB, despite the reconstruction results apparently being distorted and having many spurious scatterers just like other methods. However, the imaging quality gradually enhanced with the increase of SNR. Compared with SBL algorithm under all SNR levels, the proposed algorithm provides higher resolution results in which the scattering intensity more authentically and the target outline is clearer. For TVAL3 algorithm, the target can be perfectly reconstructed when the SNR is larger than 15 dB. However, its performance degraded dramatically as SNR go lower. In general, our performance is competitive with the typical optimization iterative algorithms when SNR is not larger than 10 dB. It is main reason that CS-based algorithms are not quite robust since some errors exist in the reference signal matrix **S**. Nonetheless, the developed network can automatically fix these errors during the training. With the increase of SNR, we also note that the imaging performance of the proposed method is ultimately bounded by the imaging network's representational error. However, as can be seen in the predicted target, the reconstructions are semantically correct.

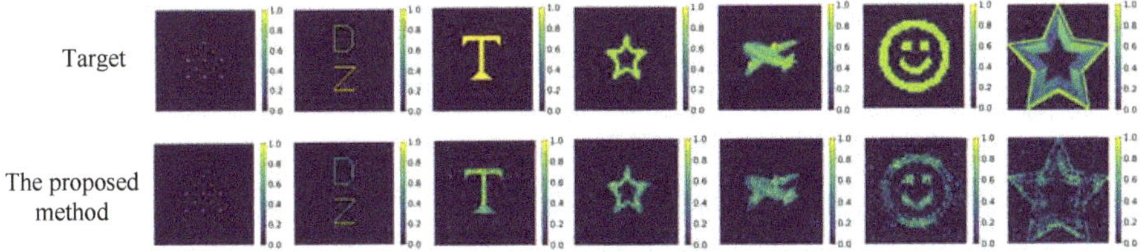

Figure 3. Reconstruction results for different ratio of T/M based on the proposed algorithm.

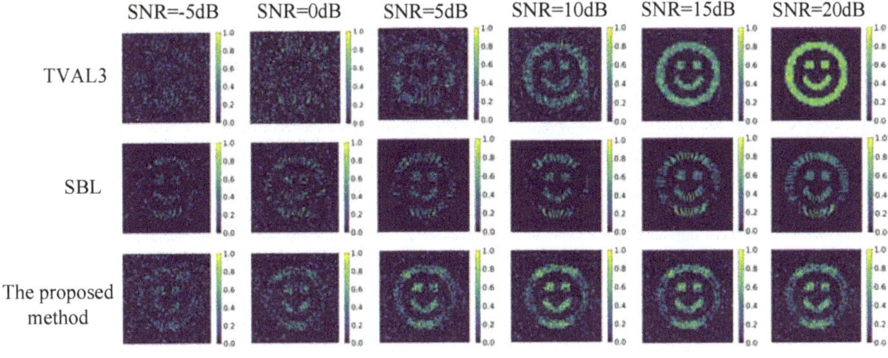

Figure 4. Comparison of reconstruction results from various methods at different signal to noise ratio (SNR).

Quantitatively, the MSEs of different methods under different SNRs are calculated and the results are drawn in Figure 5. For each SNR, the shape of the target changes randomly in each trial. As expected, the SBL algorithm is worse than the proposed method no matter for which SNR. Although the TVAL3 algorithm shows good performance in high SNR, but it is sensitive to noise and takes more time than the presented algorithm. Table 2 shows the time cost of various algorithms, each of which is the average of 100 trials. Due to the neural network-based approach can be easily parallelized, we also recorded the reconstruction time of the presented method with GPU implementation. It is easy to calculate that the imaging frame rate of our method is no less than 280 Hz. From these results, we can see that the proposed algorithm has great superiority with imaging efficiency. One explanation is that an end-to-end network can directly transform the echo signal into the target, while classic imaging techniques require a large amount of iterations to estimate a satisfactory solution. Therefore, it is unsurprising that the neural network based method is much less time consuming. Thus, experimental results above fully illustrate that the proposed method is a promising tool for fast TCAI under low SNR. More evidence is shown in Figure 6. Again, we can see the superiority of deep learning-based approach clearly.

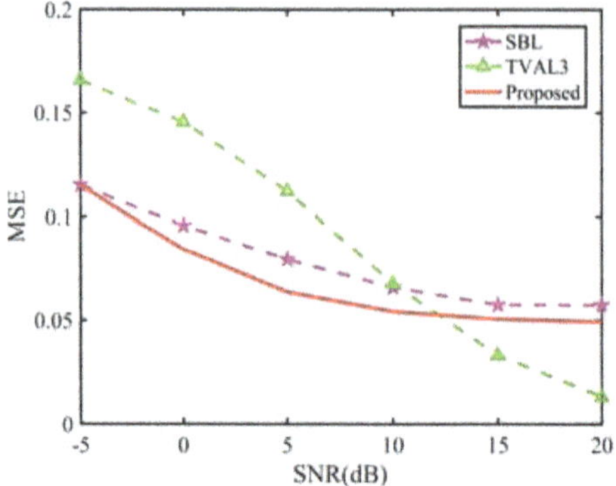

Figure 5. Performance with different SNRs.

Table 2. Comparison on elapsed time for different methods.

Parameters	Values
TVAL3 [39]	2.2499 s
SBL [38]	4.5522 s
The proposed method	0.1369 s (GPU:0.0285 s)

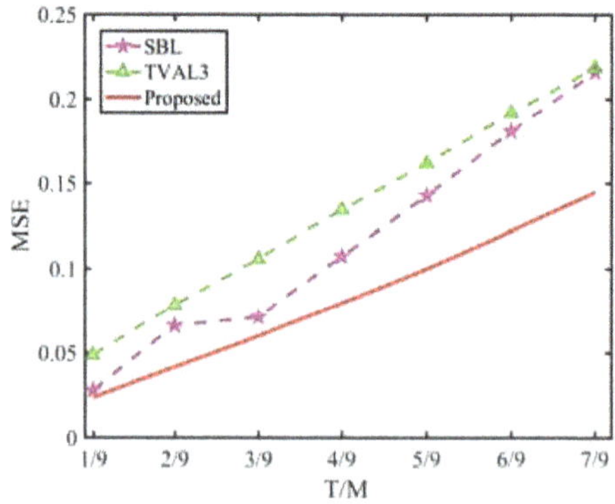

Figure 6. Performance with different ratio of T/M.

4. Discussion

Compared with the classic imaging techniques, the end-to-end neural network can adjust the imaging error adaptively, so its stability and robustness are easier to be guaranteed. Even though the reconstructed image is not as perfect as TVAL3 at high SNR, the time cost is encouraging. The dynamic response video was obtained with GPU implementation and included in Media. As aforementioned, the imaging frame rate is quite high. Therefore, we set the pause time of each frame to 0.008 s. It is less than the recovery time of a batch of images, so the video does not play very smoothly. Nonetheless, the superiority of the proposed approach on imaging efficiency can be clearly. A frame image from target video is shown in Figure 7a ($T/M = 367/3600$), and the corresponding reconstruction result is shown in Figure 7b.

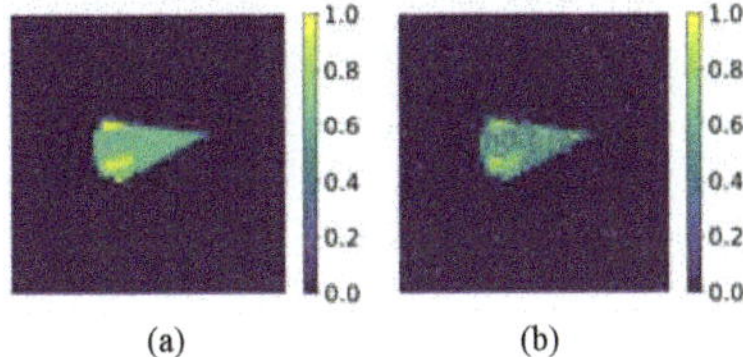

Figure 7. (**a**) Target. (**b**) The corresponding reconstruction result.

To investigate the influence of noiseless and noisy training data on network imaging performance, we trained two networks using different data sets. In test, the target size is the same as the original target in Figures 4 and 8 shows the evidence that whether the training data is noisy or not has little impact on the target restoration accuracy.

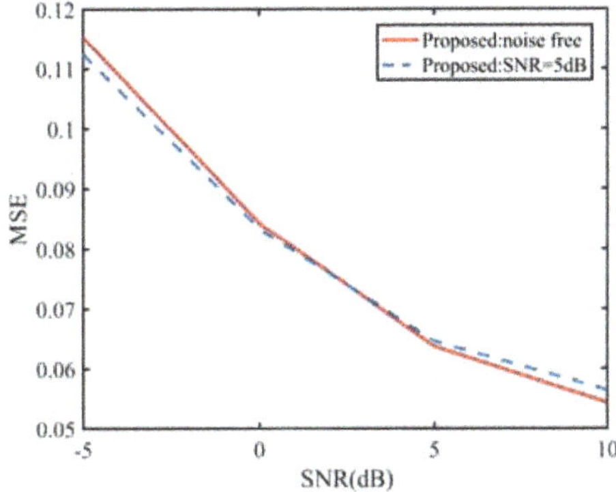

Figure 8. Compare the imaging performance of different networks.

Also, it is worth noting that the method presented here in this manuscript is especially useful for TCAI when we get the empirical data set. The large-scale reference signal matrix does not need to be estimated in advance and the total imaging complexity will be reduced significantly. In the absence of an empirical data set, the simulation data can be employed to train a practically usable imaging network as in References [29] and [30]. In conclusion, the neural network based approach is a promising tool for TCAI.

5. Conclusions

In this paper, a fast TCAI method is presented. We first introduced the TCAI system and learning-based approach. Subsequently, an end-to-end network was developed and tens of thousands of training pairs were generated to learn the mapping relationship between the echo signal and original target. The developed network includes the nonlinear part of the encoded information and the linear part of the decoded information. The experimental results of the simulation data set demonstrated that the proposed method can outperform the state-of-the-art iterative algorithms in both accuracy and efficiency. Also, it can quickly reconstruct targets of different sizes and the pixel interval that can be resolved is no larger than 0.005 m. With the advantages of the neural network based approach, it has a potential application in security screening, battlefield reconnaissance, nondestructive detection and so on. In further work, we intend to improve the network performance and experimentally verify our method with training data sets and echo signals.

Author Contributions: Conceptualization, F.G. and C.L.; methodology, F.G., C.L. and H.W.; software, F.G.; validation, F.G.; formal analysis, F.G. and C.L.; investigation, F.G., X.L. and L.P.; resources, C.L.; data curation, F.G.; writing—original draft preparation, F.G.; writing—review and editing, X.L., H.W. and L.P.; visualization, F.G.; supervision, C.L.; project administration, C.L. All authors have read and agreed to the published version of the manuscript.

Funding: This research was funded by National Natural Science Foundation of China (NSFC) (61701513, 61971427, 61871386).

Conflicts of Interest: The authors declare no conflict of interest.

Abbreviations

The following abbreviations are used in this manuscript:

TCAI	Terahertz coded-aperture imagin
SNR	Signal-to-noise ratio
THz	terahertz
CNN	Convolution neural4network
DL	Deep learning
CS	compressed sensing
MSE	Mean Square Error
SBL	Sparse Bayesian Learning

References

1. Gildenburg, V.; Vvedenskii, N. Optical-to-THz wave conversion via excitation of plasma oscillations in the tunneling-ionization process. *Phys. Rev. Lett.* **2007**, *98*, 245002. [CrossRef] [PubMed]
2. Kübler, C.; Huber, R.; Tübel, S.; Leitenstorfer, A. Ultrabroadband detection of multi-terahertz field transients with GaSe electro-optic sensors: Approaching the near infrared. *Appl. Phys. Lett.* **2004**, *85*, 3360–3362. [CrossRef]
3. Vicarelli, L.; Vitiello, M.; Coquillat, D.; Lombardo, A.; Ferrari, A.C.; Knap, W.; Polini, M.; Pellegrini, V.; Tredicucci, A. Graphene field-effect transistors as room-temperature terahertz detectors. *Nat. Mater.* **2012**, *11*, 865–871. [CrossRef] [PubMed]
4. Kono, S.; Tani, M.; Gu, P.; Sakai, K. Detection of up to 20 THz with a low-temperature-grown GaAs photoconductive antenna gated with 15 fs light pulses. *Appl. Phys. Lett.* **2000**, *77*, 4104–4106. [CrossRef]
5. Zangeneh-Nejad, F.; Safian, R. Significant enhancement in the efficiency of photoconductive antennas using a hybrid graphene molybdenum disulphide structure. *J. Nanophotonics* **2016**, *10*, 036005. [CrossRef]
6. Kim, J.H.; Polley, A.; Ralph, S.E. Efficient photoconductive terahertz source using line excitation. *Opt. Lett.* **2005**, *30*, 2490–2492. [CrossRef]
7. Zangeneh-Nejad, F.; Safian, R. A graphene-based THz ring resonator for label-free sensing. *IEEE Sens. J.* **2016**, *16*, 4338–4344. [CrossRef]
8. Rahm, M.; Li, J.S.; Padilla, W.J. THz wave modulators: A brief review on different modulation techniques. *J. Infrared Millim. Terahertz Waves* **2013**, *34*, 1–27. [CrossRef]
9. Ullmann, I.; Adametz, J.; Oppelt, D.; Vossiek, M. Millimeter Wave Radar Imaging for Non-Destructive Detection of Material Defects. In Proceedings of the AMA Conferences 2017, Nürnberg, Germany, 5 June–1 July 2017; pp. 467–472.
10. Gui, S.; Li, J.; Pi, Y. Security imaging for multi-target screening based on adaptive scene segmentation with terahertz radar. *IEEE Sens. J.* **2018**, *19*, 2675–2684. [CrossRef]
11. Semenov, A.; Richter, H.; Böttger, U.; Hübers, H.W. Imaging terahertz radar for security applications. Terahertz for Military and Security Applications VI. *Int. Soc. Opt. Photonics* **2008**, *6949*, 694902.
12. Lynch, J.J.; Herrault, F.; Kona, K.; Virbila, G.; McGuire, C.; Wetzel, M.; Fung, H.; Prophet, E. Coded aperture subreflector array for high resolution radar imaging. In Proceedings of the Passive and Active Millimeter-Wave Imaging XXII, Anaheim, CA, USA, 11 May 2017; Volume 10189, p. 101890I.
13. Cheng, Y.; Zhou, X.; Xu, X.; Qin, Y.; Wang, H. Radar coincidence imaging with stochastic frequency modulated array. *IEEE J. Sel. Top. Signal Process.* **2016**, *11*, 414–427. [CrossRef]

14. Yin, Z.; Lu, Y.; Xia, T.; Lai, W.; Yang, J.; Lu, H.; Deng, G. Electrically tunable terahertz dual-band metamaterial absorber based on a liquid crystal. *RSC Adv.* **2018**, *8*, 4197–4203. [CrossRef]
15. Hunt, J.; Driscoll, T.; Mrozack, A.; Lipworth, G.; Reynolds, M.; Brady, D.; Smith, D.R. Metamaterial apertures for computational imaging. *Science* **2013**, *339*, 310–313. [CrossRef] [PubMed]
16. Ulander, L.M.; Hellsten, H.; Stenstrom, G. Synthetic-aperture radar processing using fast factorized back-projection. *IEEE Trans. Aerosp. Electron. Syst.* **2003**, *39*, 760–776. [CrossRef]
17. Chan, W.L.; Charan, K.; Takhar, D.; Kelly, K.F.; Baraniuk, R.G.; Mittleman, D.M. A single-pixel terahertz imaging system based on compressed sensing. *Appl. Phys. Lett.* **2008**, *93*, 121105. [CrossRef]
18. Candès, E.J.; Romberg, J.; Tao, T. Robust uncertainty principles: Exact signal reconstruction from highly incomplete frequency information. *IEEE Trans. Inf. Theory* **2006**, *52*, 489–509. [CrossRef]
19. Li, Y.B.; Li, L.L.; Xu, B.B.; Wu, W.; Wu, R.Y.; Wan, X.; Cheng, Q.; Cui, T.J. Transmission-type 2-bit programmable metasurface for single-sensor and single-frequency microwave imaging. *Sci. Rep.* **2016**, *6*, 23731. [CrossRef]
20. Luo, C.G.; Deng, B.; Wang, H.Q.; Qin, Y.L. High-resolution terahertz coded-aperture imaging for near-field three-dimensional target. *Appl. Opt.* **2019**, *58*, 3293–3300. [CrossRef]
21. Xu, G.; Xing, M.; Zhang, L.; Liu, Y.; Li, Y. Bayesian inverse synthetic aperture radar imaging. *IEEE Geosci. Remote. Sens. Lett.* **2011**, *8*, 1150–1154. [CrossRef]
22. Cetin, M.; Stojanovic, I.; Onhon, O.; Varshney, K.; Samadi, S.; Karl, W.C.; Willsky, A.S. Sparsity-driven synthetic aperture radar imaging: Reconstruction, autofocusing, moving targets, and compressed sensing. *IEEE Signal Process. Mag.* **2014**, *31*, 27–40. [CrossRef]
23. Chen, S.; Luo, C.; Deng, B.; Qin, Y.; Wang, H. Study on coding strategies for radar coded-aperture imaging in terahertz band. *J. Electron. Imaging* **2017**, *26*, 053022. [CrossRef]
24. Yuan, Z.; Wang, H. Multiple Scattering Media Imaging via End-to-End Neural Network. *arXiv* **2018**, arXiv:1806.09968.
25. Rivenson, Y.; Zhang, Y.; Günaydın, H.; Teng, D.; Ozcan, A. Phase recovery and holographic image reconstruction using deep learning in neural networks. *Light. Sci. Appl.* **2018**, *7*, 17141. [CrossRef] [PubMed]
26. Sinha, A.; Lee, J.; Li, S.; Barbastathis, G. Lensless computational imaging through deep learning. *Optica* **2017**, *4*, 1117–1125. [CrossRef]
27. Lyu, M.; Wang, W.; Wang, H.; Wang, H.; Li, G.; Chen, N.; Situ, G. Deep-learning-based ghost imaging. *Sci. Rep.* **2017**, *7*, 17865. [CrossRef] [PubMed]
28. He, Y.; Wang, G.; Dong, G.; Zhu, S.; Chen, H.; Zhang, A.; Xu, Z. Ghost Imaging Based on Deep Learning. *Sci. Rep.* **2018**, *8*, 6469. [CrossRef]
29. Gao, J.; Deng, B.; Qin, Y.; Wang, H.; Li, X. Enhanced radar imaging using a complex-valued convolutional neural network. *IEEE Geosci. Remote. Sens. Lett.* **2018**, *16*, 35–39. [CrossRef]
30. Wang, F.; Wang, H.; Wang, H.; Li, G.; Situ, G. Learning from simulation: An end-to-end deep-learning approach for computational ghost imaging. *Opt. Express* **2019**, *27*, 25560–25572. [CrossRef]
31. Antholzer, S.; Haltmeier, M.; Schwab, J. Deep learning for photoacoustic tomography from sparse data. *INverse Probl. Sci. Eng.* **2019**, *27*, 987–1005. [CrossRef]
32. Chen, S.; Luo, C.; Deng, B.; Wang, H.; Cheng, Y.; Zhuang, Z. Three-dimensional terahertz coded-aperture imaging based on single input multiple output technology. *Sensors* **2018**, *18*, 303. [CrossRef]
33. McCann, M.T.; Jin, K.H.; Unser, M. Convolutional neural networks for inverse problems in imaging: A review. *IEEE Signal Process. Mag.* **2017**, *34*, 85–95. [CrossRef]
34. Jin, K.H.; McCann, M.T.; Froustey, E.; Unser, M. Deep convolutional neural network for inverse problems in imaging. *IEEE Trans. Image Process.* **2017**, *26*, 4509–4522. [CrossRef] [PubMed]
35. Ferguson, T.S. An inconsistent maximum likelihood estimate. *J. Am. Stat. Assoc.* **1982**, *77*, 831–834. [CrossRef]
36. Kingma, D.P.; Ba, J. Adam: A method for stochastic optimization. *arXiv* **2014**, arXiv:1412.6980.
37. Peng, L.; Luo, C.; Deng, B.; Wang, H.; Qin, Y.; Chen, S. Phaseless Terahertz Coded-Aperture Imaging Based on Incoherent Detection. *Sensors* **2019**, *19*, 226. [CrossRef] [PubMed]

38. Wipf, D.P.; Rao, B.D. Sparse Bayesian learning for basis selection. *IEEE Trans. Signal Process.* **2004**, *52*, 2153–2164. [CrossRef]
39. Li, C.; Yin, W.; Jiang, H.; Zhang, Y. An efficient augmented Lagrangian method with applications to total variation minimization. *Comput. Optim. Appl.* **2013**, *56*, 507–530. [CrossRef]

Article

Indirect Method for Measuring Absolute Acoustic Nonlinearity Parameter Using Surface Acoustic Waves with a Fully Non-Contact Laser-Ultrasonic Technique

Jihyun Jun [1] and Kyung-Young Jhang [2,*]

[1] Department of Mechanical Convergence Engineering, Graduate School, Hanyang University, Seoul 04763, Korea; jjhhelen@naver.com

[2] School of Mechanical Engineering, Hanyang University, Seoul 04763, Korea

* Correspondence: kyjhang@hanyang.ac.kr; Tel.: +82-2-2220-4220

Received: 23 July 2020; Accepted: 24 August 2020; Published: 26 August 2020

Abstract: This paper proposes an indirect method to measure absolute acoustic nonlinearity parameters using surface acoustic waves by employing a fully non-contact laser-ultrasonic technique. For this purpose, the relationship between the ratio of relative acoustic nonlinearity parameters measured using the proposed method in two different materials (a test material and a reference material) and the ratio of absolute acoustic nonlinearity parameters in these two materials was theoretically derived. Using this relationship, when the absolute nonlinearity parameter of the reference material is known, the absolute nonlinearity parameter of the test material can be obtained using the ratio of the measured relative parameters of the two materials. For experimental verification, aluminum and copper specimens were used as reference and test materials, respectively. The relative acoustic nonlinearity parameters of the two materials were measured from surface waves generated and received using lasers. Additionally, the absolute parameters of aluminum and copper were measured using a conventional direct measurement method, with the former being used as a reference value and the latter being used for comparison with the estimation result. The absolute parameter of copper estimated by the proposed method showed good agreement with the directly measured result.

Keywords: nondestructive evaluation; acoustic nonlinearity parameter; indirect method; laser ultrasound; fully non-contact; surface acoustic wave

1. Introduction

The acoustic nonlinearity parameter (β) is widely used for diagnosing and inferring material damage and it can be measured using displacement amplitudes of fundamental and second-order harmonics waves [1–11]. The exact value of an acoustic nonlinearity parameter is called the absolute acoustic nonlinearity parameter. To measure the absolute acoustic nonlinearity parameter, it is necessary to measure the extremely small displacement amplitudes of the second harmonic frequency component [12]. However, such experimental measurement methods are complicated and practically difficult to apply in the field. Hence, the relative acoustic nonlinearity parameter available for relative comparison is frequently measured using the voltage amplitude of the device, measuring the detected ultrasonic wave. Nevertheless, measurement of the absolute acoustic nonlinearity parameter is indispensable for quantitative characterization of materials.

There are two ways to obtain the absolute acoustic nonlinearity parameter: a direct method and an indirect method. The direct method is used to measure the displacement amplitude of the ultrasonic wave directly or to use a calibration to convert the detected voltage signal amplitude into the displacement amplitude [12]. The indirect method is used to estimate the absolute acoustic nonlinearity parameter of a test material by measuring the ratio of relative acoustic nonlinearity parameters between the test material and the reference material, where the absolute acoustic nonlinearity parameter of the reference material is known [13]. This indirect method, introduced in the aforementioned paper, is simpler than the direct method, but requires the assumption that the voltage-displacement proportionality coefficients (VDCs) of the test and reference materials are the same. The VDC indicates the proportionality coefficient between the detected voltage signal amplitude of ultrasonic wave and the displacement amplitude of that wave, which is dependent on the material properties and the sensitivity of the receiving transducer. If the experimental conditions are kept constant for two similar materials, an indirect method can be applied to them because the VDCs of the two materials are almost the same. On the other hand, if the test and reference materials are dissimilar, an indirect method cannot be applied even if the experimental conditions are kept consistent because the VDCs of the two materials are different. In particular, the material dependency of the VDC is critical in the case of contact detection. For example, in the case of contact reception of ultrasonic waves using piezoelectric transducers, energy loss due to impedance mismatching can occur when ultrasonic energy is converted into electrical energy, so that the VDC varies if the material's acoustic impedance is different. Therefore, to overcome the limitation of the previously proposed indirect method, a non-contact detection method using a laser interferometer has been proposed for longitudinal waves [14]. Since the interferometer obtains an output directly proportional to the ultrasonic displacement, its proportionality is independent of the material.

Meanwhile, many studies have investigated the acoustic nonlinearity parameter of surface acoustic waves [3,4,15–19]. In this regard, an indirect method to measure the acoustic nonlinearity parameter of surface acoustic waves has also been studied for the case of similar test and reference materials [3]. Unfortunately, however, when a surface acoustic wave is transmitted and received using wedges as a contact technique, this method cannot be applied for dissimilar materials because of the constraints mentioned above.

This paper proposes a fully non-contact surface acoustic wave technique using lasers for the indirect measurement of the acoustic nonlinearity parameter. This technique allows the application of the indirect method even when the test and reference materials are dissimilar. Here, a pulsed Nd:YAG laser was used to generate surface acoustic waves, and a laser beam was irradiated onto the specimens through a line-arrayed slit mask to generate tone burst waves. In the contact method, the initial second harmonic frequency coming from the system, i.e., electronic devices and transducers, can pose a problem. However, in the proposed technique the initial second harmonic frequency can be easily suppressed by adjusting the duty ratio of the line-arrayed slit mask [17]. This characteristic is an advantage that may be difficult to obtain in the case of using other non-contact excitations, for example EMAT (electromagnetic acoustic transducer) or ACT (air-coupled transducer). For surface acoustic wave reception, a laser interferometer produced an output signal directly proportional to the out-of-plane displacement of the surface waves and maintained a constant VDC regardless of the used material. Furthermore, we established a measurement principle that applies the proposed non-contact technique to the indirect measurement of the absolute acoustic nonlinearity parameter using surface acoustic waves. This study differs from previous studies on longitudinal waves [14] in that additional compensation is required for the difference in wavenumbers in the test and reference materials.

For experimental verification, aluminum and copper specimens were used as reference and test materials, respectively, and the relative acoustic nonlinearity parameters of the two materials were measured from the surface acoustic waves generated and received using lasers. The absolute acoustic nonlinearity parameter of copper was estimated from the ratio of relative acoustic nonlinearity parameters with the compensation of wavenumbers. Additionally, the absolute parameters of aluminum and copper were measured by longitudinal wave using a conventional direct measurement method, with the former being used as a reference value and the latter being used for comparison with the estimation result.

2. Principles

The acoustic nonlinearity parameter of a surface acoustic wave (β) can be derived in terms of the out-of-plane displacement amplitude for the fundamental and second-order harmonic frequency components of the surface acoustic wave as follows [1,6,18,19],

$$\beta = \frac{8}{k_L^2 x}\frac{A_2}{A_1^2}\frac{k_T^2\sqrt{k_S^2 - k_L^2}}{k_S(2k_S^2 - k_T^2)} \tag{1}$$

Here, A_1 and A_2 are the out-of-plane displacement amplitudes of the fundamental and second-order harmonic components of surface acoustic waves, respectively; x is the wave propagation distance; and k_L, k_T, and k_S are the wavenumbers of longitudinal, transverse, and surface acoustic waves, respectively. If the term consisting only of wavenumbers in Equation (1) is defined as a parameter F as in Equation (2),

$$F = \frac{k_T^2\sqrt{k_S^2 - k_L^2}}{k_S(2k_S^2 - k_T^2)} \tag{2}$$

Equation (1) can then be expressed as Equation (3).

$$\beta = \frac{8}{k_L^2 x}\frac{A_2}{A_1^2}F \tag{3}$$

In contrast, the relative acoustic nonlinearity parameter of surface acoustic wave β'_{SAW} is defined by the voltage amplitudes as follows.

$$\beta'_{SAW} = \frac{A'_2}{A'^2_1} \tag{4}$$

where A_1' and A_2' are the detected signal amplitudes of the fundamental and second-order harmonic components of the surface acoustic wave, respectively.

In a previous study, it has been proved that when a laser interferometer is used as a receiver, the detected signal amplitude is proportional to the displacement amplitude regardless of the difference in materials. Thus, the VDC is not dependent on the material, and the displacement amplitudes of the fundamental and second-order harmonic components can be expressed as follows [14].

$$\begin{aligned} A_{1,t} &= \alpha_1 \cdot A'_{1,t} \\ A_{1,r} &= \alpha_1 \cdot A'_{1,r} \\ A_{2,t} &= \alpha_2 \cdot A'_{2,t} \\ A_{2,r} &= \alpha_2 \cdot A'_{2,r} \end{aligned} \tag{5}$$

Here, the subscripts t and r refer to the test and reference materials, respectively. α_1 and α_2 are the VDCs of the fundamental and second-order harmonic frequencies, respectively. They are only related to the sensitivity of the laser detector, which is dependent on the frequency but not on the material.

Now, in order to apply the indirect method, we consider the relationship between the ratio of the absolute acoustic nonlinearity parameters of the two materials and the ratio of their relative acoustic nonlinearity parameters, as shown in the following equation.

$$\frac{\beta_t}{\beta_r} = \frac{\frac{8}{k_{t,L}^2 x}\frac{\alpha_2 A'_{2,t}}{\alpha_1^2 A'^{\,2}_{1,t}}F_t}{\frac{8}{k_{r,L}^2 x}\frac{\alpha_2 A'_{2,r}}{\alpha_1^2 A'^{\,2}_{1,r}}F_r} = \frac{\frac{F_t}{k_{t,L}^2}\frac{A'_{2,t}}{A'^{\,2}_{1,t}}}{\frac{F_r}{k_{r,L}^2}\frac{A'_{2,r}}{A'^{\,2}_{1,r}}} = k'\frac{\beta'_{SAW,t}}{\beta'_{SAW,r}} \tag{6}$$

The second term of Equation (6) was obtained by substituting Equation (5) into Equation (3) for the test and reference materials. This can be simplified to the third term when the propagation distance x is fixed. The VDCs of the two materials are canceled out. The fourth term is a rearrangement of the third term using Equation (4), where k' is a factor representing the ratio of F values in the two materials and is defined in Equation (7).

$$k' = \frac{F_t \cdot k_{r,L}^2}{F_r \cdot k_{t,L}^2} \tag{7}$$

Finally, the absolute acoustic nonlinearity parameter of the test material can be determined from the relationship in Equation (6).

$$\beta_t = k' \frac{\beta'_{SAW,t}}{\beta'_{SAW,r}} \beta_r \tag{8}$$

Equation (8) indicates that the absolute acoustic nonlinearity parameter of the test material can be estimated from the ratio of the relative acoustic nonlinearity parameters of the test and reference materials. The absolute acoustic nonlinearity parameter of the reference material as well as the wavenumber-dependent factor k' (referred to as the wavenumber compensation factor in this paper) should be known in advance. If the test material and the reference material are similar, k' is almost one and thus it can be ignored. However, if they are different, the wavenumbers in the two materials are different and they should be taken into account.

3. Specimens

To verify the proposed method, two kinds of specimens with different materials, pure copper and aluminum (Al2024), were prepared as shown in Figure 1. The dimensions of both the specimens were same at 120 mm × 40 mm × 20 mm. Aluminum was used as the reference material and copper was used as the test material.

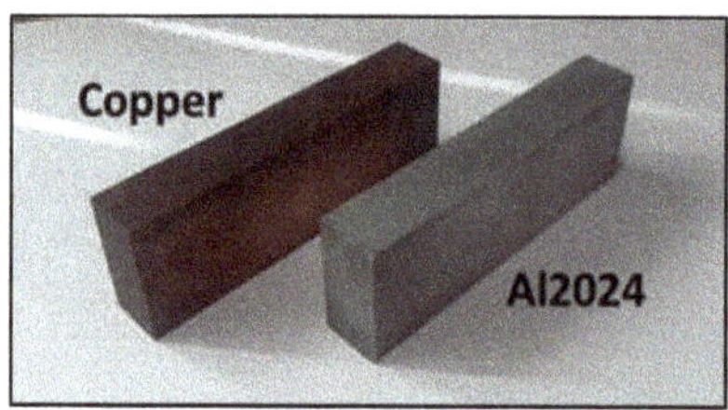

Figure 1. Specimens.

4. Experiments

4.1. Experimental Setup

The experimental scheme for the relative acoustic nonlinearity measurements using the fully non-contact surface acoustic wave method is shown in Figure 2. A pulsed Nd:YAG laser beam of 1064 nm (SL280, Spectron Laser Systems, Warwickshire, UK) was used to excite surface acoustic waves. The pulse duration of the pulsed laser is 10 ns, the repetition rate is 10 Hz, and the maximum energy is 300 mJ. A line-arrayed slit mask was designed to create surface acoustic waves with a wavelength of 2.92 mm, which corresponds to the fundamental frequency of 1 MHz in aluminum and 0.75 MHz in copper. The duty ratio of the slit mask was 50%, which theoretically does not generate a second harmonic component [12].

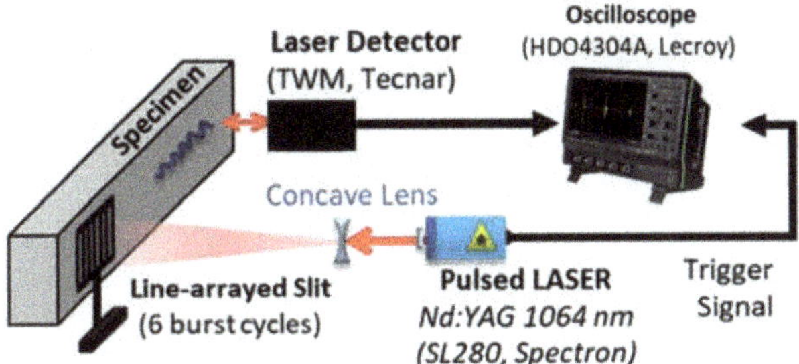

Figure 2. Experimental setup to generate and receive surface acoustic waves using the fully non-contact laser-ultrasonic technique.

The wave propagated on the specimens was received at the other point on the same side of the specimen by a photorefractive interferometer (TWM, Tecnar, Saint-Bruno-de-Montarville, QC, Canada). In general, when detecting the displacement of the surface wave, the diameter of the measuring laser beam should be smaller than the wavelength to avoid a decrease in signal sensitivity. The beam diameter of the interferometer used in the experiment was 0.8 mm, which is sufficiently small compared to the wavelength of the excited surface acoustic wave. The propagation distance was 65 mm. If the propagation distance is too long, the effects of attenuation and diffraction cannot be ignored. The received surface acoustic wave signal was captured by a digital oscilloscope (Lecroy HDO4034A, Teledyne LeCroy, Chestnut Ridge, NY, USA).

The Hanning window was applied to minimize the effect of the side lobe, and the fast Fourier transform was used to obtain the amplitudes of the fundamental frequency component A_1' and the second harmonic frequency component A_2'. The Hanning window size is the same as the data length. Sampling frequency was 10 GHz, and the number of data points was 10,000. As a result, the frequency resolution in FFT (Fast fourier transform) spectrum is 0.001 MHz. In order to obtain the magnitudes at the fundamental and second harmonic frequencies, we searched for peaks within the ±0.1 MHz range at each frequency. The frequency of the peak detected in the experimental results was within the range of ±0.02 MHz from the predicted frequency. The measurements were repeated by increasing the laser energy from 95 to 145 mJ in seven steps. The relative nonlinearity parameter β'_{SAW} was determined from the linearity slope of between $A_1'^2$ and A_2'.

4.2. Measurement of Relative Nonlinearity Parameter

The received surface acoustic wave signals, their frequency spectra, and the linear fitting plots of $A_1'^2$ and A_2' are shown in Figure 3 for aluminum and copper specimens. Figure 3a–c shows the results of the aluminum specimen and Figure 3d–f shows the results of the copper specimen. The surface acoustic wave signals shown in Figure 3a–d were obtained at the highest laser intensity, and Figure 3b–e shows their frequency spectra. The magnitudes of the fundamental and second-order harmonic frequency components are indicated by red dots. The fundamental frequency in aluminum was detected at 1.0 MHz and the second harmonic frequency was detected at 2.0 MHz, as intended. In copper, the fundamental frequency and second harmonic frequency components were detected at 0.74 MHz and 1.47 MHz, as expected.

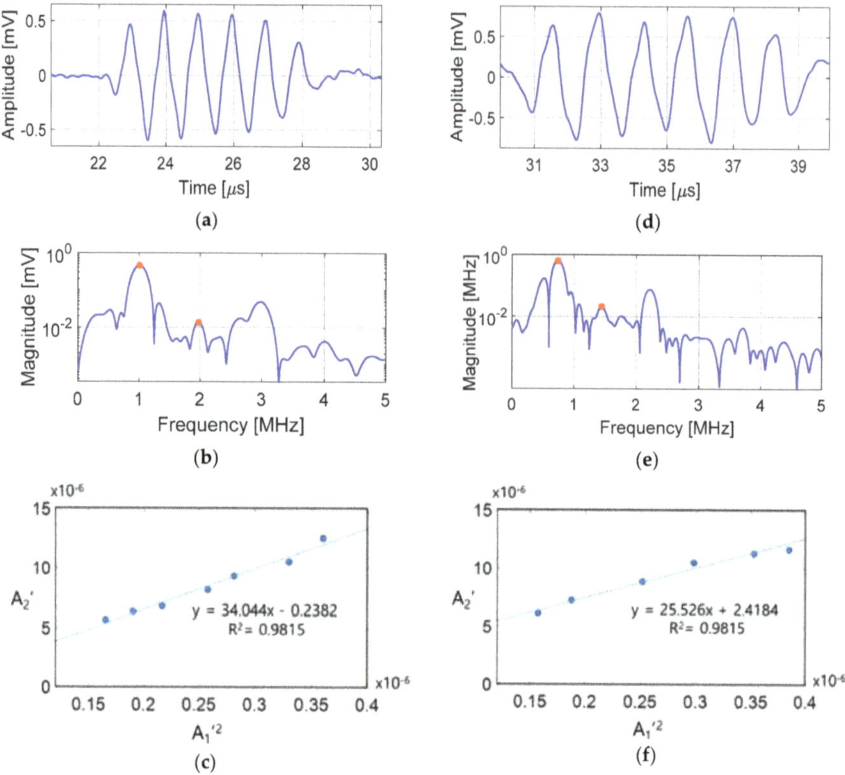

Figure 3. Experimental results: (**a–c**) represent Al2024, and (**d–f**) represent copper. (**a**,**d**): Raw signals, (**b**,**e**): frequency spectra of raw signals, (**c**,**f**): linear fitting of $A_1{'}^2$ and $A_2{'}$.

Note that a relatively large third harmonic is generated, because the surface acoustic wave close to the square wave form is generated by the line-arrayed slit mask. In this case, even-order harmonics are suppressed; however, the occurrence of odd-order harmonics is unavoidable [20]. When the third harmonic frequency is put in together with the fundamental frequency, the magnitude of the second harmonic frequency may change owing to the mixing effect. Nevertheless, as the magnitude of odd-order harmonics depends on the fundamental frequency, only the constant '8' in Equation (1) will vary. However, because this constant will be canceled in the calculation of the relative ratio in Equation (6), Equation (8) is valid as it is and does not affect the proposed measurement technique based on the relative ratio.

The linear relationship between $A_2{'}$ and $A_1{'}^2$ is shown in Figure 3c,f. In both specimens, the R-squared values are approximately 0.98, which confirms the linearity between $A_2{'}$ and $A_1{'}^2$. The measurements were repeated four times at each laser power, the deviation was so small that it cannot be seen in the figure.

Table 1 shows the relative nonlinearity parameter measurement results of each specimen and their ratio.

Table 1. Measurement values for the relative nonlinearity parameter of Al2024 and copper.

Material	Relative Nonlinearity Parameter, β'_{SAW}	$\beta'_{SAW, Copper}/\beta'_{SAW, Al2024}$
Al2024	25.71 ± 0.60	1.333 ± 0.119
Copper	34.27 ± 2.20	

4.3. Measurement of Wavenumber Compensation Factor k'

The ultrasonic velocity was measured for each specimen to determine the wavenumber of the longitudinal, transverse, and surface acoustic waves required to calculate the wavenumber compensation factor. The velocities of the longitudinal and transverse waves were obtained by measuring the time-of-flight (TOF) between the back-wall echo signals.

Pulser-receiver (Olympus 5077PR, Tokyo, Japan) and PZT transducers with main-resonance frequencies of 5.0 MHz for longitudinal waves and 2.25 MHz for transverse waves were used in the experiment. Figure 4 shows the experimental setup for wave velocity measurement. The TOF between the echo signals was measured using the auto-correlation of the received signal [21]. By using the measured TOF and thickness of the specimen, the velocities of the longitudinal and transverse waves were calculated for each specimen.

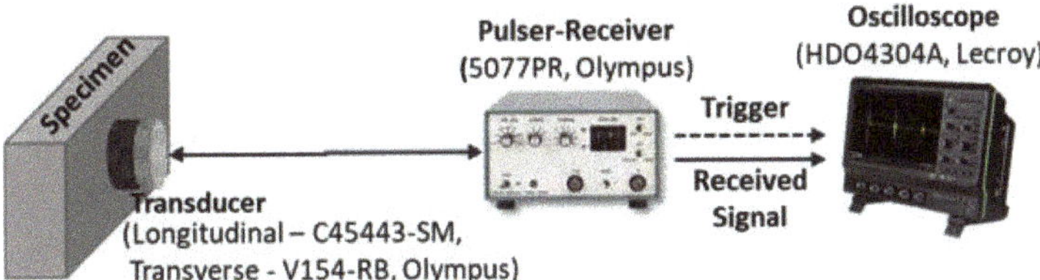

Figure 4. Experimental setup for measuring the velocities of longitudinal and transverse waves.

The surface wave velocity was measured using the same apparatus shown in Figure 2. Velocity of the surface acoustic wave is measured by non-contact method using a laser. The velocity was obtained by multiplying the fundamental frequency of the received signal by the wavelength given in the line array interval [17].

The measured longitudinal, transverse, and surface acoustic wave velocities are shown in Table 2. The F values of each material were calculated using Equation (2), and the wavenumber compensation factor k' was obtained using Equation (7).

Table 2. Measurement results for wave velocities and wavenumber compensation factor.

Material	Longitudinal Wave (m/s)	Transverse Wave (m/s)	Surface Acoustic Wave (m/s)	F	k'
Al2024	6403.2 ± 2.4	3120.9 ± 2.1	2935.0 ± 3.8	3.73 ± 0.080	0.501 ± 0.018
Copper	4373.0 ± 1.3	2266.8 ± 1.2	2150.8 ± 4.7	4.00 ± 0.001	

4.4. Estimation of Absolute Nonlinearity Parameter

In order to verify the validity of the proposed method, the absolute nonlinearity parameters of the two materials were measured first using the conventional calibration method [12], which uses a pre-measured calibration function converting the electrical output of the receiving transducer into the displacement amplitude. The results are shown in Table 3. The absolute parameter of aluminum was used as a reference value and the absolute parameter of copper was used for comparison with the result estimated by the proposed technique.

Table 3. Absolute acoustic nonlinearity parameters of Al2024 and copper measured by the direct method.

Material	Absolute Nonlinearity Parameter (β)	$\beta_{Copper}/\beta_{Al2024}$
Al2024	6.93 ± 0.30	0.671 ± 0.068
Copper	4.65 ± 0.25	

Next, the absolute nonlinearity parameter of copper was estimated by substituting the measured wavenumber compensation factor k' and the relative nonlinearity parameters of Al2024 and copper in Equation (8), in which the absolute nonlinearity parameter of Al2024 shown in Table 3 was used as the reference. The result is shown in Table 4. The absolute nonlinearity parameter measured using the direct method is shown for comparison. Considering the deviation, the estimated value agrees well with the direct measurement result.

Table 4. Absolute nonlinearity parameter of copper measured by the proposed method and the direct measurement method.

	Proposed Method	Direct Method (Calibration)
Copper	4.62 ± 0.24	4.65 ± 0.25

The results verify that the proposed indirect method using surface acoustic waves with a fully non-contact laser-ultrasonic technique is effective for estimating acoustic nonlinearity parameters.

5. Conclusions

This paper proposed a novel indirect method to measure the absolute acoustic nonlinearity parameter using surface acoustic waves with a fully non-contact laser-ultrasonic technique. The relationship between the ratio of relative nonlinearity parameters of two different materials (a test material and a reference material) measured using the proposed method and the ratio of absolute nonlinearity parameters of these two materials was theoretically derived. Using this relationship, when the absolute nonlinearity parameter of the reference material is known, the absolute nonlinearity parameter of the test material can be obtained from the ratio of the measured relative parameters of the two materials. The effectiveness of the proposed method was verified using the experimental results; the absolute nonlinearity parameter of copper measured by the proposed method was in good agreement with that obtained from direct measurement, in which aluminum (Al2024) was used as the reference material. The proposed technique avoids the inconvenience of direct measurement, maintains the advantage of surface waves, and can be applied even when the test material and the reference material are dissimilar, which is difficult to investigate using conventional contact techniques.

Author Contributions: Conceptualization, J.J. and K.-Y.J.; methodology, J.J.; software, J.J.; validation, J.J. and K.-Y.J.; formal analysis, J.J.; investigation, J.J.; resources, K.-Y.J.; data curation, J.J.; writing—original draft preparation, J.J.; writing—review and editing, K.-Y.J.; visualization, J.J.; supervision, K.-Y.J.; project administration, K.-Y.J.; funding acquisition, K.-Y.J. All authors have read and agreed to the published version of the manuscript.

Funding: This research was supported by the Nuclear Power Research and Development Program through the National Research Foundation of Korea (NRF) funded by the Ministry of Science, ICT and Future Planning (NRF-2013M2A2A9043241).

Conflicts of Interest: The authors declare no conflict of interest.

References

1. Thiele, S.; Kim, J.Y.; Qu, J.; Jacobs, L.J. Air-coupled detection of nonlinear Rayleigh surface waves to assess material nonlinearity. *Ultrasonics* **2014**, *54*, 1470–1475. [CrossRef] [PubMed]
2. Deng, M.X.; Pei, J.F. Assessment of accumulated fatigue damage in solid plates using nonlinear Lamb wave approach. *Appl. Phys. Lett.* **2007**, *90*, 121902. [CrossRef]
3. Herrmann, J.; Kim, J.Y.; Jacobs, L.J.; Qu, J.M.; Littles, J.W.; Savage, M.F. Assessment of material damage in a nickel-base superalloy using nonlinear Rayleigh surface waves. *J. Appl. Phys.* **2006**, *99*, 124913. [CrossRef]
4. Seo, H.; Jun, J.; Jhang, K.Y. Assessment of Thermal Aging of Aluminum Alloy by Acoustic Nonlinearity Measurement of Surface Acoustic Waves. *Res. Nondestruct. Eval.* **2017**, *28*, 3–17. [CrossRef]
5. Xiang, Y.; Zhu, W.; Liu, C.-J.; Xuan, F.-Z.; Wang, Y.-N.; Kuang, W.-C. Creep degradation characterization of titanium alloy using nonlinear ultrasonic technique. *NDT E Int.* **2015**, *72*, 41–49. [CrossRef]

6. Kim, J.Y.; Jacobs, L.J.; Qu, J.M.; Littles, J.W. Experimental characterization of fatigue damage in a nickel-base superalloy using nonlinear ultrasonic waves. *J. Acoust. Soc. Am.* **2006**, *120*, 1266–1273. [CrossRef]
7. Liu, M.; Kim, J.-Y.; Jacobs, L.; Qu, J. Experimental study of nonlinear Rayleigh wave propagation in shot-peened aluminum plates—Feasibility of measuring residual stress. *NDT E Int.* **2011**, *44*, 67–74. [CrossRef]
8. Nagy, P.B. Fatigue damage assessment by nonlinear ultrasonic materials characterization. *Ultrasonics* **1998**, *36*, 375–381. [CrossRef]
9. Walker, S.V.; Kim, J.Y.; Qu, J.M.; Jacobs, L.J. Fatigue damage evaluation in A36 steel using nonlinear Rayleigh surface waves. *NDT E Int.* **2012**, *48*, 10–15. [CrossRef]
10. Kim, J.-Y.; Jacobs, L.; Qu, J. Nonlinear Ultrasonic Techniques for Material Characterization. In *Nonlinear Ultrasonic and Vibro-Acoustical Techniques for Nondestructive Evaluation*; Springer: Berlin/Heidelberg, Germany, 2019; pp. 225–261.
11. Xiang, Y.; Deng, M.; Xuan, F.-Z. Thermal degradation evaluation of HP40Nb alloy steel after long term service using a nonlinear ultrasonic technique. *J. Nondestruct. Eval.* **2014**, *33*, 279–287. [CrossRef]
12. Kim, J.; Song, D.G.; Jhang, K.Y. Absolute Measurement and Relative Measurement of Ultrasonic Nonlinear Parameters. *Res. Nondestruct. Eval.* **2017**, *28*, 211–225. [CrossRef]
13. Ren, G.; Kim, J.; Jhang, K.Y. Relationship between second- and third-order acoustic nonlinear parameters in relative measurement. *Ultrasonics* **2015**, *56*, 539–544. [CrossRef] [PubMed]
14. Park, S.H.; Kim, J.; Jhang, K.Y. Relative measurement of the acoustic nonlinearity parameter using laser detection of an ultrasonic wave. *Int. J. Precis. Eng. Manuf.* **2017**, *18*, 1347–1352. [CrossRef]
15. Gross, J.; Kim, J.-Y.; Jacobs, L.; Kurtis, K.; Qu, J. Evaluation of near surface material degradation in concrete using nonlinear Rayleigh surface waves. In *AIP Conference Proceedings*; American Institute of Physics: College Park, MD, USA, 2013; pp. 1309–1316.
16. Guo, S.F.; Zhang, L.; Mirshekarloo, M.S.; Chen, S.T.; Chen, Y.F.; Wong, Z.Z.; Shen, Z.; Liu, H.; Yao, K. Method and analysis for determining yielding of titanium alloy with nonlinear Rayleigh surface waves. *Mater. Sci. Eng. A* **2016**, *669*, 41–47. [CrossRef]
17. Choi, S.; Seo, H.; Jhang, K.Y. Noncontact Evaluation of Acoustic Nonlinearity of a Laser-Generated Surface Wave in a Plastically Deformed Aluminum Alloy. *Res. Nondestruct. Eval.* **2015**, *26*, 13–22. [CrossRef]
18. Jun, J.; Seo, H.; Jhang, K.-Y. Nondestructive Evaluation of Thermal Aging in Al6061 Alloy by Measuring Acoustic Nonlinearity of Laser-Generated Surface Acoustic Waves. *Metals* **2020**, *10*, 38. [CrossRef]
19. Matlack, K.H.; Kim, J.Y.; Jacobs, L.J.; Qu, J. Review of Second Harmonic Generation Measurement Techniques for Material State Determination in Metals. *J. Nondestruct. Eval.* **2015**, *34*, 273. [CrossRef]
20. Choi, S.; Nam, T.; Jhang, K.-Y.; Kim, C.S. Frequency response of narrowband surface waves generated by laser beams spatially modulated with a line-arrayed slit mask. *J. Korean Phys. Soc.* **2012**, *60*, 26–30. [CrossRef]
21. Cutard, T.; Fargeot, D.; Gault, C.; Huger, M. Time-Delay and Phase-Shift Measurements for Ultrasonic Pulses Using Autocorrelation Methods. *J. Appl. Phys.* **1994**, *75*, 1909–1913. [CrossRef]

Article

Proposal of UWB-PPM with Additional Time Shift for Positioning Technique in Nondestructive Environments

Nguyen Thi Huyen [1], Nguyen Le Cuong [2] and Pham Thanh Hiep [1,*]

1 Faculty of Radio-Electronic Engineering, Le Quy Don Technical University, 236 Hoang Quoc Viet, Ha noi 100000, Vietnam; nguyenhuyenhvktqs@gmail.com

2 Faculty of Electronics and Telecommunications, Electric Power University, 235 Hoang Quoc Viet, Ha noi 100000, Vietnam; cuongnl@epu.edu.vn

* Correspondence: phamthanhhiep@gmail.com; Tel.: +84-982-535-203

Received: 20 July 2020; Accepted: 27 August 2020; Published: 30 August 2020

Abstract: The ultra-wide band (UWB) technology has many advantages in positioning and measuring systems; however, powers of UWB signals rapidly reduce while traveling in propagation environments, hence detecting UWB signals are difficult. Various modulation techniques are applied for UWB signals to increase the ability for detecting the reflected signal from transmission mediums, such as pulse amplitude modulation (PAM), pulse position modulation (PPM), and so on. In this paper, we propose an ultra-wide band pulse position modulation technique with optimized additional time shift (UWB-PPM-ATS) to enhance the accuracy in locating buried object in nondestructive environments. Moreover, the Levenberg–Marquardt Fletcher algorithm (LMFA) is applied to determine the medium parameters and buried object location simultaneously. The influences of proposed modulation technique on determining system's parameters, such as a propagation time, distance, and properties of the medium are analyzed. Calculation results indicate that the proposed UWB-PPM-ATS gives higher accuracy than the conventional one such as UWB-OOK and UWB-PPM in both homogeneous and heterogeneous environments. Furthermore, the LMFA with the proposed UWB-PPM-ATS outperforms the LMFA with the traditional modulation method, especially for unknown propagation environment.

Keywords: UWB-PPM; UWB-OOK; buried objects; nondestructive environment; Levenberg–Marquardt method

---

1. Introduction

With ultra-wide bandwidth, the ultra-wide band (UWB) signal is considered as an ideal locating technique in a short-range with high spatial resolution. As defined by the Federal Communications Commission (FCC), UWB technology has a center frequency higher than 2.5 GHz, or if less than 2.5 GHz, there must be a minimum bandwidth ratio of 0.2 [1], or the minimum bandwidth must reach 500 MHz [2]. To avoid affecting other narrow band systems, the rules of FCC allow the effective isotropic radiated power (EIRP) level of UWB devices to be lower than −41.3 dBm/MHz in the frequencies range of 3.1 to 10.6 GHz [3], so the UWB devices can work for more extended periods than narrow band systems with the same battery power, and due to the use of very narrow pulses, UWB signals are better able to penetrate in the nondestructive environments. General modulation techniques are used for UWB signals such as pulse amplitude modulations (PAMs), On-Off Keying (OOK), and pulse position modulation (PPM) [4]. One can use time-hopping (TH) in UWB systems to create TH-PPM, TH-BPSK signal types [5], or design a generator circuit which generates the 4-th and 5-th order derivative of Gaussian pulses in TH-QPSK system applied to multipath channels [6].

Each modulation technique has a different application range. The choice of the right modulation configuration not only increases the efficiency of system implementation but also maximizes the benefits of ultra-wide bandwidth and reduces the complexity of device hardware. In [7], a simple peak detection based on noncoherent UWB receiver is proposed for low data rate wireless sensor networks (WSN) and Internet of things (IoT) applications. It has improved receiver performance with TH-PPM UWB signal. In [8], to reduce the complexity of the TH-UWB receiver, a channel shortening equalizer design method is proposed based on an eigen filter using a new objective function, whereby the proposed system has dramatically reduced the power of channel impulse response, spectral distortion, multiaccess interference, and noise power. Therefore, different UWB signal modulation types have affected the quality and application of the UWB system.

In those modulation techniques, the PPM technique is one of the widely used configurations in UWB systems. Studies on UWB-PPM in wireless communication networks mainly focus on solutions to reduce the conflicts in multiuser access systems; for example, [9] proposed an M-ary PPM modulation configuration for the UWB (M-PPM) system and indicated that the proposed system significantly improved performance compared to systems using direct spreading sequences in the environments with a low signal to noise (SNR) ratio. In [10], Vinod Venkatesan et al. proposed the application of a direct spreading sequence with the optimized UWB-PPM technique for multiaccess systems. The proposed method reduced the impact of multiaccess interference (MAI) and significantly reduced the floor error compared to the orthogonal signal configuration at a large SNR ratio. Besides, there are several studies on improving the quality of the receiver for UWB-PPM signals [11], determining the optimal integration time for the energy detector of the UWB-PPM system [12], and developing a measurement matrix combined with randomly Fourier transform converters for UWB-PPM signals [13]. The combination of PPM symbols and time of arrival (TOA) estimation algorithm using the Sub-Nyquist IR-UWB signal in the IR-UWB device is discussed in [14]. Turbo codes for PPM-IR UWB signals to improve the power spectral density (PSD) power signal density [15] and randomizing the pulses to improve the UWB system [16] were proposed. The noncoherent modulation techniques based on the use of the receiver adaptive thresholds applied to enhanced PPM in the IR-UWB and the direct chaotic communication UWB (DCC-UWB) systems were proposed to improve the bit error rate (BER) performance of the system in a multipath transmission environment [17].

For testing purposes, material penetrating systems using UWB technology to examine nondestructive environments are discussed in [18]; the result indicated that this system can detect imperfect structures of metal. Besides that, the estimation of the layer's thickness based on the processing of the GPR's data with the optimized techniques (such as neural networks) is discussed in [19,20]. When using UWB technology in the testing, positioning, or another application in communication to improve the resolution of the systems, one of the main problems is choosing the appropriate modulation technique combined with the receiver's signal processing methods. The selection of a modulation scheme based on determination distance was discussed in [21]. From those results, we can recognize that the correct detection of UWB pulse signals is one of the essential factors which affect the accuracy of the distance estimation technique. In [22], an UWB indoor positioning system is presented to exploit two-way flight time to calculate range measurements to determine the transceiver location based on Pozyz inner algorithm with a range accuracy of 320 ± 30 mm. In [23], to locate underground personnel in coal mines, an UWB wireless sensor network and time difference of arrival (TDOA) algorithm was proposed, and this system can achieve high-precision positioning in real-time. Furthermore, in the nondestructive environment, direct sequence ultra-wide band (DS-UWB) transmission system with an adaptive pseudo random sequence length is proposed in [24] to reduce processing time and increase positioning accuracy. As mentioned above, the PPM modulation is widely applied to the UWB system, especially for detecting the location of objects; however, the accuracy of estimated results is still low. In this paper, we focus on proposing a new modulation scheme based on PPM modulation to improve the resolution of the estimated distance. The main contributions of this research are listed as follows.

- Firstly, the UWB-PPM signal which is used in determining the distance under the homogeneous and heterogeneous propagation conditions is mathematically analyzed.
- Secondly, based on the analysis of received signal, the delay time and then the propagation distances are estimated.
- Thirdly, to increase the accuracy of the estimated distance, an enhanced UWB-PPM modulation technique, called ultra-wide band pulse position modulation with additional time shift (UWB-PPM-ATS) is proposed and compared with other techniques such as conventional UWB-PPM, UWB- OOK with the properties of transmission environment is known.
- Finally, the Levenberg–Marquardt nonlinear estimation algorithm is applied to estimate the system parameter and the target when the propagation environment is unknown.

The remainder of this paper is organized as follows. Section 2 describes the system model for estimating distances in nondestructive environments. The proposed system model and the parameters are presented in Section 3. The simulation results are provided in Section 4, and finally, conclusions and further work are discussed in Section 5.

2. System Model

A distance measurement and positioning system using UWB technology is illustrated in Figure 1, where d_i and ε_i are the thickness and the relative permittivity of the ith layer in the nondestructive environment, respectively. $s(t)$ denotes the transmitted pulse signals, and $r(t)$ is the received signal.

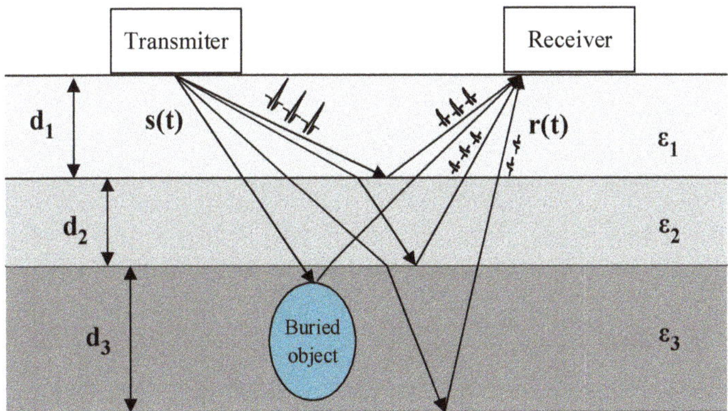

Figure 1. The distance measurement and positioning system in the nondestructive environments.

For the environment with three layers and one buried object, the $r(t)$ is represented by

$$\begin{aligned} r(t) &= \sum r_i(t) + r_{ob}(t) + n(t), \\ r_1(t) &= A_1 s(t-\tau_1), \\ r_2(t) &= A_1 A_2 s(t-\tau_1-\tau_2), \\ r_3(t) &= A_1 A_2 A_3 s(t-\tau_1-\tau_2-\tau_3), \\ r_{ob}(t) &= A_{ob} s(t-\tau_{ob}), \end{aligned} \quad (1)$$

where A_i denotes the amplitude factor which represents the reflection and transmission properties of the propagation environment, and its value depends on the properties of the ith layers, τ_i is the traveling time of UWB pulse in the ith layer. A_{ob} and τ_{ob} are respectively the amplitude and traveling time of the signal which is reflected from the buried object. $n(t)$ represents the additive white Gaussian noise.

At the receiver side, the procedure for determining the position of buried object is indicated in Figure 2. Accordingly, the position of buried object is determined by the distance from it to the transceiver. The LMFA method is applied to calculate the relative permittivity of environment and these distances.

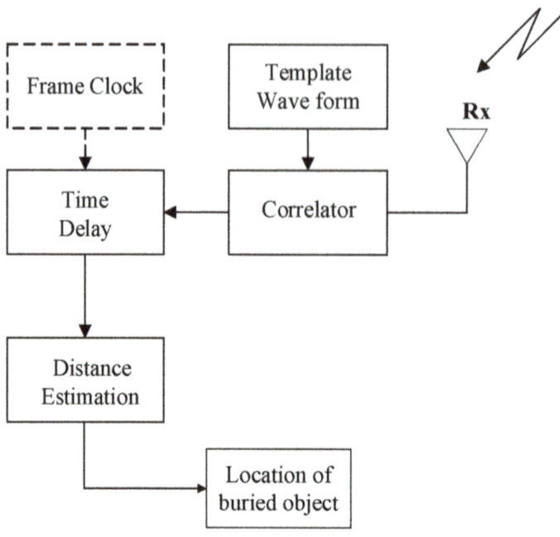

Figure 2. The procedure for determining the position of buried object.

In the UWB system, $s(t)$ is generated based on Gaussian [25] or Hermite [26] functions and their derivatives. Compared to Gaussian pulses, the Hermite pulse is useful for parallel data transmission with high data rates, but it is hard to achieve in the real world [27]. Therefore, the Gaussian pulses UWB signals are applied in this work. A typical Gaussian pulse usually takes the form [25]:

$$g(t) = A_p e^{-2\pi(\frac{t}{\mu_p})^2}, \tag{2}$$

where A_p denotes the amplitude of pulse, μ_p is a factor which influences the amplitude and the width of Gaussian pulse, also called time normalization factor, the width of a pulse becomes narrower when the μ_p is reduced. The nth derivative of Gaussian pulse, named n^{th}-order monocycle, is

$$g_n(t) = A_{np} \frac{d^n}{dt^n} e^{-2\pi(\frac{t}{\mu_p})^2}, \tag{3}$$

where A_{np} is the amplitude of the nth-order monocycle. The shapes of different types of Gaussian monocycles are indicated in Figure 3.

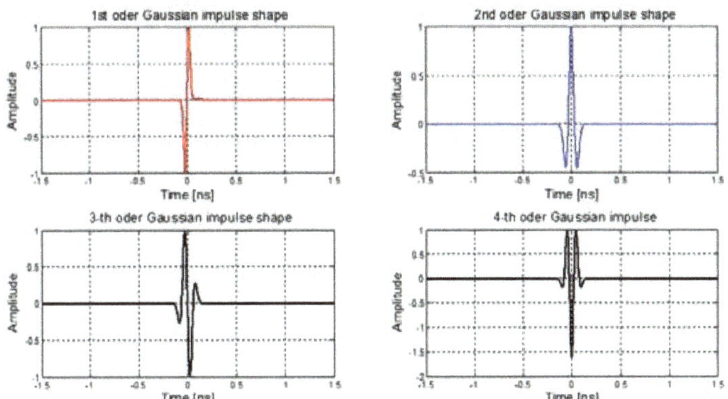

Figure 3. The different types of Gaussian monocycles.

The selected Gaussian pulse shape must meet FCC requirements about power and frequency range used. According to Figure 4, it is observed that for the power spectral density (PSD), only the first-order Gaussian monocycle does not comply with the FCC requirements about the effective isotropic radiated power (EIRP) level. In this paper, we restrict our performance analysis of UWB system to the fourth-order Gaussian monocycle in (4). However, our method can be applied for any Gaussian pulse shape.

$$g_4(t) = A_p\left[-12\pi + 96\pi^2\left(\frac{t}{\mu_p}\right)^2 - 64\pi^3\left(\frac{t}{\mu_p}\right)^4\right] e^{-2\pi.\left(\frac{t}{\mu_p}\right)^2}. \tag{4}$$

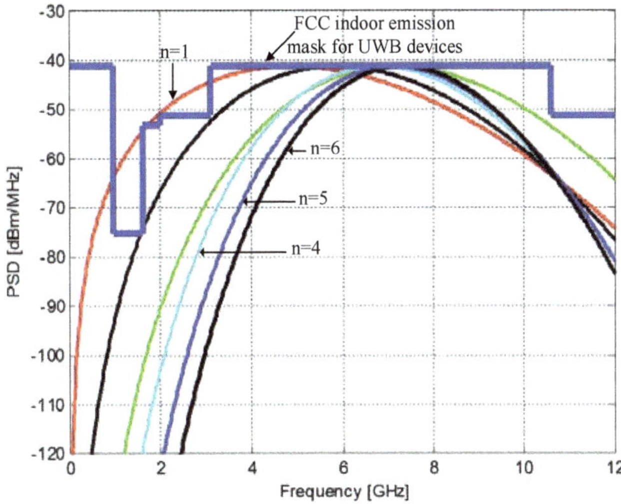

Figure 4. The power spectral density (PSD) of the different derivatives of Gaussian pulses.

It is observed that with impulse radio UWB (IR-UWB) signal, the correlation and therefore the propagation time can be calculated. However, the IR-UWB signal is not strong enough to be processed at the receiver in the case of multiple layers medium (the value of cross-correlation is lower than the noise lever). Hence, an orthogonal pseudo sequence $a(n)$ is applied to the UWB signal in pulse position modulation (PPM) to ensure the cross-correlation is obtained when receiving many reflected signals from many surfaces and buried objects. The conventional UWB-PPM signal with the fourth-order Gaussian monocycle is given as:

$$s(t) = \sum_{n=1}^{N} g_4(t - nT_r - a_n T_{\text{PPM}}), \tag{5}$$

where T_r is pulse repetition period and T_{PPM} is the time shift associated with binary PPM, $a_n \in \{0, 1\}$, and N is the length of UWB pulses. In this modulation technique, when a_n is 0, there is no additional time shift that modulates the pulse $g_4(t)$ and a time shift T_{PPM} is added to $g_4(t)$ when a_n is 1. In our system model, a_n does not carry information, it only indicates whether the corresponding pulse is shifted or not.

The propagation distance d in the UWB-PPM system used for positioning the buried object in nondestructive medium is computed by:

$$d = \frac{c\tau}{2\sqrt{\varepsilon}}, \tag{6}$$

where c is the velocity of light, and τ, ε are the traveling time of UWB pulses and the relative permittivity of the nondestructive medium, respectively.

The traveling time τ is calculated by the correlation between the received signal from the ith layer, $r_i(t)$, and the template waveform $\omega(t)$:

$$R_i(\tau) = \int_{-\infty}^{\infty} r_i(t)\omega(t)dt. \tag{7}$$

With the conventional UWB-PPM systems, the template waveform at the receiver is $(g_4(t) - g_4(t - T_{\text{PPM}}))$ and the correlation of this system is denoted by R_{0i} and has the form:

$$R_{0i}(\tau) = \int_{-\infty}^{\infty} r_i(t)[g_4(t) - g_4(t - T_{\text{PPM}})]dt. \tag{8}$$

Define the autocorrelation function of the fourth-order Gaussian monocycle in (4) as $R_{G4}(\tau)$, we have

$$R_{G4}(\tau) = \int_{-\infty}^{\infty} g_4(t)g_4(t - \tau)dt, \tag{9}$$

$$R_{0i}(\tau) = R_{G4}(\tau) - R_{G4}(\tau - T_{\text{PPM}}). \tag{10}$$

The traveling time of UWB signal is the value that makes the correlation function hit its maximum; it corresponds to twice the transmission time from the transceiver to the reflective surface of the ith layer as shown in Figure 5 and given as

$$\tau = arg_{\tau}(max(R_i)), \tag{11}$$

where $max(R_i)$ is the local extreme value corresponding to the reflected signals from the surface of the layers or object.

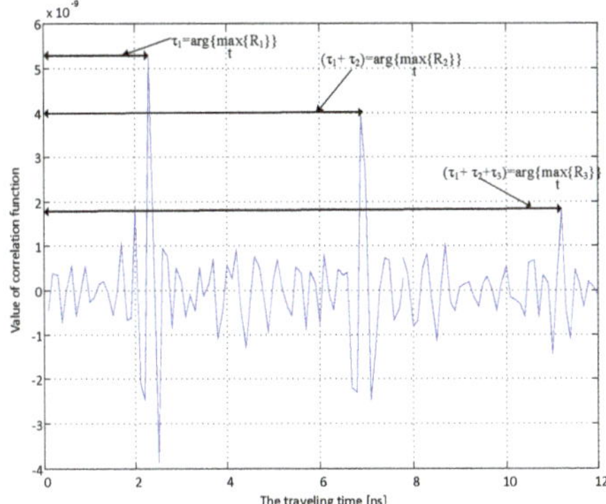

Figure 5. The local extreme values of correlation output corresponding to propagation distances in Figure 1.

According to (6), τ is a function of ε and the distance d, due to both d and ε are unknown; they should be estimated by large enough values of τ using LMFA. To increase the accuracy in the τ determination procedure, and in estimation of d, ε, the proposed UWB-PPM-ATS technique is applied.

3. Proposal of Positioning Approach in Nondestructive Environments

3.1. Proposal of UWB-PPM-ATS

From (1), (5), and (6), it can be seen that the time shift T_{PPM} directly affects the quality of UWB-PPM systems in positioning applications. The performance of the system can be improved by selecting the appropriate value of the time shift. To increase the accuracy in the estimated distance and the position of buried objects by using the UWB-PPM system, the time shift in this system should be selected so that the ability to detect the received UWB pulse is the best. To accomplish this task, in UWB-PPM systems with the time shift T_{PPM} is invariant, we recommend adjusting this time shift with a certain time constant to achieve its optimal value. The optimal value of T_{PPM} defined in this paper is the value at which the UWB-PPM system gives the smallest error in distance estimation. So we propose an UWB pulse position modulation with an additional time shift (UWB-PPM-ATS). In the proposed technique, the pulse position will be changed with a time constant denoted by ζ. The signal of UWB-PPM-ATS is given as:

$$s(t) = \sum_{n=1}^{N} g_4(t - nT_r - a_n.(T_{PPM} + \zeta)). \tag{12}$$

The "+" sign in (12) means that algebraic additions, so ζ, can take either positive or negative values. The effect of ζ on the quality of the system will be evaluated later.

An example of the conventional UWB, UWB-PPM, and UWB-PPM-ATS signal shapes are illustrated in Figure 6. In Figure 6, there is an example of a seven pulses sequence carrying seven bits 1010101. The pulse width is 0.28 ns, and it adopts the fourth derivative of the Gaussian function described by the blue line; the conventional PPM modulated pulses sequence is denoted by the red line. Whereby, when the bit value is 0, the transmitter will send out a pulse $g_{40}(t)$ as the original pulse (without any change from the unmodulated pulse) when the value bit is 1, the transmitter will send out a pulse $g_{41}(t)$ which is the original pulse with a time shift of T_{PPM}. With the proposed UWB-PPM-ATS (black line), instead of the normal T_{PPM} time shift, the new shift level is set to (T_{PPM} + ζ).

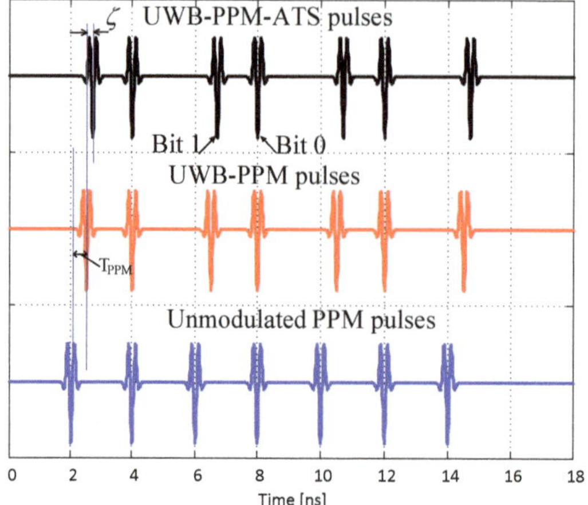

Figure 6. The ultra-wide band (UWB) pulse shapes with pulse position modulation (PPM) and PPM-ATS.

The procedure for estimating distances in UWB-PPM-ATS system is the same as in the conventional system (UWB-PPM) and in (7); the template waveform is $(g_4(t) - g_4(t - T_{\mathrm{PPM}} - \zeta))$. The correlation of the proposed system denoted by R_{1i} and determined as:

$$R_{1i}(\tau) = \int_{-\infty}^{\infty} r_i(t)[g_4(t) - g_4(t - T_{\mathrm{PPM}} - \zeta)]dt. \tag{13}$$

So, we have:

$$R_{1i}(\tau) = R_{G4}(\tau) - R_{G4}(\tau - T_{\mathrm{PPM}} - \zeta)]dt. \tag{14}$$

The shapes of $R_{G4}(\tau)$,$R_{0i}(\tau)$ and $R_{1i}(\tau)$ are shown in Figure 7.

Figure 7 indicates the autocorrelation function of the fourth-order Gaussian monocycle R_{G4} and correlation functions of UWB-PPM R_{0i} and UWB-PPM-ATS R_{1i} systems (with additional time $\zeta = -0.08$ ns and $\zeta = 0.08$ ns); those functions are compared at the value of $\tau = 0$. In the figure below inside Figure 7, the blue line with "+" sign denotes the value of $R_{0i}(\tau)$, the red line denotes the value of $R_{1i}(\tau)$ with ζ gets a negative value of -0.08 ns, and the dashed black line denotes the value of $R_{1i}(\tau)$ with ζ gets a positive value of 0.08 ns. We can observe that the choice of a negative value of ζ makes $R_{1i}(\tau)$ get the maximum value at $\tau = 0$ and this value is greater than $R_{1i}(0)$- with a positive value of ζ and also $R_{0i}(0)$ (both of these functions are not maximized at $\tau = 0$). Thus ζ will be selected that makes the value of $R_{G4}(\tau - T_{\mathrm{PPM}} - \zeta)$ reaches to the minimum point of the R_{G4} function which is denoted by $R\tau_{opt}$ as shown in Figure 7. It is clear that, in the UWB-PPM-ATS scheme, the negative values of ζ give the better correlation function than positive values. The variation of the correlation function with the different of ζ values is illustrated in Figure 8. In Figure 8, the different values of ζ lead to the different shapes of the correlation function; the optimal value of ζ in this case is $\zeta_{opt} = -0.08$ ns which makes $R_{1i}(0)$ have the maximum value. When the magnitude of the ζ increases close to the pulse width ($\zeta = -0.16$ ns, -0.25 ns in Figure 8), the corresponding correlation functions will not reach the global maximum near the value of $\tau = 0$ and its local maximum points have the values close to the global maximum. This leads to the higher error in determining the global maximum point of these functions. Therefore, depending on the specific parameter configuration of each UWB-PPM system, the value of ζ will be chosen such that $R_{1i}(0)$ in the (14) has the largest value.

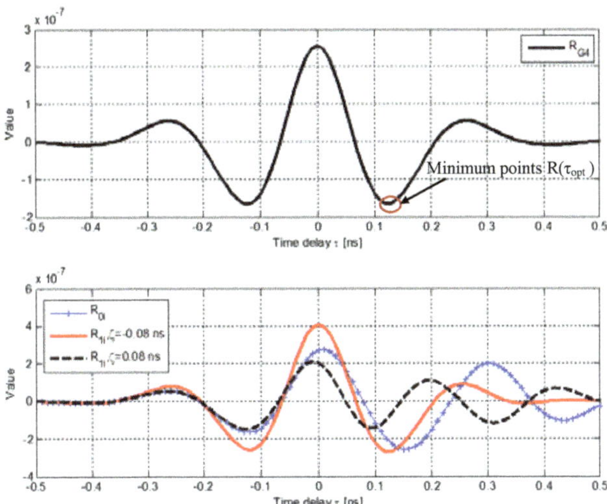

Figure 7. Correlation functions of the conventional UWB-PPM and the proposed UWB-PPT-ATS systems with different time shift.

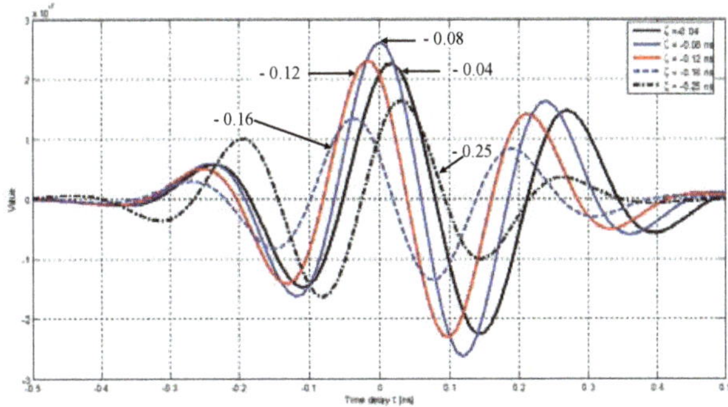

Figure 8. Correlation functions of proposed UWB-PPM-ATS with different additional time shift.

The procedure for estimating distances is based on the traveling time mentioned above and can be applied to a well-known propagation environment; however, it is unavailable for an unknown propagation environment. Consequently, we propose to apply LMFA [28] into the calculation of distances, relative permittivities, and determination of buried object location (depth and horizontal direction) in the case of the unknown propagation environment. The proposed method comes from the fact that the traveling time depends on the relative permittivities, buried object location, and transceiver position. So in order to find the value of these parameters, we change the position of the receiver horizontally and calculate the traveling time corresponding to each position of the transceiver. Based on the transceiver location and computed traveling time, the remaining parameters will be estimated.

3.2. Estimation Algorithm

Let us consider a specific system model of UWB-PPM systems illustrated in Figure 9, in which the environment is assumed to have a nondestructive structure that has two layers with relative permittivities of ε_1 and ε_2.

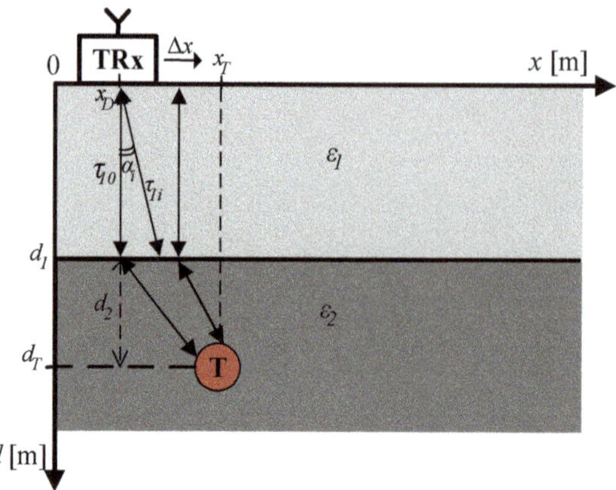

Figure 9. Target system model in an environment with nondestructive structure of two layers.

To determine the environmental parameters and also the location of the target "T", the transceiver generates UWB pulses with rotation angles of the transceiver antenna are α_i and get the reflected pulses. The traveling time τ_{1i} corresponding to the ith time of emission is calculated according to (11), the geometric relationship between τ_{1i}, d_1, and ε_1 is indicated in (15).

$$\tau_{1i} = \frac{2d_1\sqrt{\varepsilon_1}}{c.cos\alpha_i}, \tag{15}$$

where α_i is the rotation angle of antenna and:

$$\alpha_i = i.\Delta\alpha; \tag{16}$$

where $\Delta\alpha$ is the minimum rotation angle of the transceiver antenna. Thus the values of ε_1 and d_1 in (15) are estimated based on known pairs (α_i, τ_{1i}). Then, to locate buried objects, the device is moved horizontally, and the transmitter emits pulses in the perpendicular direction after every movement step of Δx. Denote τ_{2i} is the traveling time of the transmitted signals, reflected from the buried object and returned to the transceiver at position of x_{Di}, it is computed by:

$$\tau_{2i} = \tau_{10} + \frac{\sqrt{\varepsilon_2(d_2^2 + (x_T - x_{Di})^2)}}{c}, \tag{17}$$

where x_T is the horizontal coordinates of the buried object and τ_{10} is the traveling time of reflected signal at the interface between two layers of the environment in *d* direction. τ_{2i} also computed according to (11), the values of ε_2, d_2, and x_T in (17) are estimated based on known pairs (τ_{2i}, τ_{10}, x_{Di}), and

$$x_{Di} = i.\Delta x. \tag{18}$$

The proposed algorithm to estimate the values of those parameters is indicated in Figure 10 and described in detail as below

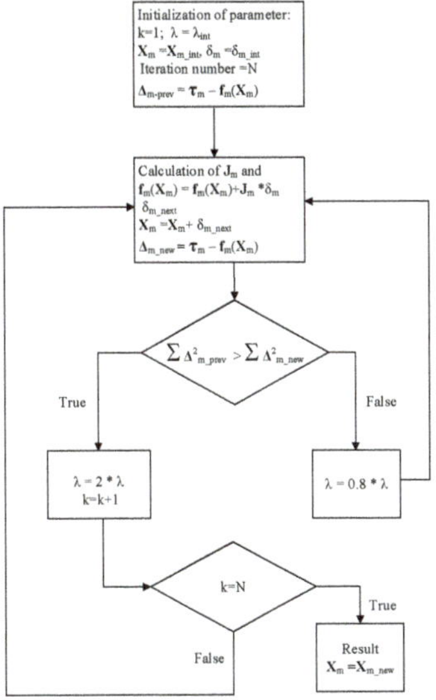

Figure 10. Flowchart of estimated algorithms.

In Figure 10, the unknown parameter vectors are denoted by:

$$\mathbf{X}_1 = (\hat{\varepsilon}_1, \hat{d}_1); \mathbf{X}_2 = (\hat{\varepsilon}_2, \hat{d}_2, \hat{x}_T), \tag{19}$$

and the index m in Figure 10 is used to replace for 1, 2. With the known pairs (τ_{1i}, α_i) and (τ_{2i}, x_{Di}), the unknown parameter vectors $\mathbf{X}_1$ and $\mathbf{X}_2$ are determined such that the sum of the squares of the deviations $S_i(\mathbf{X}_i)$ are minimized:

$$X_1 = arg\min_{\mathbf{X}_1} S_1(\mathbf{X}_1) = arg\min_{\mathbf{X}_1} \sum_{i=1}^{M} [\tau_{1i} - f_1(\alpha_i, \mathbf{X}_1)]^2, \tag{20}$$

$$X_2 = arg\min_{\mathbf{X}_2} S_2(\mathbf{X}_2) = arg\min_{\mathbf{X}_2} \sum_{i=1}^{M} [\tau_{2i} - f_2(x_{Di}, \mathbf{X}_2)]^2, \tag{21}$$

where $f_1(\alpha_i, \mathbf{X}_1)$ and $f_2(x_{Di}, \mathbf{X}_2)$ are:

$$f_1(\alpha_i, \mathbf{X}_1) = \frac{\hat{d}_1.\sqrt{\hat{\varepsilon}_1}}{c.cos\alpha_i}, \tag{22}$$

$$f_2(x_{Di}, \mathbf{X}_2) = \frac{\sqrt{\hat{\varepsilon}_2(\hat{d}_2^2 + (\hat{x}_T - x_{Di})^2)}}{c}, \tag{23}$$

where $\hat{}$ denotes the estimated values of the parameters; the vectors $\mathbf{X}_1$ and $\mathbf{X}_2$ were estimated according to the steps as shown in Figure 10 as follows:

+ Assign to $\mathbf{X}_1$, $\mathbf{X}_2$ any arbitrary initialization values, denoted by $\mathbf{X}_{1-int}$, $\mathbf{X}_{2-int}$;

+ In each iteration step, the parameter vectors $\mathbf{X}_1$, $\mathbf{X}_2$ are replaced by new estimates ($\mathbf{X}_1 + \delta_1$), ($\mathbf{X}_2 + \delta_1$) with δ_1, δ_2 are updated step vectors and we have:

$$\begin{aligned} \mathbf{S}_1(\mathbf{X}_1 + \delta_1) \approx\ & [\tau_1 - \mathbf{f}_1(\mathbf{X}_1)]^T[\tau_1 - \mathbf{f}_1(\mathbf{X}_1)] \\ & -2[\tau_1 - \mathbf{f}_1(\mathbf{X}_1)]^T\mathbf{J}_1\delta_1 + \delta_1^T\mathbf{J}_1^T\mathbf{J}_1\delta_1, \\ \mathbf{S}_2(\mathbf{X}_2 + \delta_2) \approx\ & [\mathbf{x}_D - \mathbf{f}_2(\mathbf{f}_2)]^T[\mathbf{x}_D - \mathbf{f}_2(\mathbf{X}_2)] \\ & -2[\mathbf{x}_D - \mathbf{f}_2(\mathbf{X}_2)]^T\mathbf{J}_2\delta_2 + \delta_2^T\mathbf{J}_2^T\mathbf{J}_2\delta_2, \end{aligned}$$

with $\mathbf{J}_{1,2}$ is the Jacobian matrix, whose i^{th} row equals $\mathbf{J}_{1i}$ and $\mathbf{J}_{2i}$, respectively:

$$\mathbf{J}_{1i} = \frac{\partial f_1(\alpha_i, \mathbf{X}_1)}{\partial \mathbf{X}_1}, \tag{24}$$

$$\mathbf{J}_{2i} = \frac{\partial f_2(x_{Di}, \mathbf{X}_2)}{\partial \mathbf{X}_2}. \tag{25}$$

To get the minimum value of the sums $\mathbf{S}_1, \mathbf{S}_2$, the update step vectors $\delta_{1,2}$ is calculated such that the derivative of $\mathbf{S}_{1,2}(\mathbf{X}_{1,2} + \delta_{1,2})$ with respect to $\delta_{1,2}$ has a result of zero, so, $\delta_{1,2}$ can be determined satisfying:

$$[\mathbf{J}_1^T \mathbf{J}_1 + \lambda \mathrm{diag}(\mathbf{J}_1^T \mathbf{J}_1)]\delta_1 = \mathbf{J}_1^T [\tau_1 - \mathbf{f}_1(\mathbf{X}_1)], \tag{26}$$

$$[\mathbf{J}_2^T \mathbf{J}_2 + \lambda \mathrm{diag}(\mathbf{J}_2^T \mathbf{J}_2)]\delta_2 = \mathbf{J}_2^T [\mathbf{x}_D - \mathbf{f}_2(\mathbf{X}_2)], \tag{27}$$

where the damping factor λ (non-negative) is adjusted at each iteration. If $\mathbf{S}_{1,2}$ is reduced rapidly, a smaller value of λ can be used, whereas if in an iteration does not reduce the residual, λ can be increased. The update step vectors are computed as follows:

$$\delta_1 = [\mathbf{J}_1^{\mathbf{T}} \mathbf{J}_1 + \lambda \mathrm{diag}(\mathbf{J}_1^T \mathbf{J}_1)]^{-1} \mathbf{J}_1^T [\tau_1 - \mathbf{f}_1(\mathbf{X}_1)], \tag{28}$$

$$\delta_2 = [\mathbf{J}_2^T \mathbf{J}_2 + \lambda \mathrm{diag}(\mathbf{J}_2^T \mathbf{J}_2)]^{-1} \mathbf{J}_2^T [\mathbf{x}_D - \mathbf{f}_2(\mathbf{X}_2)]. \tag{29}$$

After a certain number of iterations, the output of LMFA is the final estimated values of system parameters that meet the constraint condition in (21), as shown in Figure 10.

4. Simulation Results and Discussion

4.1. Simulation Parameters

The accuracy of methods for determining the distance and characteristics of a multilayered reflective environments using UWB pulses is strongly dependent on UWB signal processing techniques. UWB-PPM is one of the candidates for positioning technology in a nondestructive environment with multiple reflective layers; with the proposed UWB-PPM-ATS, our trials indicated that the UWB-PPM pulses shifted with a certain time constant can be used to improve the precision in estimating the distance. All the numerical results in this paper were computed using Matlab.

First, we compare the exact estimation errors of the UWB-OOK, UWB-PPM and UWB-PPM-ATS systems with the results obtained from the simulations and the actual value of the parameters use (6), (8), (11), (13) and (30). Here OOK is one of the UWB pulse amplitude modulation techniques and has two level modulation with the bits are 0 and 1. When sending a bit 0, the transmitter will not send anything. When sending a bit 1, the transmitter will send a pulse [29], and the template waveform $\omega(t)$ at the receiver of UWB-OOK system is $g_4(t)$. In addition, the performance of UWB-OOK, UWB-PPM and UWB-PPM-ATS systems were evaluated for an environment with known characteristics (for example, here are three layers as indicated in Figure 1). Finally, the location determination technique in a unknown environment with multiple reflection layers using the LMFA based on the estimated parameters of those UWB systems is presented with the system model as shown in Figure 9. The comparisons between the above systems are evaluated in term of errors between the estimated values from the considered system and the true values. Based on the PSD plot of the Gaussian impulses shown in Figure 4, most of the numerical results presented in this section are based on analysis using the fourth-order Gaussian monocycle. The parameters of the example UWB systems are listed in Table 1 and follow [5].

To assess the performance of those UWB systems, the error of estimated distance is defined as follows:

$$\delta_d = \left| \hat{d}_i - d_{tri} \right|, \tag{30}$$

where $\hat{d}_i$ denotes the estimated distance and d_{tri} is the true value.

Table 1. Simulation parameters.

Parameter	Notation	Value
Time normalization factor	μ_p	0.2877 ns
Transmitted power	P_{Tx}	−35.4 dB
The amplitude factors	A_1, A_2, A_3	0.33, 0.13, 0.14
Number of pulses	N	100
Noise power	$N_0/2$	−102 dB
Time shift of PPM	T_{PPM}	0.2 ns
Additional time shift	ζ	−0.08 ns, −0.16 ns
Relative permittivity with heterogeneous medium	$\varepsilon_1, \varepsilon_2, \varepsilon_3$	4; 3; 5

In Figure 1, without generality, we assume that the propagation environment is heterogeneous with three layers: sand (dry), sandy soil (dry) and granite (dry) and the relative permittivities $\varepsilon_1, \varepsilon_2, \varepsilon_3$ are 4, 3, 5, respectively. Those layers are assumed to be dry to reduce the attenuation of environment, so propagation velocity in those layers are 15 cm/ns, 17.32 cm/ns, 13.42 cm/ns and the attenuation is 0.01 dB/m [30]. With a known transmission environment, those distance can be estimated using (6), where τ_i is computed according to (8), (11) and (13). Then the errors of estimated distances are calculated according to (30) and illustrated in Figure 11 for OOK, PPM, ATS systems with $\zeta = -0.08$ ns and $\zeta = -0.16$ ns. In Figure 11, the black line denotes the average error of the OOK system, the blue line denote the error of the PPM system and the red lines denote the error of the ATS system, all with the same system parameters. Observe that the relative error of the UWB-OOK system is about 24%, of UWB-PPM is about 11%, of UWB-PPM-ATS with $\zeta = -0.08$ ns is about 7%, with $\zeta = -0.16$ ns is about 13%; so the UWB-PPM- ATS with $\zeta = -0.08$ ns performs better for all distance values. This can be explained by comparing the correlation functions which are given in (8), (13) and Figures 7 and 8; we observe that $R_{1i}(0)$ is significantly greater than $R_{0i}(0)$ for $\zeta = -0.08$ ns and $R_{0i}(0)$ is not the maximum value of $R_{0i}(\tau)$, which leads to a smaller than average error in the PPM-ATS with $\zeta = -0.08$ ns when compared to PPM scheme; but with $\zeta = -0.16$ ns, $R_{1i}(0)$ is less than $R_{0i}(0)$ thus the PPM outperforms the PPM-ATS. Besides that, the UWB-OOK is one of the amplitude modulation techniques, thus it is affected by the transmission environment and more difficult to separate the received signal from noise interference in comparison with the PPM techniques which have a constant amplitude. The results in Figure 11 indicate that the time shift ζ directly affects the performance of PPM systems. The value of ζ is selected so that $(T_{\text{PPM}} + \zeta)$ achieves the optimal value at which $R_{G4}(\tau - T_{\text{PPM}} - \zeta)$ gets the minimum value. From (6) and (13), we also know that ζ is determined by the pulse shape and the pulse width normalization factor μ_p. Due to the influence of the UWB-PPM system to the choice of the time shift, a suitable value of ζ should be chosen according to the specific pulse shape and pulse width employed in a particular UWB application.

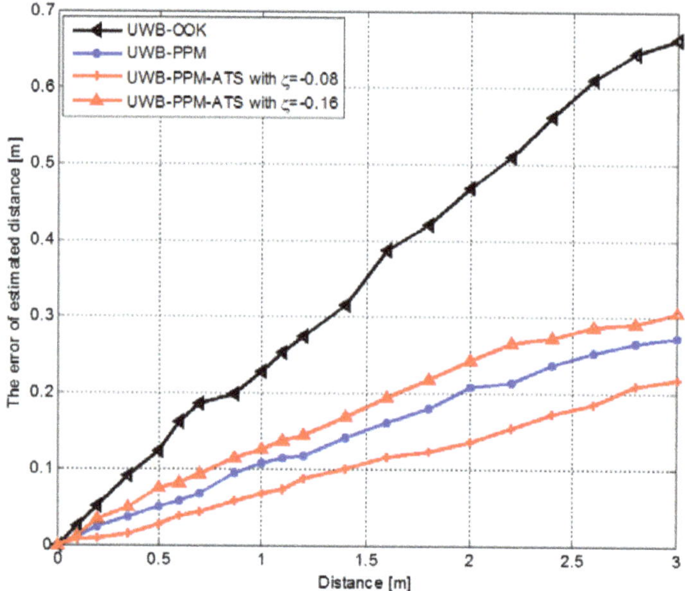

Figure 11. Comparison of distance estimation errors between OOK, PPM and the proposed UWB-PPM-ATS modulation techniques.

4.2. Determine the Location of Buried Object by LMFA

As mentioned above, the propagation distance can only be determined using the correlation function if the characteristic of the environment is well-known; however, this is impossible with unknown environment. Moreover, the position of buried object cannot be determined by only the propagation distances; it must be determined in both x and d directions (x_T, d_T in Figure 9). The parameters of each layer in the system model in Figure 9 are estimated sequentially using LMFA as presented in Section 3.2. The parameters of system model and the initialization vectors of the LMFA algorithm are listed in Table 2 in which two layers of environment are assumed to be dry sand and concrete with the corresponding propagation velocity of 16 cm/ns and 14.14 cm/s.

Table 2. Initialization parameters of the model.

Parameter	Notation	Value
The thickness of the 1st layer	d_1	0.63 m
Depth of 'T' in the 2nd layer	d_2	0.42 m
X-coordinate of 'T'	x_T	1.6 m
Relative permittivity	$\varepsilon_1, \varepsilon_2$	3.5, 4.5
Damping factor	λ	3
Movement step of the device	Δx	20 cm

First the parameters ε_1, d_1 are estimated based on the propagation time τ_{1i} and rotation angle α_i with both UWB-PPM and UWB-PPM-ATS systems. Then, in a similar way, the parameters d_2, ε_2, and x_T are estimated based on $\mathbf{ø_2}$ and $\mathbf{x_D}$; therefore, the depth of buried object $d_T = d_1 + d_2$ is also determined. The estimated results after 30 iterations and errors of UWB systems are listed in Table 3 and Figures 12–14. Figures 12 and 13 show the relationship between τ_{1i} and α_i in (15). In those figures, the dashed black line denote the curve of τ_{1i} vs. α_i with the true values of d_1 and ε_1; the blue dots are τ_{1i} values measured by simulated UWB systems and the red line denote the curve of τ_{1i} vs. α_i with the estimated values $\widehat{d_1}$ and $\widehat{\varepsilon_1}$. Figure 14 shows the curve of $(\tau_{2i} - \tau_{10})$ depends on x_{Di} according to (17).

In those figures, the dashed black line denotes the curve of $(\tau_{2i} - \tau_{10})$ vs. x_{Di} with the true values of d_2, ε_2 and x_T; the blue dots are $(\tau_{2i} - \tau_{10})$ values measured by simulated UWB systems and the red line denotes the curve of $(\tau_{2i} - \tau_{10})$ vs. x_{Di} with the estimated values $\widehat{d_2}$, $\widehat{\varepsilon_2}$ and $\widehat{x_T}$.

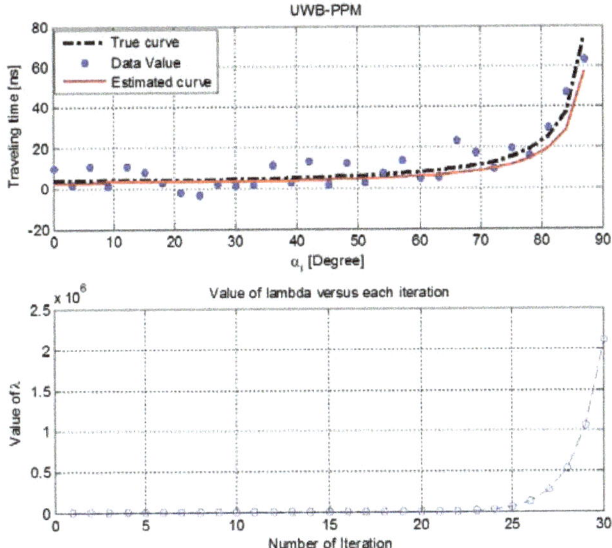

Figure 12. The results of estimating d_1 and ε_1 of the conventional UWB-PPM based on Levenberg–Marquardt Fletcher algorithm (LMFA).

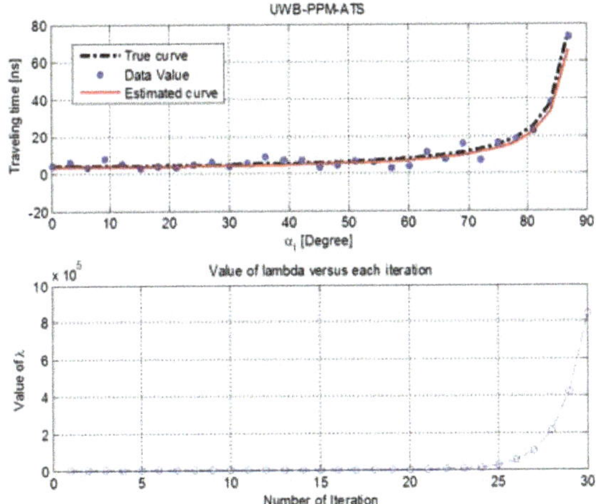

Figure 13. The results of estimating d_1 and ε_1 of the proposed UWB-PPM-ATS based on LMFA.

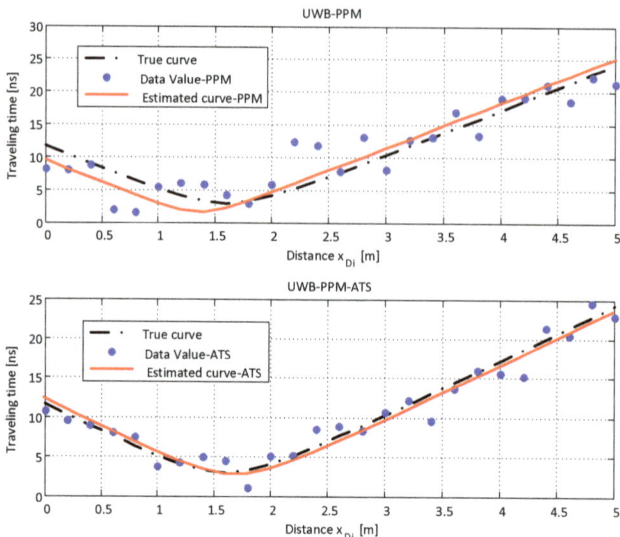

Figure 14. The results of estimating d_2, ε_2 and x_T of the conventional UWB-PPM and the proposed UWB-PPM-ATS based on LMFA.

Table 3 and Figure 15 show the estimated results by the UWB systems. As seen, the modulation techniques for UWB pulses combine with nonlinear estimation method LMFA can be used to determine the thickness of different layers in the nondestructive environments, their relative permittivities and also the position of buried object. We observe that the UWB-PPM-ATS system outperforms the UWB-PPM system for all parameters of the model. This behavior results directly from the features of the correlation functions of different signals shown in Figures 7 and 8. When using UWB-PPM-ATS technique with the optimal value of ζ, the estimated traveling time values have a smaller error than using UWB-PPM technique, so the results from the LMFA of the PPM-ATS system give the higher accuracy than conventional PPM system.

Table 3. The results of the estimated parameters.

Parameter/Notation	d_1	d_2	x_T	ε_1	ε_2
True value [m]	0.63	0.42	1.6	3.5	4.5
By UWB-OOK [m]	0.9418	0.6623	2.1362	4.3216	5.7261
By UWB-PPM [m]	0.7923	0.5742	1.9316	2.7109	3.3590
By UWB-PPM-ATS [m]	0.6523	0.3899	1.7107	3.2829	4.8167
Error of UWB-OOK	0.3118	0.2423	0.5362	0.8216	1.2261
Error of UWB-PPM	0.1623	0.1542	0.3316	0.7891	1.141
Error of UWB-PPM-ATS	0.0223	0.0301	0.1107	0.2171	0.3167

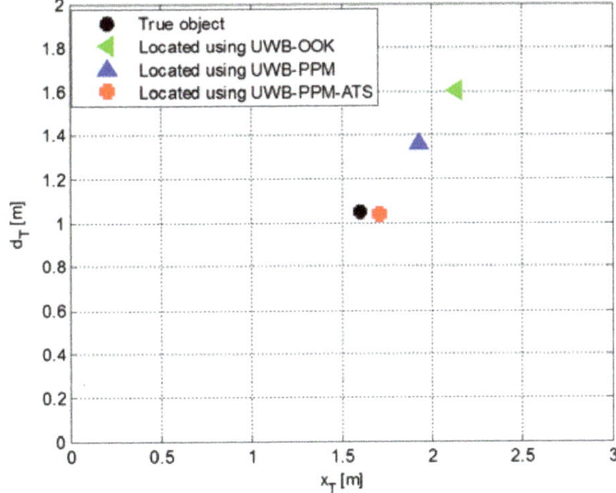

Figure 15. The true and estimated locations of buried object using the conventional UWB-PPM and the proposed UWB-PPM-ATS.

5. Conclusions

In this paper, the authors have proposed an UWB pulse position modulation with an additional time-shift technique for positioning applications in nondestructive environments of the UWB systems. The proposed method is utilized to determine the distances and the position of a buried object in heterogeneous environments. The proposed technique's quality is assessed based on the calculation errors compared to the actual values and comparison with the conventional modulation methods. Combining with the Levenberg–Marquardt Fletcher algorithm to determine the buried object location when the propagation environment is unknown, the efficiency of the proposed technique is better than the PPM-UWB system with the same model. Evaluating different UWB systems based on the calculation of errors provides reliable information for choosing modulation schemes with suitable parameters in the design of UWB systems for positioning applications, especially in nondestructive environments.

However, in this work, only one buried object location is determined; multiple buried object estimation algorithm will be researched in our future works. Additionally, the research on parameters of the buried object, such as shape, size, permittivity and so on, is left for the future work.

Author Contributions: Conceptualization, N.T.H. and P.T.H.; methodology, N.T.H. and N.L.C.; software, N.T.H.; validation, N.T.H. and P.T.H.; formal analysis, P.T.H. and N.L.C.; investigation, N.T.H.; resources, N.T.H.; data curation, P.T.H.; writing—original draft preparation, N.T.H.; writing—review and editing, N.T.H. and N.L.C.; visualization, N.L.C.; supervision, P.T.H.; project administration, P.T.H. All authors have read and agreed to the published version of the manuscript.

Funding: This research received no external funding.

Conflicts of Interest: The authors declare that there is no conflict of interest regarding the publication of this paper.

Abbreviations

The following abbreviations are used in this manuscript:

UWB	Ultra-wide band
UWB-PPM	Ultra-wide band pulse position modulation
UWB-OOOK	Ultra-wide band On-Off Keying
UWB-PPM-ATS	Ultra-wide band pulse position modulation with optimized additional time shift
LMFA	Levenberg–Marquardt Fletcher algorithm

References

1. Goyal, V. Pulse generation and analysis of ultra wide band system model. *GESJ Comput. Sci. Telecommun.* **2012**, *2*, 3–6. ISSN 1512-1232.
2. U. F. C. Commission. *Revision of Part 15 of the Commission's Rules Regarding Ultra-Wideband Transmission Systems*; First report and order, technical report, Feb, Tech. Rep.; FCC: Washington, DC, USA, 2002.
3. Goyal, V.; Dhaliwal, B. Optimal pulse generation for the improvement of ultra wideband system performance. In Proceedings of the 2014 Recent Advances in Engineering and Computational Sciences (RAECS), Chandigarh, India, 6–8 March 2014; IEEE: Piscataway, NJ, USA, 2014; pp. 1–6.
4. Welborn, M.L. System considerations for ultra-wideband wireless networks. In Proceedings of the RAWCON 2001, 2001 IEEE Radio and Wireless Conference (Cat. No. 01EX514), Waltham, MA, USA, 19–22 August 2001; IEEE: Piscataway, NJ, USA, 2001; pp. 5–8.
5. Hu, B.; Beaulieu, N.C. Accurate evaluation of multiple-access performance in th-ppm and th-bpsk uwb systems. *IEEE Trans. Commun.* **2004**, *52*, 1758–1766. [CrossRef]
6. Khalesi, H.; Ghods, V. QPSK Modulation Scheme Based on Orthogonal Gaussian Pulses for IR-UWB Communication Systems. *J. Circuits Syst. Comput.* **2019**, *28*, 1950008. [CrossRef]
7. Sharma, S.; Gupta, A.; Bhatia, V. A simple modified peak detection based UWB receiver for WSN and IoT application. In Proceedings of the 2017 IEEE 85th Vehicular Technology Conference (VTC Spring), Sydney, NSW, Australia, 4–7 June 2017; IEEE: Piscataway, NJ, USA, 2017; pp. 1–6.
8. Benotmane, N.B.; Elahmar, S.A.; Dayoub, I.; Hamouda, W. Improved eigenfilter design method for channel shortening equalizer in TH-UWB. *IEEE Trans. Veh. Technol.* **2018**, *67*, 7749–7753. [CrossRef]
9. Zhao, L.; Haimovich, A.M. Capacity of m-ary ppm ultra-wideband communications over awgn channels. In Proceedings of the IEEE 54th Vehicular Technology Conference, VTC Fall 2001, Proceedings (Cat. No. 01CH37211), Atlantic City, NJ, USA, 7–11 October 2001; IEEE: Piscataway, NJ, USA, 2001; Volume 2, pp. 1191–1195.
10. Venkatesan, V.; Liu, H.; Nilsen, C.; Kyker, R.; Magana, M.E. Performance of an optimally spaced ppm ultra-wideband system with direct sequence spreading for multiple access'. In Proceedings of the 2003 IEEE 58th Vehicular Technology Conference, VTC 2003-Fall (IEEE Cat. No. 03CH37484), Orlando, FL, USA, 6–9 October 2003; IEEE: Piscataway, NJ, USA, 2003; Volume 1, pp. 602–606.
11. Wu, J.; Xiang, H.; Tian, Z. Weighted noncoherent receivers for uwb ppm signals. *IEEE Commun. Lett.* **2006**, *10*, 655–657.
12. Almodovar-Faria, J.M.; McNair, J. Optimal integration time for energy-detection ppm uwb systems. In Proceedings of the 2012 IEEE Global Communications Conference (GLOBECOM), Anaheim, CA, USA, 3–7 December 2012; IEEE: Piscataway, NJ, USA, 2012; pp. 4054–4059.
13. Yin, H.B.; Yang, J.A.; Wang, W.D. A randomly permuted fourier measurement matrix for uwb-ppm signals. *Appl. Mech. Mater.* **2014**, *556*, 2646–2649. [CrossRef]
14. Yajnanarayana, V.; Händel, P. Joint estimation of toa and ppm symbols using sub-nyquist sampled ir-uwb signal. *IEEE Commun. Lett.* **2017**, *21*, 949–952. [CrossRef]
15. Jabbar, M. Improving PSD of PPM-IR for UWB Signal Using Turbo Encoder. *J. Ind. Eng. Manag. S* **2018**, *3*, 2169-0316. [CrossRef]
16. Goyal, V.; Dhaliwal, B. Improving ultra wideband (uwb) system by modified random combination of pulses. *Eng. Rev.* **2018**, *38*, 189–203. [CrossRef]
17. Maali, A.; Boukhelifa, A.; Mesloub, A.; Sadoudi, S.; Benssalah, M. An enhanced pulse position modulation (ppm) for both ir-uwb and dcc-uwb communication. In Proceedings of the 2019 13th European Conference on Antennas and Propagation (EuCAP), Krakow, Poland, 31 March–5 April 2019; IEEE: Piscataway, NJ, USA, 2019; pp. 1–5.
18. Daniels, D.J. Ground penetrating radar. In *Encyclopedia of RF and Microwave Engineering*; John Wiley & Sons, Inc.: New York, NY, USA, 2005.
19. Cao, Y.; Labuz, J.; Guzina, B. Evaluation of pavement system based on ground-penetrating radar full-waveform simulation. *Transp. Res. Rec.* **2011**, *2227*, 71–78. [CrossRef]
20. Karim, H.H.; Al-Qaissi, A.M. Assessment of the accuracy of road flexible and rigid pavement layers using gpr. *Eng. Technol. J. Part (A) Eng.* **2014**, *32*, 788–799.

21. Nakayama, Y.; Kohno, R. Novel variable spreading sequence length system for improving the processing speed of ds-uwb radar. In Proceedings of the 2008 8th International Conference on ITS Telecommunications, Phuket, Thailand, 24 October 2008; IEEE: Piscataway, NJ, USA, 2008; pp. 357–361.
22. Dabove, P.; Di Pietra, V.; Piras, M.; Jabbar, A.A.; Kazim, S.A. Indoor positioning using Ultra-wide band (UWB) technologies: Positioning accuracies and sensors performances. In Proceedings of the 2018 IEEE/ION Position, Location and Navigation Symposium (PLANS), Monterey, CA, USA, 23–26 April 2018; IEEE: Piscataway, NJ, USA, 2018; pp. 175–184.
23. Xie, W.; Li, X.; Long, X. Underground operator monitoring platform based on ultra-wide band WSN. *Int. J. Online Biomed. Eng. (iJOE)* **2018**, *14*, 219–229. [CrossRef]
24. Huyen, N.T.; Hiep, P.T. Proposing adaptive PN sequence length scheme for testing non-destructive structure using DS-UWB'. In Proceedings of the 2019 3rd International Conference on Recent Advances in Signal Processing, Telecommunications & Computing (SigTelCom), Hanoi, Vietnam, 21–22 March 2019; IEEE: Piscataway, NJ, USA, 2019; pp. 10–14.
25. Chen, X.; Kiaei, S. Monocycle shapes for ultra wideband system. In Proceedings of the 2002 IEEE International Symposium on Circuits and Systems, Proceedings (Cat. No. 02CH37353), Phoenix-Scottsdale, AZ, USA, 26–29 May 2002; IEEE: Piscataway, NJ, USA, 2002; Volume 1, pp. I-597–I-600.
26. De Abreu, G.T.F.; Mitchell, C.J.; Kohno, R. On the orthogonality of hermite pulses for ultra wideband communications systems. In Proceedings of the 6th International Symposium on Wireless Personal Multimedia Communications (WPMC), Yokosuka, Japan, 21–22 October 2003.
27. Gao, X. Uwb Indoor Localization System. Ph.D. Thesis, The George Washington University, Washington, DC, USA, 2018.
28. Kanzow, C.; Yamashita, N.; Fukushima, M. Withdrawn: Levenberg–marquardt methods with strong local convergence properties for solving nonlinear equations with convex constraints. *J. Comput. Appl. Math.* **2005**, *173*, 321–343. [CrossRef]
29. Benedetto, D. Understanding ultra wide band radio fundamentals. In *Understanding Ultra Wide Band Radio Fundamentals*; Pearson Education India: Delhi, India, 2008.
30. Hubbard, S.S.; Peterson, J.E., Jr.; Majer, E.L.; Zawislanski, P.T.; Williams, K.H.; Roberts, J.; Wobber, F. Estimation of permeable pathways and water content using tomographic radar data. *Lead. Edge J.* **1997**, *16*, 1623–1630. [CrossRef]

Article

An Attention-Based Network for Textured Surface Anomaly Detection

Gaokai Liu, Ning Yang * and Lei Guo

School of Automation, Northwestern Polytechnical University, Xi'an 710129, China; lgk@mail.nwpu.edu.cn (G.L.); lguo@nwpu.edu.cn (L.G.)

* Correspondence: ningyang@nwpu.edu.cn

Received: 27 July 2020; Accepted: 3 September 2020; Published: 8 September 2020

Abstract: Textured surface anomaly detection is a significant task in industrial scenarios. In order to further improve the detection performance, we proposed a novel two-stage approach with an attention mechanism. Firstly, in the segmentation network, the feature extraction and anomaly attention modules are designed to capture the detail information as much as possible and focus on the anomalies, respectively. To strike dynamic balances between these two parts, an adaptive scheme where learnable parameters are gradually optimized is introduced. Subsequently, the weights of the segmentation network are frozen, and the outputs are fed into the classification network, which is trained independently in this stage. Finally, we evaluate the proposed approach on DAGM 2007 dataset which consists of diverse textured surfaces with weakly-labeled anomalies, and the experiments demonstrate that our method can achieve 100% detection rates in terms of TPR (True Positive Rate) and TNR (True Negative Rate).

Keywords: textured surface anomaly detection; computer vision; deep learning; attention mechanism; adaptive fusion

---

1. Introduction

Automatic surface-anomaly detection is one of the most vital tasks in manufacturing processes to guarantee that the end product is visually free of anomalies. It is mostly relevant in various domains of industrial production, such as steels [1,2], fibers [3], and plastics [4]. As a matter of fact, surface anomaly detection is usually carried out manually due to the constraints of technical conditions, which is very inefficient and errors are apt to occur due to fatigue. Over the past two decades, automated surface inspection approaches based on computer vision have been proven to be very effective and are attracting more research attentions. In particular, deep learning techniques have achieved great success in the domain of visual inspections.

In the early years, classical image processing approaches were often applied to controlled environments, such as stable lighting conditions. Sanchez-Brea et al. [5] put forward a thresholding technique to detect the anomaly according to the intensity variations of rings which is caused when laser beams illuminate the wire. However, such methods are no longer applicable for complex backgrounds or the strong interference of noises. More appropriate methods should be designed for these challenging tasks. Later approaches can be mainly divided into two categories: models based on selective features, and deep learning-based methods [6,7]. Feature-based approaches such as visual saliency map [8], gray level co-occurrence matrix [9], and statistical projection [10] are usually appropriate for specific tasks. These features are not only hard to design, but being hand-crafted features, they are also not useable for other applications, which causes the extension of development cycles to adapt different products. The emergence of deep learning-based models has significantly improved this issue, as such methods are data-driven, and can automatically seek optimal features which avoids the special feature-design processes for different applications.

In recent years, there have emerged more and more excellent convolutional neural network models, such as FPN [11], ResNet [12], and SegNet [13]. Apart from these models mentioned above, FCN [14] is the first network applied to semantic segmentation tasks in an end-to-end manner. In this method, an encoder–decoder module and skip connections are used to combine deep with more shallow features. Attention U-Net [15] highlights the foreground via the supplement of more semantic information in the encoder parts. Hi-Net [16] utilizes more information from different modalities via the fusion of each learned feature representations. Liu et al. [17] present a sample balancing strategy via the assignment different weights to the edge and background pixels to further improve the extraction accuracy.

Early work on textured surface anomaly detection where deep learning is utilized can be found in Ref. [18], which investigated the performance differences generated by different hyper-parameter settings. Racki et al. [19] presented a compact convolutional neural architecture for the detection of surface anomalies. This network firstly acquires good features via segmentation network, then all the parameters are frozen, and only the classification network is trained. Mei et al. [20] proposed an unsupervised algorithm for fabric anomaly detection. It reconstructs image patches via convolutional denoising autoencoder networks under multi-scale gaussian pyramid levels, and the residual maps of each image patch are used for pixel-wise prediction.

In order to further improve the surface anomaly detection performance on the DAGM 2007 dataset, this paper presents an attention-based network inspired by the works mentioned above. On one hand, feature extraction module be used to capture detailed information. On the other hand, anomaly attention module is designed to strengthen the potential objects and simultaneously weaken the background noises. The validity of the proposed method is confirmed by a series of experiments.

The remainder of this paper is organized as follows. The proposed model is elaborately described in Section 2. The experiment and discussion are presented in Section 3. Finally, Section 4 draws the conclusion of this paper.

2. Materials and Methods

For the dataset with limited samples, overfitting is prone to occur when detection or classification approaches are employed directly. However, segmentation-based two-stage models can settle this issue, and the methods of this type generally follow the same paradigm, i.e., segmentation network is applied to extract good feature representations, then the classification network is trained upon these features. The validity of this mode can be explained that for segmentation tasks, overfitting problems can be largely improved as image segmentation belongs to pixel-level classifications, enabling effective samples to be added in the training process [21].

2.1. Segmentation Model

An encoder–decoder architecture is adopted to capture the detailed information [22] as much as possible, especially for the small anomaly structure used in this paper. However, unlike U-Net network [22], we also design an attention-based fusion module to focus on the potential objects.

The overall segmentation part is shown in Figure 1, which consists of the encoder, decoder, skip connections, and the proposed fusion block. Specifically, pool and corresponding transpose convolution operations divide the whole process into four stages, and in the second and third stages, the relevant layers from encoder and decoder are combined via skip connections. As a whole, we integrate encoder and decoder information of the first stage via the attention-based fusion module to realize background weakness and anomaly reinforcement.

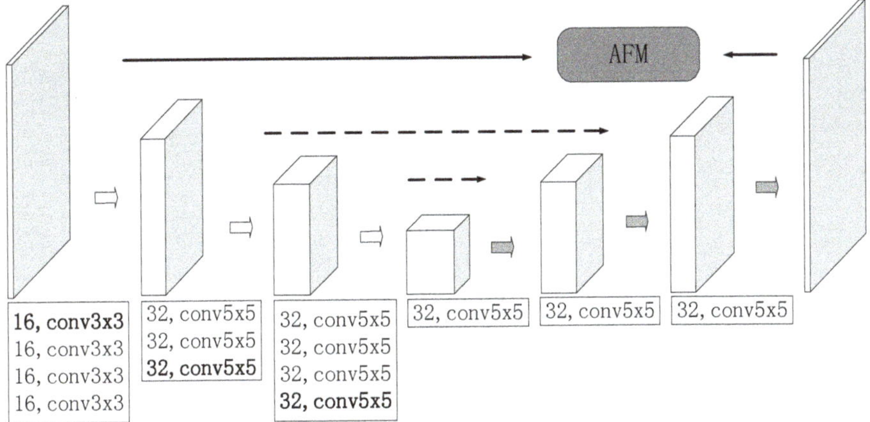

Figure 1. Segmentation network (AFM: Attention-based fusion module).

2.2. Attention-Based Fusion Module

The attention-based fusion module is designed to capture the detailed information, meanwhile strengthen the salient features and weaken irrelevant and noisy responses as well. The detailed procedure is presented in Figure 2.

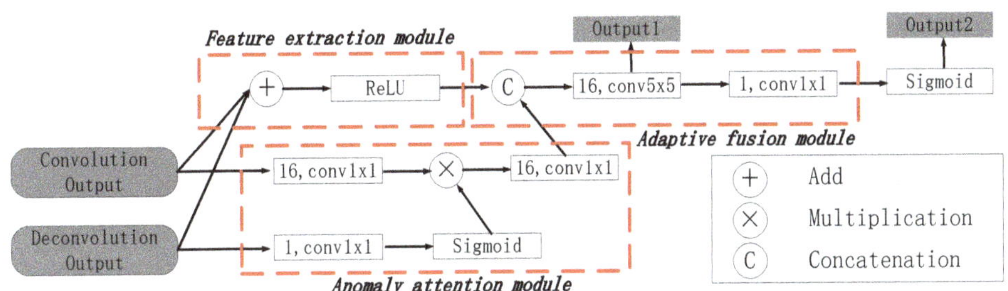

Figure 2. Attention-based fusion module.

In the feature extraction module, the information from encoder and decoder are merged by add operation and ReLU activation function, which ensures that detailed features can be retained, and visual saliency maps are highlighted to some extent. Moreover, the semantic gap between the feature extraction module and following anomaly attention module is narrowed when semantic information is supplemented in this section, which is more conducive to the training process.

Anomaly attention module provides a more comprehensive perspective to focus on the potential objects and weaken background information. Specifically, on one side, the information of the encoder from the first stage is input into a convolution layer to increase nonlinearity and feature depth. On the other side, the corresponding part from the decoder is exerted by the convolution operation and sigmoid function to obtain an attention coefficient. Then the output of these two sides are integrated via pixel-wise multiplications, which are followed by convolution operation to further increase nonlinearity and feature depth.

In order to weigh the detailed information and visual saliency extraction, an adaptive fusion module is designed to combine the two information from feature extraction and anomaly attention modules in a learning manner, and the weight parameters can be updated adaptively to maximally meet the demands of the different applications. As shown in Figure 2, the output of the feature extraction and anomaly attention modules above are concatenated firstly, then the result is executed by 5×5 and

1×1 convolution calculations respectively, which can learn the weight parameters from channel inners and inters. The output 1 and output 2 are reserved for the following classification module.

In addition, for the imbalance issue of samples, the loss of each pixel is formulated as Equation (1) to attach more weight for positive samples.

$$\ell(X_i) = -\frac{1}{N}\sum_{i=1}^{N}(\alpha log(1 - P(y_i = 0|X_i)) + \beta log P(y_i = 1|X_i)) \tag{1}$$

Here X_i, y_i denote feature vector and label at pixel i respectively. P represents the sigmoid activation function, and $\alpha = 1, \beta = 3$ in this paper.

2.3. Classification Network

The classification part relies on the outputs of segmentation network where all the parameters are frozen. As shown in Figure 3, we introduce this module according to Ref. [21]. The main difference is that a mergence operation of multiple dilated convolutions similar to deeplabv3+ [23] is employed to acquire enough receptive field and mitigate the loss of detailed information as much as possible. Moreover, the ReLU activation function after add operation [15] is utilized to be conductive to increasing the network sparsity and alleviating the overfitting issue, in the same manner as in the segmentation model.

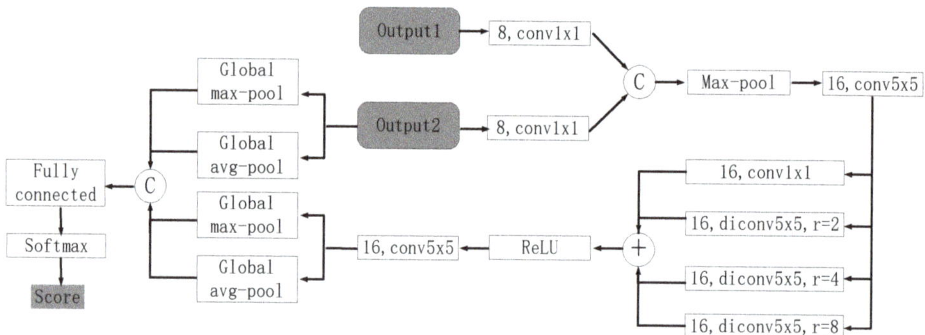

Figure 3. Classification network (diconv: dilated convolution).

Similarly, the loss of each sample in classification network can be calculated by softmax cross-entropy function as follows,

$$\ell(S_j) = -\frac{1}{M}\sum_{j=1}^{M}(S_j + log\sum_{k=1}^{C} S_{jk}) \tag{2}$$

where S_j indicates the input of softmax function for sample j. C, M refer to the number of categories and samples.

2.4. DAGM Textured Dataset

The proposed approach is evaluated on the open textured surface dataset DAGM 2007 (https://hci.iwr.uni-heidelberg.de/node/3616) for industrial optical inspection. It consists of 10 sub-datasets with different classes of anomalies, the distribution condition of training and testing samples with the size of 512×512 is listed in Table 1, and the positive means the textured samples with defects, while the negative represents the defect-free samples. All the defective areas are roughly labeled with an encircling ellipse.

Table 1. Sample distribution of the DAGM dataset.

	Train		Test	
Class	Positive	Negative	Positive	Negative
1	79	496	71	504
2	66	509	84	491
3	66	509	84	491
4	82	493	68	507
5	70	505	80	495
6	83	492	67	508
7	150	1000	150	1000
8	150	1000	150	1000
9	150	1000	150	1000
10	150	1000	150	1000

Considering the imbalance of positive and negative samples in training set, we generate another three samples for each positive training example via rotating with 180 degree, mirroring along horizontal and vertical axis in the same manner as in Ref. [19], and a series of relevant experiments for the augmented dataset are also carried out for further analysis.

3. Result and Discussion

All experiments are implemented using Tensorflow [24] and the process is divided into two steps. Firstly, the segmentation network is trained independently, then these optimized parameters are frozen and only classification network are trained in the second stage. Batch normalization [25] is implemented in each convolutional layer. The Adam [26] optimizer and a learning rate of 0.1 are used in this paper.

Firstly, the effectiveness of the anomaly attention module is verified from the views that the segmentation results should provide good interpretability as human experts and the optimal performance is liable to achieve. Figure 4a illustrates two examples when the anomaly attention module is used or not used, and the relevant variations of classification score on the test dataset is shown in Figure 4b.

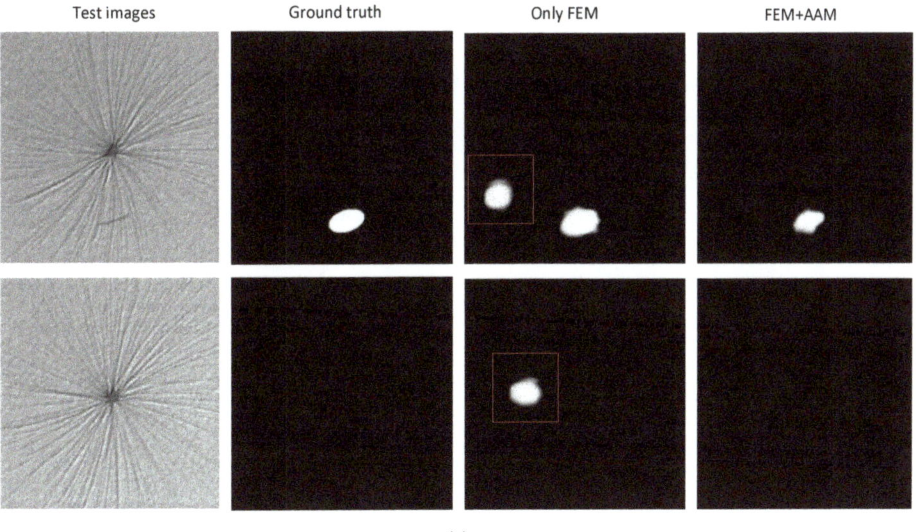

(a)

Figure 4. *Cont.*

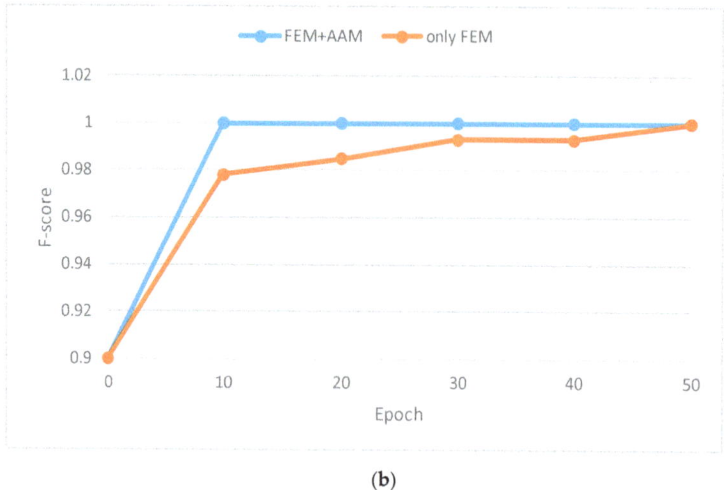

(b)

Figure 4. (a) Test samples of segmentation network (FEM: feature extraction module; AAM: anomaly attention module). The red squares represent the areas erroneously detected due to background interference; (b) Relevant test f-score variations in training process of classification network (the initial value is set as 0.9).

It can be explained that on one side, it can better focus on the object regions for the anomaly attention module than that when only the feature extraction module is applied, and improve the condition that background noises are apt to be erroneously identified as anomalies. On the other hand, due to the interference of the false-alarm blocks, the difficulty in differentiating anomalies from background noises is increased for classification model, which causes us to conclude that the classification network is harder to be optimize as Figure 4b.

Then, we report the results of comparative experiments with Compact CNN in Ref. [19] on the original and augmented datasets as depicted in Table 2.

Table 2. Comparative experiments with Compact CNN on original and augmented datasets (Our results are marked with square brackets. TPR: True Positive Rate; TNR: True Negative Rate).

	Original		**Augmented**	
Class	**TPR**	**TNR**	**TPR**	**TNR**
1	100[100]	96.4[100]	100[100]	98.8[100]
2	98.8[100]	99.6[100]	100[100]	99.8[100]
3	100[100]	97.1[100]	100[100]	96.3[100]
4	77.9[100]	95.7[100]	98.5[100]	99.8[100]
5	100[100]	99.6[100]	100[100]	100[100]
6	100[100]	100[100]	100[100]	100[100]
7	100[100]	98.9[100]	100[100]	100[100]
8	100[100]	99.9[100]	100[100]	100[100]
9	100[100]	100[100]	100[100]	99.9[100]
10	100[100]	99.7[100]	100[100]	100[100]

From Table 2 we can see that Compact CNN is apt to be affected by the imbalance of positive and negative samples, while our model is not very sensitive to this quantity difference and can work well under the two conditions. Therefore, the proposed approach is quite applicable to practical industrial scenarios where defective samples are hard to acquire but numerous defect-free samples are usually available.

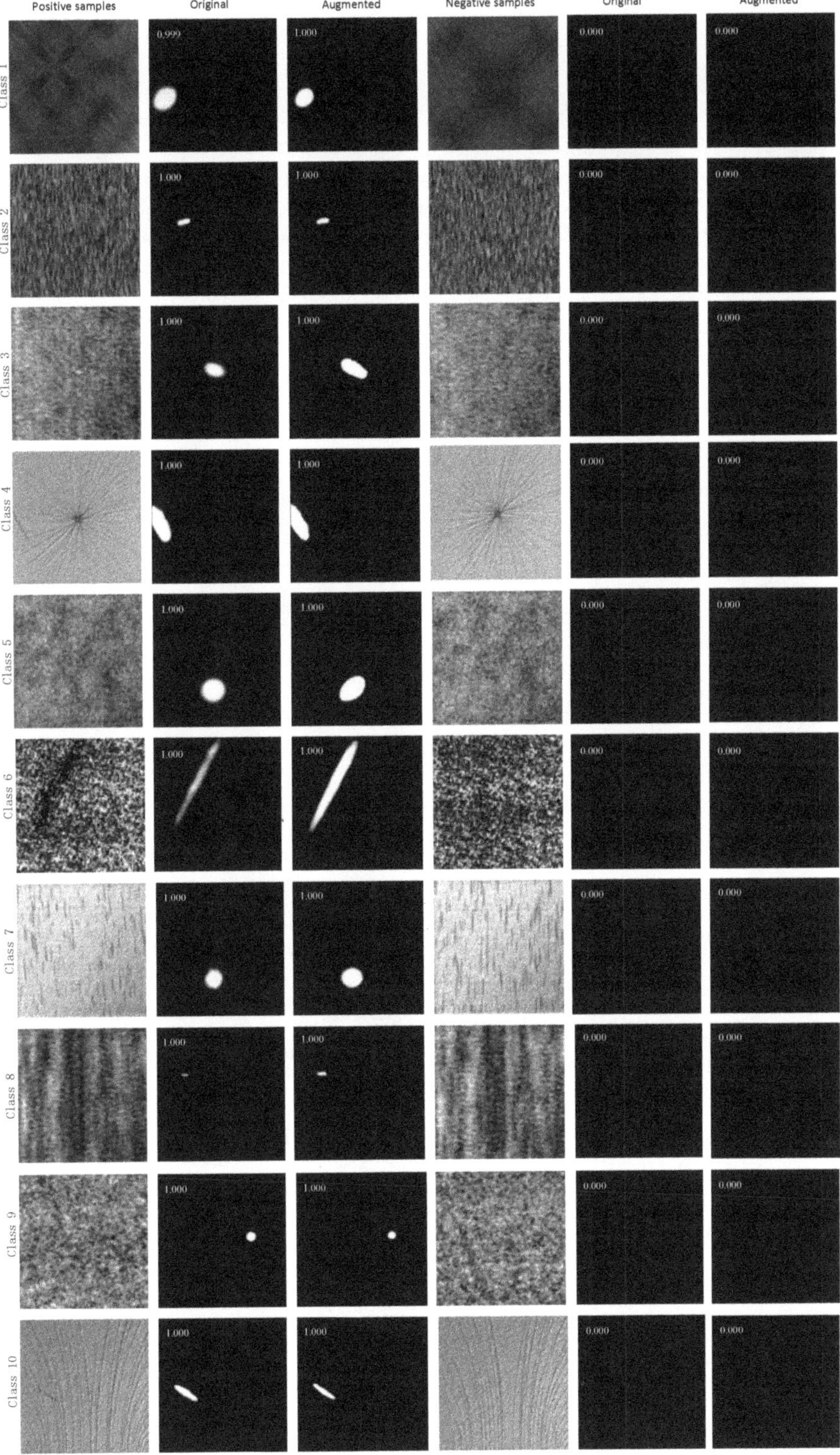

Figure 5. Examples of segmentation and classification.

Finally, more comparative results are listed in Table 3 for widespread comparison and comprehensive analysis. It is clear that deep learning-based approaches bring about significant performance improvements relative to feature selection methods, which can be explained by the fact that deep learning models can extract more high-level feature representations which are similar to the intrinsic properties of defects and backgrounds, while hand-crafted features merely describe the coarse or middle-level information, and the ability of feature expression is largely limited. Moreover, the proposed model can achieve better results than previous deep learning-based works on DAGM 2007. We hold that the proposed attention-based fusion module plays a crucial role in it. Specifically, it is known that the classification network is highly dependent on the frozen segmentation parameters, and the presented attention-based fusion approach can further optimize them to improve segmentation outputs in the ways of highlighting the potential anomalies and weakening background noises. A number of examples are shown in Figure 5. Figure 6 shows some samples compared with Compact CNN.

Table 3. Classification performance of the proposed model vs. others (TPR: True Positive Rate; TNR: True Negative Rate).

	Proposed	Compact CNN [19]	FC-CNN [18]	SIF [27]	Weibull [28]
Class			**TPR(TNR)**		
1	100(100)	100(98.8)	100(100)	98.9(100)	87.0(98.0)
2	100(100)	100(99.8)	100(97.3)	95.7(91.3)	-
3	100(100)	100(96.3)	95.5(100)	98.5(100)	99.8(100)
4	100(100)	98.5(100)	100(98.7)	-	-
5	100(100)	100(100)	98.8(100)	98.2(100)	97.2(100)
6	100(100)	100(100)	100(99.5)	99.8(100)	94.9(100)
7	100(100)	100(100)	-	-	-
8	100(100)	100(100)	-	-	-
9	100(100)	100(99.9)	-	-	-
10	100(100)	100(100)	-	-	-

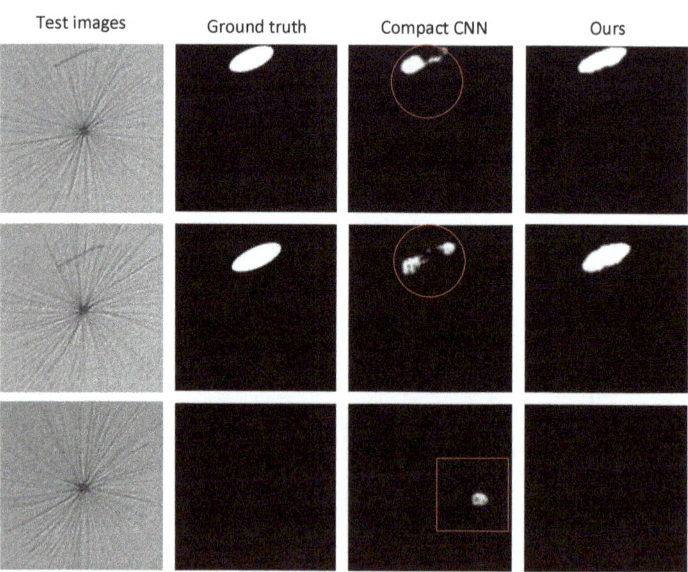

Figure 6. Examples compared with Compact CNN. The red circles and squares denote information loss owing to the lack of detailed features and the areas erroneously detected due to background interference, respectively.

4. Conclusions

We put forward an attention-based approach to improve textured surface anomaly detection. A number of experiments demonstrate that our approach is quite insensitive to the imbalance of positive and negative samples; meanwhile, 100% detection results can be achieved without false alarms and missing detections on the original as well as the augmented DAGM 2007 dataset. Consequently, it can be expected that the proposed model will be further applied in the practical industrial scenes where the quantity of anomaly samples is usually limited. Finally, how to implement the quantitative comparison for the segmentation result under weak supervised labels will be the focus of our next work.

Author Contributions: Conceptualization, G.L. and N.Y.; methodology, G.L.; software, G.L.; resources, L.G.; writing-original draft preparation, G.L.; project administration and writing–review & editing, N.Y. All authors have read and agreed to the published version of the manuscript.

Funding: This research received no external funding.

Acknowledgments: We sincerely appreciate the contribution of the DAGM and GNSS institutions for the open dataset to promote the development of textured surface anomaly detection.

Conflicts of Interest: The authors declare no conflict of interest.

References

1. Masci, J.; Meier, U.; Ciresan, D.; Schmidhuber, J.; Fricout, G. Steel Defect Classification with Max-Pooling Convolutional Neural Networks. In Proceedings of the IEEE International Joint Conference on Neural Networks (IJCNN), Brisbane, Australia, 10–15 June 2012; pp. 1–6.
2. Ghorai, S.; Mukherjee, A.; Gangadaran, M.; Dutta, P.K. Automatic defect detection on hot-rolled flat steel products. *IEEE Trans. Instrum. Meas.* **2013**, *62*, 612–621. [CrossRef]
3. Napoletano, P.; Piccoli, F.; Schettini, R. Anomaly detection in nanofibrous materials by cnn-based self-similarity. *Sensors* **2018**, *18*, 209. [CrossRef] [PubMed]
4. Liu, G.K.; Yang, N.; Guo, L.; Guo, S.P.; Chen, Z. A One-Stage Approach for Surface Anomaly Detection with Background Suppression Strategies. *Sensors* **2020**, *20*, 1829. [CrossRef] [PubMed]
5. Sanchez-Brea, L.M.; Siegmann, P.; Rebollo, M.A.; Bernabeu, E. Optical technique for the automatic detection and measurement of surface defects on thin metallic wires. *Appl. Opt.* **2000**, *39*, 539–545. [CrossRef] [PubMed]
6. LeCun, Y.; Bengio, Y.; Hinton, G. Deep learning. *Nature* **2015**, *521*, 436. [CrossRef] [PubMed]
7. Dumoulin, V.; Visin, F. A guide to convolution arithmetic for deep learning. *arXiv* **2016**, arXiv:1603.07285.
8. Song, K.; Yan, Y.H. Micro surface defect detection method for silicon steel strip based on saliency convex active contour model. *Math. Probl. Eng.* **2013**, *2013*, 1–13. [CrossRef]
9. Shanmugamani, R.; Sadique, M.; Ramamoorthy, B. Detection and classification of surface defects of gun barrels using computer vision and machine learning. *Measurement* **2015**, *60*, 222–230. [CrossRef]
10. Gong, R.; Chu, M.; Wang, A.; Yang, Y. A fast detection method for region of defect on strip steel surface. *Isij Int.* **2015**, *55*, 207–212. [CrossRef]
11. Lin, T.Y.; Dollár, P.; Girshick, R.; He, K.M.; Hariharan, B.; Belongie, S. Feature pyramid networks for object detection. In Proceedings of the IEEE Conference on Computer Vision and Pattern Recognition (CVPR), Honolulu, HI, USA, 21–26 July 2017; pp. 2117–2125.
12. He, K.M.; Zhang, X.Y.; Ren, S.Q.; Sun, J. Deep residual learning for image recognition. In Proceedings of the IEEE Conference on Computer Vision and Pattern Recognition (CVPR), Las Vegas, NV, USA, 26 June–1 July 2016; pp. 770–778.
13. Badrinarayanan, V.; Kendall, A.; Cipolla, R. SegNet: A deep convolutional encoder-decoder architecture for image segmentation. *IEEE Trans. Pattern Anal. Mach. Intell.* **2017**, *39*, 2481–2495. [CrossRef]
14. Long, J.; Shelhamer, E.; Darrell, T. Fully convolutional networks for semantic segmentation. In Proceedings of the IEEE Conference on Computer Vision and Pattern Recognition (CVPR), Boston, MA, USA, 7–12 June 2015; pp. 3431–3440.

15. Oktay, O.; Schlemper, J.; Folgoc, L.L.; Lee, M.; Heinrich, M.; Misawa, K.; Mori, K.; McDonagh, S.; Hammerla, N.Y.; Kainz, B.; et al. Attention U-Net: Learning Where to Look for the Pancreas. *arXiv* **2018**, arXiv:1804.03999.
16. Zhou, T.; Fu, H.; Chen, G.; Shen, J.; Shao, L. Hi-net: Hybrid-fusion network for multi-modal MR image synthesis. *IEEE Trans. Med. Imaging* **2020**. [CrossRef] [PubMed]
17. Liu, Y.; Cheng, M.M.; Hu, X.; Wang, K.; Bai, X. Richer convolutional features for edge detection. In Proceedings of the IEEE Conference on Computer Vision and Pattern Recognition (CVPR), Honolulu, HI, USA, 21–26 July 2017; pp. 3000–3009.
18. Weimer, D.; Scholz-Reiter, B.; Shpitalni, M. Design of deep convolutional neural network architectures for automated feature extraction in industrial inspection. *CIRP Ann. Manuf. Technol.* **2016**, *65*, 417–420. [CrossRef]
19. Racki, D.; Tomazevic, D.; Skocaj, D. A compact convolutional neural network for textured surface anomaly detection. In Proceedings of the IEEE Winter Conference on Applications of Computer Vision (WACV), Lake Tahoe, CA, USA, 12–15 March 2018; pp. 1331–1339.
20. Mei, S.; Wang, Y.; Wen, G. Automatic fabric defect detection with a multi-scale convolutional denoising autoencoder network model. *Sensors* **2018**, *18*, 1064. [CrossRef] [PubMed]
21. Tabernik, D.; Šela, S.; Skvarč, J.; Skočaj, D. Segmentation-based deep-learning approach for surface-defect detection. *J. Intell. Manuf.* **2019**, *31*, 759–776. [CrossRef]
22. Ronneberger, O.; Fischer, P.; Brox, T. U-net: Convolutional networks for biomedical image segmentation. In Proceedings of the International Conference on Medical Image Computing and Computer-Assisted Intervention, Munich, Germany, 5–9 October 2015; pp. 234–241.
23. Chen, L.C.; Zhu, Y.; Papandreou, G.; Schroff, F.; Adam, H. Encoder-decoder with atrous separable convolution for semantic image segmentation. In Proceedings of the European Conference on Computer Vision, (ECCV), Munich, Germany, 8–14 September 2018; pp. 801–818.
24. Abadi, M.; Barham, P.; Chen, J.; Chen, Z.; Davis, A.; Dean, J.; Devin, M.; Ghemawat, S.; Irving, G.; Isard, M. Tensorflow: A system for large-scale machine learning. In Proceedings of the Symposium on Operating Systems Design and Implementation (OSDI), Savannah, GA, USA, 2–4 November 2016; pp. 265–283.
25. Ioffe, S.; Szegedy, C. Batch Normalization: Accelerating Deep Network Training by Reducing Internal Covariate Shift. *arXiv* **2015**, arXiv:1502.03167.
26. Kingma, D.P.; Ba, J. Adam: A method for stochastic optimization. *arXiv* **2014**, arXiv:1412.6980.
27. Scholz-Reiter, B.; Weimer, D.; Thamer, H. Automated surface inspection of cold-formed micro-parts. *CIRP Ann. Manuf. Technol.* **2012**, *61*, 531–534. [CrossRef]
28. Timm, F.; Barth, E. Non-parametric texture defect detection using Weibull features. In Proceedings of the SPIE—The International Society for Optical Engineering, San Francisco, CA, USA, 25–27 January 2011; p. 78770J.

Article

A Comparison of Power Quality Disturbance Detection and Classification Methods Using CNN, LSTM and CNN-LSTM

Carlos Iturrino Garcia [1,2,*,†], Francesco Grasso [1,2,†], Antonio Luchetta [1,2,*,†], Maria Cristina Piccirilli [1,2,†], Libero Paolucci [1,2,*,†] and Giacomo Talluri [1,2,†]

1 Smart Energy Lab, University of Florence, 50139 Florence, Italy; francesco.grasso@unifi.it (F.G.); mariacristina.piccirilli@unifi.it (M.C.P.); giacomo.talluri@unifi.it (G.T.)
2 Department of Information Engineering, University of Florence, 50139 Florence, Italy
* Correspondence: carlos.iturrinogarcia@unifi.it (C.I.G.); antonio.luchetta@unifi.it (A.L.); libero.paolucci@unifi.it (L.P.)
† These authors contributed equally to this work.

Received: 30 July 2020; Accepted: 21 September 2020; Published: 27 September 2020

Abstract: The use of electronic loads has improved many aspects of everyday life, permitting more efficient, precise and automated process. As a drawback, the nonlinear behavior of these systems entails the injection of electrical disturbances on the power grid that can cause distortion of voltage and current. In order to adopt countermeasures, it is important to detect and classify these disturbances. To do this, several Machine Learning Algorithms are currently exploited. Among them, for the present work, the Long Short Term Memory (LSTM), the Convolutional Neural Networks (CNN), the Convolutional Neural Networks Long Short Term Memory (CNN-LSTM) and the CNN-LSTM with adjusted hyperparameters are compared. As a preliminary stage of the research, the voltage and current time signals are simulated using MATLAB Simulink. Thanks to the simulation results, it is possible to acquire a current and voltage dataset with which the identification algorithms are trained, validated and tested. These datasets include simulations of several disturbances such as Sag, Swell, Harmonics, Transient, Notch and Interruption. Data Augmentation techniques are used in order to increase the variability of the training and validation dataset in order to obtain a generalized result. After that, the networks are fed with an experimental dataset of voltage and current field measurements containing the disturbances mentioned above. The networks have been compared, resulting in a 79.14% correct classification rate with the LSTM network versus a 84.58% for the CNN, 84.76% for the CNN-LSTM and a 83.66% for the CNN-LSTM with adjusted hyperparameters. All of these networks are tested using real measurements.

Keywords: power quality disturbances; long short term memory; convolutional neural network; short time Fourier transform

---

1. Introduction

The wide diffusion of electronic loads in the industrial, household, commercial and public sectors has improved many aspects of everyday life. In other words, power electronics technologies have made life easier and more comfortable. On the other hand electronic devices have a nonlinear behavior that disturbs the power grid through voltage and current waveform distortions. The number of power electronics devices that are connected to the grid is constantly increasing; as a consequence, the waveform distortion levels have also increased in the last decades causing a degradation of the Power Quality (PQ) levels on the grid. Ideally, grid voltages and currents should have a purely sinusoidal behavior. If distorting components are injected, power losses can occur as well as malfunctioning of electric devices. This can severely damage industrial equipment, household appliances and commercial business plants. They also cause disturbance to other consumers and interference in nearby communication networks [1]. Besides that, the energy providers can sanction the injection of such disturbances on the grid. These disturbances concern frequency, amplitude, waveform and—in three-phase systems—symmetry of voltages and currents. Moreover, high energy demanding companies are becoming more sensitive to loss of profit margins due to power losses and plant shoutdowns caused by low PQ levels [2]. In order to adopt countermeasures and to define the origin of the phenomena, it is important to detect and classify these disturbances. This information indeed can be used to define the PQ level of a grid, to understand their behavior and to assess the responsibilities. The operation can be carried on by acquiring and processing the voltage and current signals of a power line. To do this, several techniques are currently exploited using machine learning algorithms [3]. Among the extensive inventories of deep learning algorithms, for the present work, the Long Short Term Memory (LSTM) and the Convolutional Neural Networks (CNN) are being used to detect and classify these disturbances [4–6]. Other algorithms are being used to address these problems like the Kalman Filter, Wavelet Transform and the Support Vector Machine (SVM).

In [7], a Kalman Filter is used in an UPQC to extract the state components of the distorted supply voltage and load current. The algorithm can classify PQD internally enabling the conditioning of the PQ signals for power factor correction. The technique seems to work well with the detection of sag, swell and harmonic distortion, however it shows a certain lag between the disturbance starting condition and the detection [8]; furthermore, the algorithm is usually applied to a restricted number of disturbances. On the other hand, the wavelet transform is used as a tool for analyzing PQD as shown in [9]. The tool is very useful for the extraction of the signals features for learning algorithms like the SVM as shown in [10–12]. However, it does not perform disturbances detection by itself. The SVM showed interesting performances for the detection of a wide range of PQ disturbances and it is often used as a benchmark to assess the performances of other algorithms. The main disadvantage of the PQD detection techniques mentioned above is that, once the voltage and current waveforms are acquired, a preprocessing of the signal must be performed before feeding it to the algorithm. This usually consists of a signal features extraction. Deep learning algorithms solves this problem by implicitly applying a feature extraction for the classification of the signal. In other words, these algorithms could be fed with raw data and still make accurate classifications. This can help to speed up the identification and classification process especially in real time applications.

For the training and the validation of deep learning algorithms, it is necessary an extensive dataset in order to avoid overfitting and obtain generalization. Unfortunately, it is not easy to obtain such datasets with experimental data. One reason is that performing on-field data sampling through measurement campaigns is time consuming, many of these disturbances indeed are sporadic, and it is not always possible to record an event with a desired amplitude and duration. For that reason, simulated voltage disturbances are used in order to create the dataset for training and validation. For further generalizing the dataset, data augmentation is used, since it has proven to be efficient in improving accuracy by reducing overfitting [13,14].

This work explores different deep learning architectures which were trained and validated using simulated data and tested using experimental data. Once the simulated data are generated, it is then augmented, in order to obtain a generalized result and overcome any sampling discrepancy and phase difference between simulated data and measured data. The signals are pre-processed in order to compare the accuracy of each architecture in their proven classification tasks. With respect to other works in which the training, validation and testing steps are performed using purely simulated data or purely experimental data [4,15,16], in the present work the training and validation steps are performed with simulated datasets, while the testing one is performed with experimental datasets that were acquired on the field.

2. Power Quality Disturbance Simulations and Dataset Acquisition

The definition of Power Quality by the IEEE is: Power Quality is the concept of powering and grounding sensitive equipment in a matter that is suitable to the operation of that equipment [17]. The definition given by the International Electrotechnical Commission is: The characteristics of the electricity at a given point on an electrical system, evaluated against a set of reference technical parameters. These parameters might, in some cases, relate to the compatibility between electricity supplied on a network and the loads connected to that network. From these definitions it can be said that PQ always includes voltage quality and supply reliability. The Power Quality Disturbances (PQD) are broadly classified into three categories: magnitude variations, sudden transients and steady-state harmonics, as said in [18].

As a preliminary stage of this work, the voltage and current time signals were simulated using MATLAB Simulink. By doing this, it has been possible to recreate the disturbances on the line and see how they interact with the other devices connected to the grid. Thanks to the simulation results, it has been possible to acquire a current and voltage dataset with which the identification algorithms were trained and validated. This dataset includes simulations of several disturbances such as Sag, Swell, Harmonics, Transient, Notch and Interruption. After that, the deep learning algorithms were tested with an experimental dataset of voltage and current field measurements containing the disturbances mentioned above and it has been possible to perform a performance comparison.

For the generation of the PQD dataset a MATLAB/Simulink model of a micro grid has been implemented. The model is shown in Figure A1 and it includes several different industrial loads. It is possible to identify a three-phase dynamic load which could be associated to an electrical motor with variable load, a linear load and a nonlinear load which injects disturbances on the net [18]. These disturbances include: Sag, Swell, Harmonics, Transient, Notch and Interruption.

The Simulink schematic for the PQD simulation consist of a three-phase voltage source connected to a fault block, then to a line impedance, to a transformer and finally ending in a linear load. The voltages and currents are measured after the transformer which are going to be used for the classification. Inside the fault block, there are different types of disturbance generation blocks which were listed above. The sag block reduces the voltage from its nominal value. The swell block rises the voltage from its nominal value, and it is modeled with a switch that connects the grid to a capacitor bank. The harmonic distortion is modeled as a resistor in parallel with a capacitor in parallel with a free willing diode. The transient is modeled with an impulse generator. The notch block is modeled by a thrysistor in parallel with a resistance and an inductance. Finally the interruption block is simply a switch. The simulink schematic for the simulation of the PQD along with each disturbance block are shown in Appendix A.

3. Machine Learning Algorithms

There is a wide literature concerning the feature extraction and classification of PQD exploiting different types of Machine Learning architectures. In [19], Borges implements a feature extraction using statistical data of PQD for classification. Shen in [20], uses an Improved Principal Component Analysis (IPCA) to extract the statistical features of the PQD followed by a 1-dimensional Convolutional Neural Network (1-D-CNN) to extract other features of the signal and for the classification. The results of the classification in this work were compared to a Support Vector Machine (SVM) In terms of accuracy, which is the ratio between correct classification and total classifications, the SVM gave 98.55% accuracy while the 1-D-CNN scored 99.75%. These results proved that the 1-D-CNN is slightly superior to the SVM in classifying PQD.

In [4], Mohan explores the potential of deep learning architectures for PQ disturbances classification. In this work the convolutional neural network (CNN), recurrent neural network (RNN), identity-recurrent neural network (I-RNN), long short-term memory (LSTM), gated recurrent units (GRU) and convolutional neural network-long short-term memory (CNNLSTM) are studied and compared in order to find the best architecture for PQD data. The accuracy of each deep learning architectures is: 98% for the CNN, 91.5% for RNN, 93.6% for the I-RNN, 96.7% for the LSTM, 96.4% for the GRU and 98.4% for the hybrid CNN-LSTM. These results proved that the hybrid CNN-LSTM is superior at classifying PQD.

The hallmark of deep learning architectures is that they are able to perform a feature extraction and classification by processing raw data. However, many other works proved that these techniques are also successful for the signal feature extraction for different applications including PQD [21–23].

3.1. Long Short Term Memory

A recurrent neural network (RNN) is a neural network that simulates a discrete-time dynamical system that has an input x_t, an output y_t and a hidden state h_t as defined in [24]. A drawback of the RNNs is that they suffer from vanishing or exploding gradient. By truncating the gradient where it does not harm, LSTM can learn to bridge minimal time lags in excess of 1000 discrete-time steps by enforcing constant error flow through constant error carousels within special units [25].

The LSTM has three states that help the network to reduce the long term dependency of the data. These states are called the Forget State, the Input State and the Output State. The Forget State eliminates redundant or useless data. The Input State processes the new data and finally the Output State processes the input data with the cell state. A block diagram of a LSTM cell is shown in Figure 1. In the following subsections a focus on the three steps is presented.

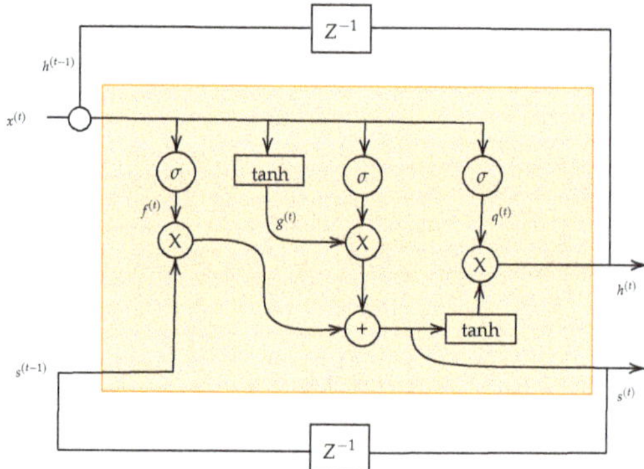

Figure 1. Block diagram of one cell of a long short term memory architecture.

3.1.1. Forget State

The forget state controls the state parameter $s^{(t)}$ via a sigmoid function σ. This state controls what the cell should remember through time and what to forget. The equation of the forget state is shown in Equation (1), where $f^{(t)}$ is the forget vector, x^t and $h^{(t-1)}$ are the input and previous output respectively. The input and the previous output are multiplied by the trained weights U and W with bias b. This result is then truncated between 0 and 1 via a sigmoid function. Basically the idea is to have an input vector added with the previous output vector passed through a neural network which outputs the values to keep with a 1 and the values to forget with a 0.

$$f_i^{(t)} = \sigma\left(b_i^f + \sum_j U_{i,j}^f x_j^t + \sum_j W_{i,j}^f h_j^{(t-1)}\right) \tag{1}$$

3.1.2. Input State

The new state of the cell is defined in the input state where the previous state is multiplied by the forget state dropping off irrelevant information. This is shown in $f_i^{(t)} s_i^{(t-1)}$ of Equation (2). Now the relevant information gets updated in $g_i^{(t)}\sigma\left(b_i + \sum_j U_{i,j}x_j^t + \sum_j W_{i,j}h_j^{(t-1)}\right)$ which is the product between the input and the previous neural network output times $g^{(t)}$ which is the candidate for the next time step of the cell state. The equation that generates the vector that contains the candidates for the next cell state is shown in Equation (3).

$$s_i^{(t)} = f_i^{(t)} s_i^{(t-1)} + g_i^{(t)} \sigma\left(b_i + \sum_j U_{i,j} x_j^t + \sum_j W_{i,j} h_j^{(t-1)}\right) \tag{2}$$

$$g_i^{(t)} = \tanh\left(b_i^g + \sum_j U_{i,j}^g x_j^t + \sum_j W_{i,j}^g h_j^{(t-1)}\right) \tag{3}$$

3.1.3. Output State

The output state decides what should be the output of the cell and of the new cell state. The output of the cell is shown in Equation (4) where the cell state goes through a hyperbolic tangent and it is then multiplied by the output of another hidden layer as shown in Equation (5).

$$h_i^t = \tanh\left(s_i^{(t)}\right) q_i^{(t)} \tag{4}$$

$$q_i^{(t)} = \sigma\left(b_i^o + \sum_j U_{i,j}^o x_j^t + \sum_j W_{i,j}^o h_j^{(t-1)}\right) \tag{5}$$

3.2. *Convolutional Neural Networks (CNN)*

Convolutional neural networks or CNN, are a particular type of neural network for data processing that has a grid-like topology. Convolutional networks proved to be successful in several practical applications. They essentially consist of neural networks that use convolution in place of general matrix multiplication in at least one of their layers [26]. The convolutional layer is accompanied by a pooling layer which is a type of under sampling that helps with processing speed.

Since the signals of interest are 1-D signals and the CNN processes a 2-D signal, a pre-processing of each signals is necessary. Hence, the Short Time Fourier Transform is performed on each signal before feeding it to the CNN; by doing this, an image containing the spectral components and amplitude of the signal of interest is generated. The characteristics of this processing technique are highlighted in Section 4.1.

3.2.1. Convolutional Layer

Convolution leverages three important ideas that can help improve a machine learning system: sparse interactions, parameter sharing and equivariant representations. Moreover, convolution provides a means for working with inputs of variable size [26]. The convolution layer of a convolutional neural network operates by applying a convolution to each dataset. Since the hallmark of the CNN is image classification, the 2 dimensional version of the discrete convolution is used. To serve as a reminder, the 2 dimensional discrete convolution operation is shown in Equation (6). The convolutional layer works by adjusting the parameter ω for each backpropagation in order to maximize the features extracted by minimizing the error in classification. This in turn creates a set of filters which are the key of the feature extraction of the data.

$$S[n_1, n_2] = \sum_{m=1}^{M_1} \sum_{m=1}^{M_2} x[m_1, m_2] \omega[n_1 - m_1, n_2 - m_2] \tag{6}$$

3.2.2. Pooling Layer

A pooling function replaces the output of the layer with a summary statistic of the previous layer outputs [26]. The most popular pooling functions include the max of a rectangular neighborhood, the average, the L2 norm, or a weighted average based on the distance from the central datum.

This layer in the architecture speeds up the training and classification since it undersamples the dataset and helps the network to obtain a more generalized result. An illustration of the pooling function is shown in Figure 2 where $\mathcal{X}_n$ is the vector containing the pooled data of the dataset as shown in Equation (7). In Equation (8) the pooling function is shown.

$$\mathcal{X}_n = \{x_j, ..., x_N\} \tag{7}$$

$$\hat{x}_n = f(x_n, x_{n+1}, x_{n+2}) = f(\mathcal{X}_n) \tag{8}$$

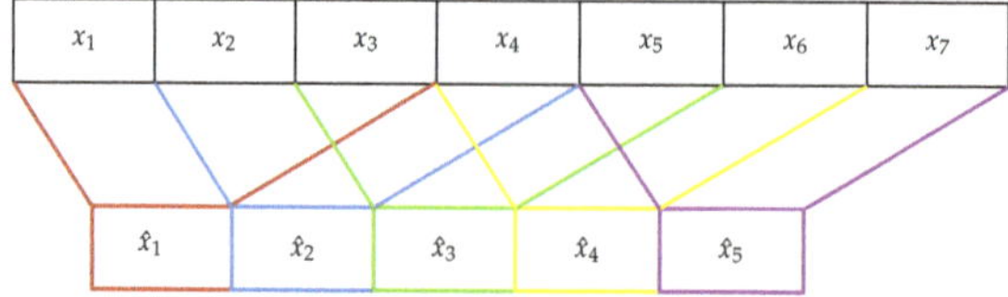

Figure 2. Pooling illustration for the CNN.

The pooling layer can contain either the maximum of the set, the average, the $\mathbf{L}_2$ norm or the weighted average of the pool. In this work the max pooling is used as shown in Equation (9).

$$f(\mathcal{X}_n) = argmax(\mathcal{X}_n) \tag{9}$$

4. Methodology

In this work, the end goal is to compare different Machine learning architectures in order to determine which is the best suited for the Power Quality disturbances identification task. The LSTM, the CNN, CNN-LSTM and the CNN-LSTM with adjusted hyperparameters were chosen for the comparison based on their performances on previous works [4]. The LSTM was designed specifically for time series data which is well suited for this task. The CNN has proven to be very effective in image classification tasks, hence it is necessary to treat the signal as an image. Since the success of CNN is attributed to its superior multi-scale high-level image representations [27], a time frequency analysis is executed to the time series data as shown in [28] so the signal can be treated as an image. To obtain this image, the Short-Time Fourier Transform is used. It has been selected because of its simplicity of implementation but others, like the wavelet transform, can be considered as well. Data Augmentation is executed in order to obtain a more generalized result for the training and validation. The different techniques are trained with simulated data and then tested with both simulated and experimental datasets. The datasets consist of three phase voltage signals which have one sinusoidal waveform shifted 120° with respect to the other; both in the experimental and in the simulated datasets, PQ disturbances are superposed to the three signals. The identification algorithms are fed with 1000 samples of the time series per classification. A rectangular sliding window is used with an overlap of 500 samples.

4.1. Short Time Fourier Transform (STFT)

As said in [29], one of the approaches for using Machine Learning techniques in the frequency domain is to transform the time series into the frequency domain first, and then apply conventional neural network components. As specified before, the transformation chosen for this task was the Short Time Fourier Transform. It is characterized by a Fourier transform executed in a fixed windowed interval. The window function $g(n)$ is called Blackman's window function and it is used to multiply a short segment of the signal by the window function. This avoids sharp sections and redundant information. A Fourier transform of this small windowed section $X_n(j\omega)$ is calculated and stacked up to form a matrix. The STFT equation is shown in Equation (10). An illustration of the Blackman's window and of the algorithm can be shown in Figures 3 and 4, respectively. The resulting matrix can now be treated as an image to train a Convolutional Neural Network which is capable of feature extraction of the frequency components of the signal. The Fourier Transform equation is shown in Equation (10) and the Blackman Window Function is shown in Equation (11).

$$X(j\omega) = \sum x(n)g(n - mR)e^{-2j\pi fn} \tag{10}$$

$$w[n] = a_0 + a_1 cos\left(\frac{2\pi n}{N}\right) + a_2 cos\left(\frac{4\pi n}{N}\right) \tag{11}$$

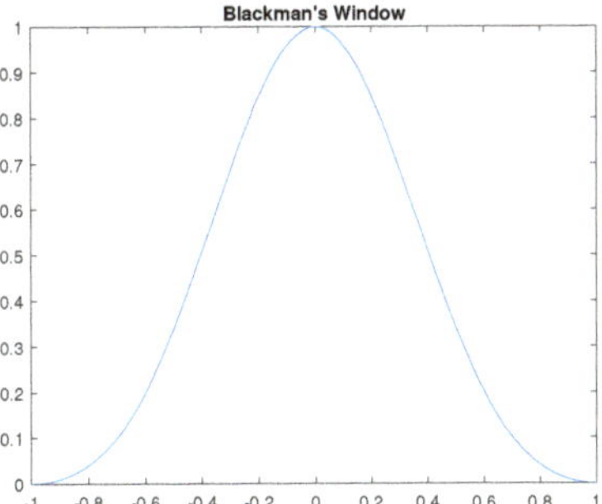

Figure 3. Blackman's window.

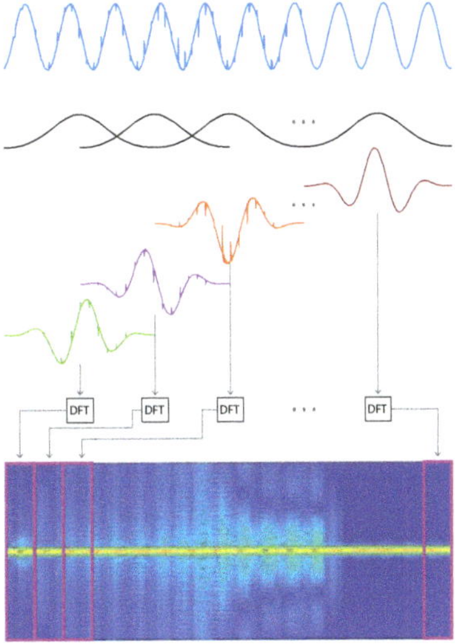

Figure 4. Short time Fourier transform illustration.

4.2. Data Augmentation

Although training and validation of the network using simulated data can be easily implemented and it is time efficient, it can often result in overfitting. To solve this issue and in order to obtain a more generalized result, data augmentation is necessary. Data augmentation consists of manipulating the training and validation set to obtain more data with small variations. The PQD waveforms obtained through simulations are limited by sampling time and starting time. In a simulated voltage signal the disturbance which is superposed to the signal is characterized by a fixed time interval. Through data augmentation, the PQD can be shifted of 0°, 60°, 120°, 180°, 240°, and 300°. These signals are then oversampled by 2 and 4 to deal with possible sampling discrepancies. In other words, for each of the three voltage waveforms, after data augmentations, six more signals are generated returning a total of 18 new waveforms.

$$\mathbf{Voltage}_{3\Phi}^{T} = \begin{bmatrix} \mathbf{V}_{\Phi_1}^{T} \\ \mathbf{V}_{\Phi_2}^{T} \\ \mathbf{V}_{\Phi_3}^{T} \end{bmatrix} \quad \mathbf{Voltage}_{3\Phi}^{0.5T} = \begin{bmatrix} \mathbf{V}_{\Phi_1}^{0.5T} \\ \mathbf{V}_{\Phi_2}^{0.5T} \\ \mathbf{V}_{\Phi_3}^{0.5T} \end{bmatrix} \quad \mathbf{Voltage}_{3\Phi}^{0.25T} = \begin{bmatrix} \mathbf{V}_{\Phi_1}^{0.25T} \\ \mathbf{V}_{\Phi_2}^{0.25T} \\ \mathbf{V}_{\Phi_3}^{0.25T} \end{bmatrix} \tag{12}$$

$$\mathbf{Class}_{d} = \begin{bmatrix} \mathbf{Voltage}_{3\Phi}^{T} \\ \mathbf{Voltage}_{3\Phi}^{0.5T} \\ \mathbf{Voltage}_{3\Phi}^{0.25T} \\ -\mathbf{Voltage}_{3\Phi}^{T} \\ -\mathbf{Voltage}_{3\Phi}^{0.5T} \\ -\mathbf{Voltage}_{3\Phi}^{0.25T} \end{bmatrix} \tag{13}$$

4.3. Implementation of the Simulink Schematic for the Generation of the Simulated Dataset

For the generation of the simulated dataset, a Simulink model of the grid has been used [18]. With this model it has been possible to generate several PQ disturbances and organize it in a cell array. The disturbances that were implemented in the simulink model are the sag, the voltage rise, the harmonic distortion, the transient, the notch and the interruption [30]. After the Simulink simulation is completed, a dataset is generated. Hence the data are gathered with a script which generates a structured cell array. Once the process is completed, each fault is labeled with a target number as shown in Equation (14). To train the neural network, the data for the Class and Target are shuffled together in order to obtain a generalized solution for the network.

$$Class = \begin{bmatrix} \textbf{Normal} \\ \textbf{Sag} \\ \textbf{Swell} \\ \textbf{Harmonics} \\ \textbf{Transient} \\ \textbf{Notch} \\ \textbf{Interruption} \end{bmatrix} \quad Target = \begin{bmatrix} 0 \\ 1 \\ 2 \\ 3 \\ 4 \\ 5 \\ 6 \end{bmatrix} \tag{14}$$

After the time series signals are stored in a structured array they are shuffled. The next step is to train the network, validate and test it. For this task, the dataset containing the time series signals is separated in two groups: one containing 75% of the signals, used for the training and the other group containing the remaining 25%, used for the validation. These networks are trained with a batch size of 20 data points with a maximum of 30 epoch making a total of 15 iterations per epoch. After training is completed for all architectures, the performance of the networks are evaluated using a confusion matrix. With this, the precision and recall are calculated for each class.

4.4. Deep Learning Architectures

Deep learning in neural networks is the approach of composing networks into multiple layers of processing with the aim of learning multiple levels of abstraction [26]. By doing so, the network can adaptively learn low-level features from raw data and higher-level features from low-level ones in a hierarchical manner, nullifying the over-dependence of shallow networks on feature engineering [31]. The two most popular types of networks are the feed forward networks and the recurrent networks. Both have evolved in what is known today as deep learning architectures. Concerning the recurrent networks the LSTM is considered, it is mostly used for time series data; with regards to feed forward networks the CNN is considered, it is used mostly for image classification. Both are found to be successful in classification problems.

4.4.1. Long-Short Term Memory

Concerning the architecture under evaluation, 100 hidden units were used, that is, 100 LSTM blocks were used for the classification of time series data. Since in this architecture it is not possible to use a pooling layer, a drop out layer is used in order to achieve generalization. After that, a fully connected layer, a soft max layer and a classification layer, which outputs the final result, are added to the architecture. The architecture is shown in Figure 5.

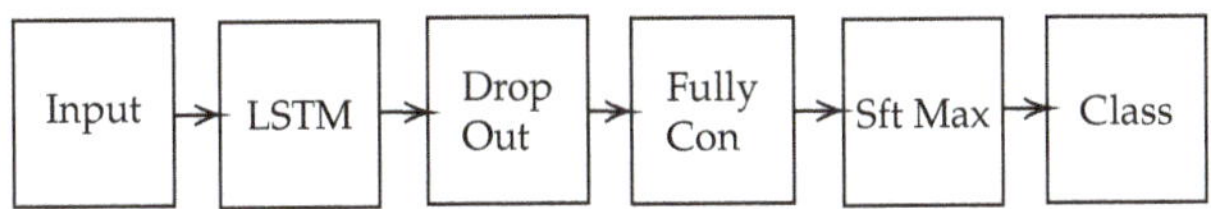

Figure 5. LSTM architecture block diagram.

4.4.2. Convolutional Neural Networks

The architecture used in this experiment has 3 stages of convolution, that is, 3 convolutional layers with 3 batch normalizations, 3 rectifier linear units or ReLU's and 2 under samplings using the max pooling layers. After the three stages of convolution, the network has a fully connected layer, a softmax layer and finally the output with the classification layer. Concerning the input, as specified before, it is necessary to first preprocess the training data using the STFT. The architecture can be seen in Figure 6 with all the details of the hyperparameters. Examples of each disturbance STFT are shown in Figure A2a–l.

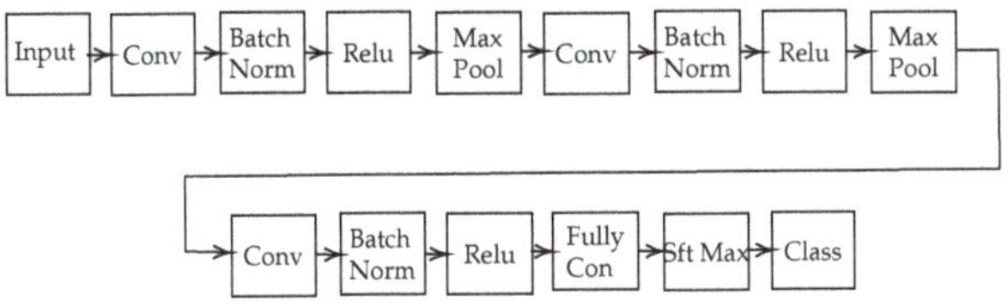

Figure 6. CNN architecture block diagram.

4.4.3. Convolutional Neural Networks—Long-Short Term Memory

This architecture mixes the CNN with the LSTM. In order to do this a sequence folding layer right after the input layer is added. The sequence folding layer converts a batch of data sequences to a batch of data. After this layer, the CNN comes into play. After the CNN, there is a sequence unfolding layer used to convert the batch of data in a batch of sequenced data. The sequence data are the input to the LSTM. Before the LSTM layer, there is a flattening layer that reshapes the input data to the input of the LSTM layer. Then a fully connected layer, a soft max and finally a classification layer are added respectively. The block diagram of the architecture is shown in Figure 7. By adjusting the parameters of the above mentioned architecture the CNN-LSTM with adjusted hyperparameters is obtained.

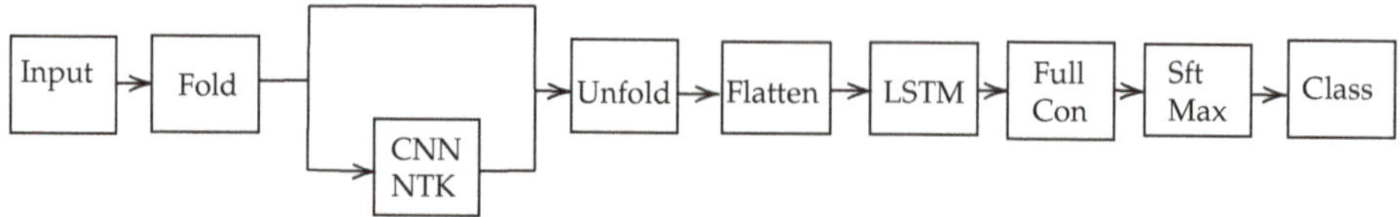

Figure 7. LSTM-CNN architecture block diagram.

5. Experimental Setup

To provide an experimental testbed for the validation of the simulation results, it has been necessary to test the identification techniques with experimental datasets. It is possible to extract this data by acquiring voltages and currents directly from the grid; however, to operate a comprehensive testing, it is necessary to reproduce different fault types by varying the amplitude, duration and intensity of the phenomena. Hence an experimental setup has been designed and implemented with which it is possible to reproduce and record several real-time PQDs. The system has been designed by the Smart Energy Lab of the University of Florence [18] and it is shown in Figure 8. To simulate the grid with PQD, a Chroma 61,500/61,600 series programmable AC source is used. To perform the measurement, a Fluke 435 Series II Power Quality Energy Analyzer (Leasametric, Villebon-sur-Yvette, France) and two Yokogawa PX8000-F-HE/G5/M1 (Yokogawa, Musashino, Japan). The first instrument was used to monitor the loading of the loads and check for any malfunctions, while with the Yokogawa instruments it was possible to obtain measures of the electrical quantities of the system recording up to 100,000 samples per second (100 kS/s) and averaging the measurement over a set period (minimum 10 ms). The load connected is given by a 1 kW linear load, a group of switching power supplies for a total of 2 kW and a three-phase inverter of 2 kW. The signal generated by the Chroma 61,500/61,600 is acquired by the Yokogawa PX8000 via the PowerViewerPlus software and is sent to the remote PC for data storage. By doing this, it is possible to test and compare the performances of different algorithms and identify strengths and weaknesses while assuring repeatability of the experiments.

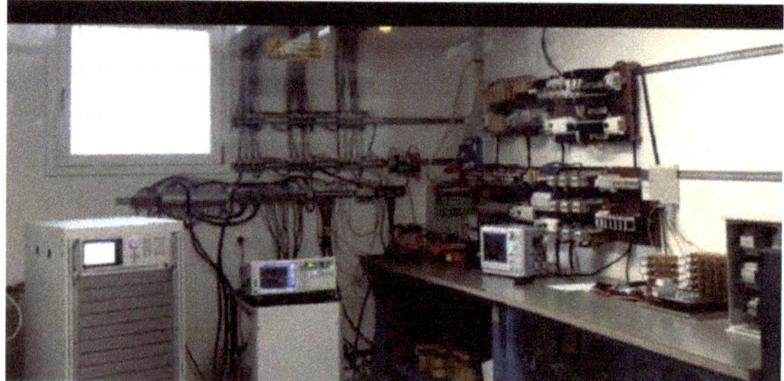

Figure 8. Experimental setup for power quality disturbances generation.

6. Testing Results

6.1. Testing of the Detection Techniques Using Simulated Signals

Simulations of the six different PQDs were made using the MATLAB Simulink model shown in Figure A1. In total, 300 simulations of each PQD were generated to create the dataset, each one containing different start times and duration of the disturbance. This signals were stacked together in a cell array along with its corresponding labels. Data augmentation was applied as explained in Section 4.2. The cell array was shuffled and divided into 75% for training and 25% for testing. This architecture had a training time of 671 min or 11.18 h with an accuracy of 79.14% for the validation.

The confusion matrix of the LSTM for training and validation is shown in Figure 9a,b along with the testing in Figure 9c. This architecture had a training time of 443 min or 7.38 h with an accuracy of 84.58% for the validation. The confusion matrix of the CNN for training and validation is shown in Figure 10a,b along with the testing in Figure 10c. This architecture had a training time of 751 min or 12.52 h with an accuracy of 83.66% for the validation. The confusion matrix of the CNN-LSTM for training and validation is shown in Figure 11a,b along with the testing in Figure 11c. This architecture had a training time of 51 min or 0.85 h with an accuracy of 84.78% for the validation. The confusion matrix of the CNN-LSTM for training and validation is shown in Figure 12a,b along with the testing shown in Figure 12c.

On each of the confusion matrix of the compared architectures, the precision and the recall was calculated. The precision of a classifier is defined as the number of retrieved relevant items as a proportion of the number of retrieved items for any given class [32]. In other words, it is the ratio between the positive identifications that are actually correct and the entire set of positive identifications of any given class. Recall, on the other hand, is defined as the number of retrieved relevant items as a proportion of all the relevant items, for any given retrieved set [32]. In other words, the proportion of the actual positives that where identified correctly. The comparison of precision and recall of different architectures are shown in Figure 13. The precision and recall of the LSTM-CNN with adjusted hyperparameters had superior results with respect to the other architectures due to the fact that it had better scores for classifying the transient in both training and testing.

The LSTM training gave an accuracy of 79.14% where most of the problems were found in the Transient disturbance as shown in the precision and recall plots. The architecture was not able to detect the transient disturbance either in the training or the validation signals, that is, the LSTM classified the Transient signal as a No Fault in 100% of the cases. Concerning the other classes it resulted in 10.2%–14.6% misclassification.

The CNN needs to be fed with an image which, in this case, represents the spectral components and amplitude of the signal of interest; hence a STFT was done on the augmented dataset. The training of the CNN gave an accuracy of 84.58% which is an improvement with respect to the LSTM. The problem with this architecture is that it classified 89% of the No Fault as a Transient from the training dataset and an 88.6% from the validation. Again, it showed confusion between the two classes as shown in the precision and recall plots.

As regards to the the hybrid CNN-LSTM, two similar strategies were tested. The first strategy consisted in joining the LSTM and CNN that were used in the previous experiment and the second was using the combined architectures while adjusting the hyperparameters. The first strategy, exploiting the CNN-LSTM, resulted in an improvement of the classification performances of almost all of the the disturbances except for the Transient which resulted in 100% misclassification as with the LSTM. The other disturbances misclassifications ranged between 2%–10%, which was an improvement. Concerning the second strategy exploiting the hybrid architecture, a significant improvement on the Transient response recognition was reached resulting in a 51.1% misclassification with the No Fault condition. In both, precision and recall, it showed more or less a 50% chance of miss-classification. For the other disturbances the misclassification ranges between 1.3% and 4.8% which is also an improvement. The hybrid CNN-LSTM with adjusted hyperparameters was able to detect the Transient disturbance in 48.9% of the signals where the transient was present. The other architectures failed completely in this task. Furthermore, this architecture obtained better results on the other disturbance classifications.

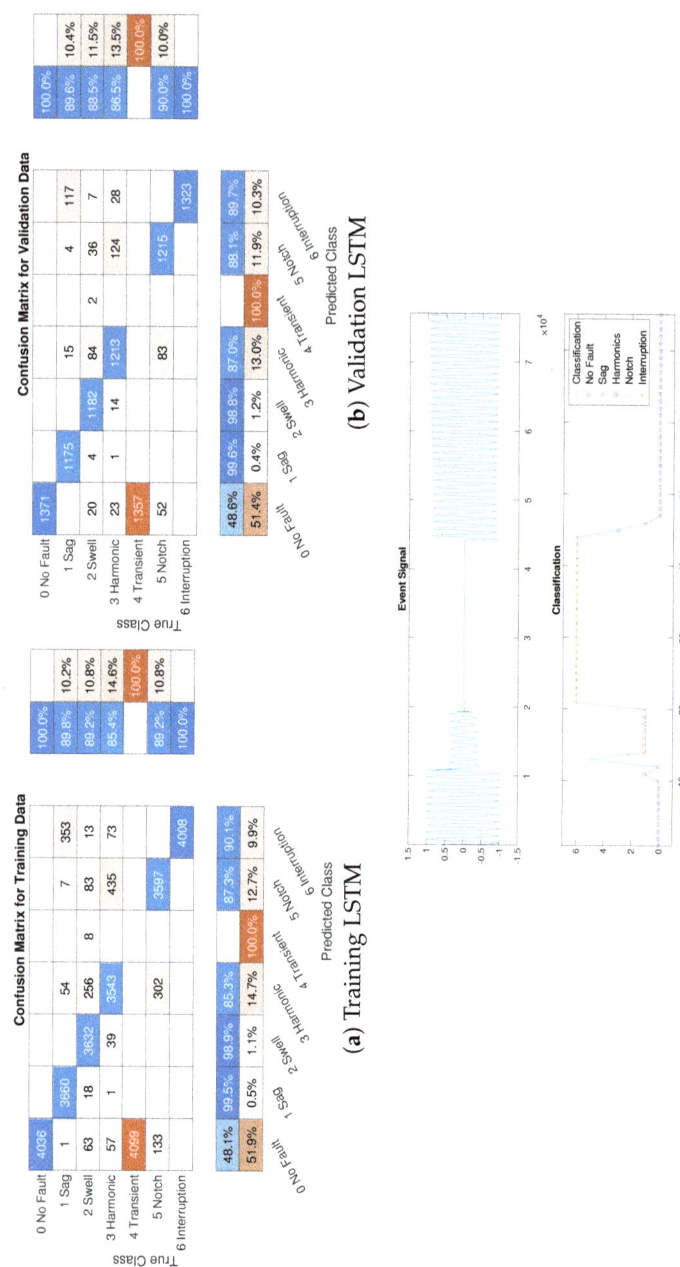

Figure 9. (**a**) Confusion matrix of the LSTM training, (**b**) Confusion matrix of the LSTM validation, (**c**) Testing of the LSTM using a simulated voltage waveform.

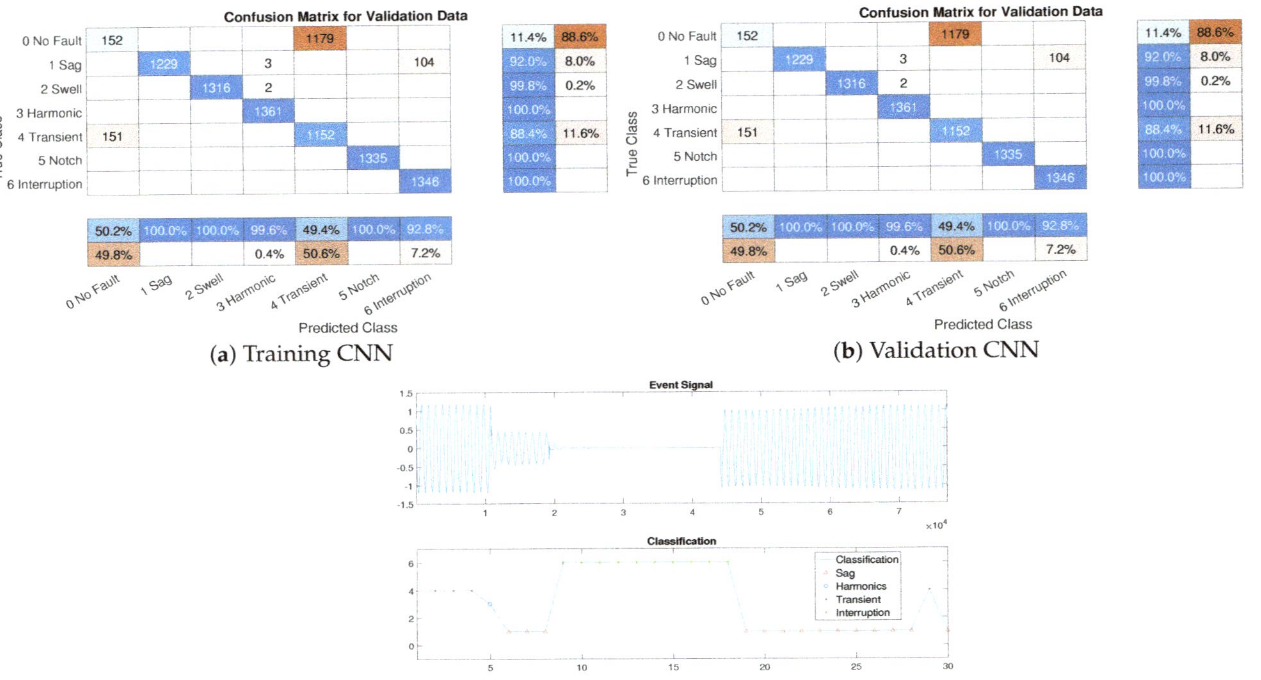

Figure 10. (**a**) Confusion matrix of the CNN training, (**b**) Confusion matrix of the CNN validation, (**c**) Testing of the CNN using a simulated voltage waveform.

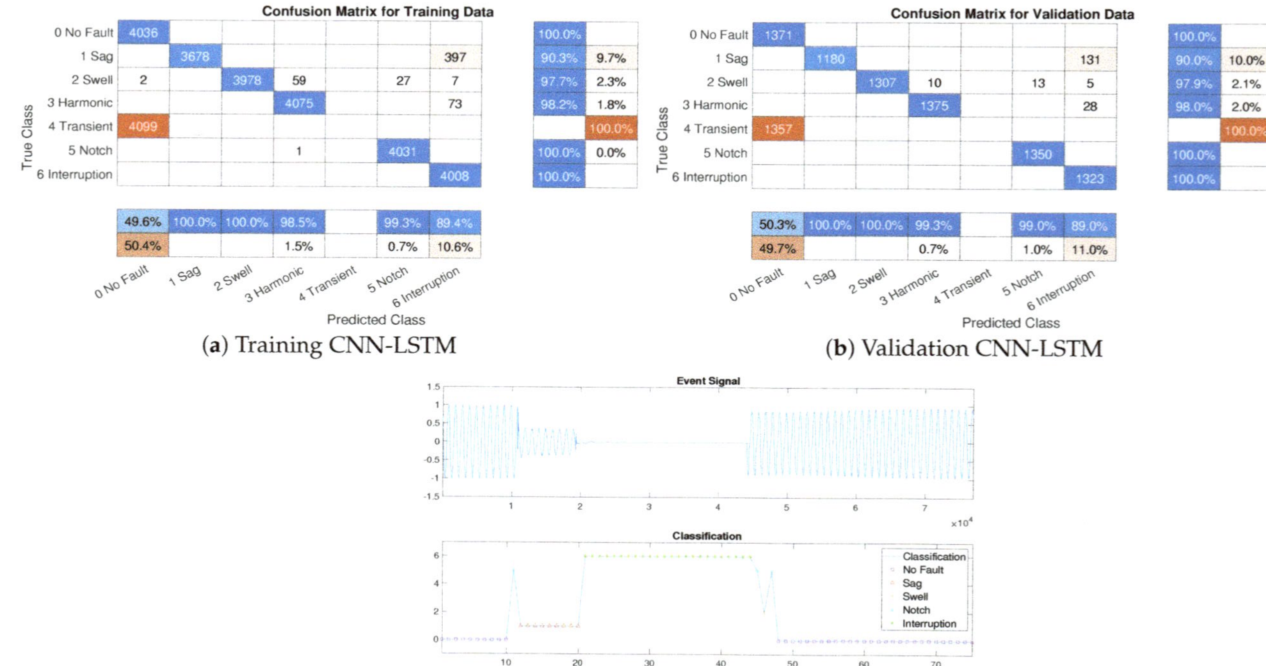

Figure 11. (**a**) Confusion matrix of the CNN-LSTM training, (**b**) Confusion matrix of the CNN-LSTM validation, (**c**) Testing of the CNN-LSTM using a simulated voltage waveform.

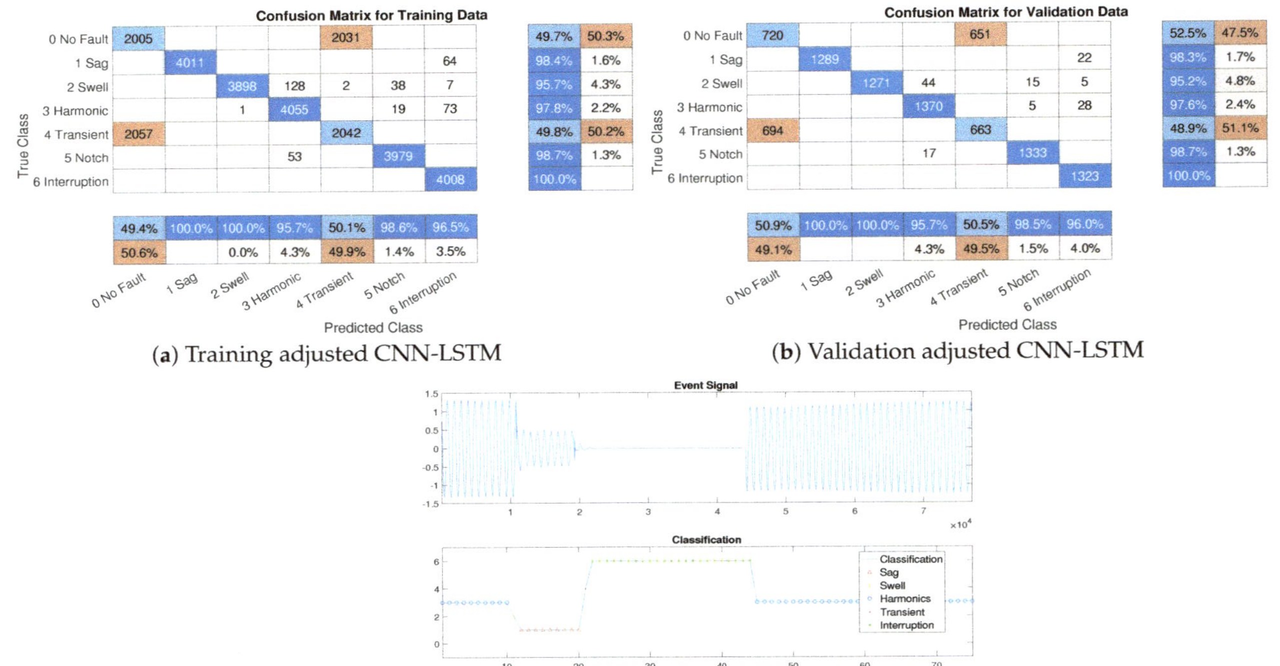

(a) Training adjusted CNN-LSTM

(b) Validation adjusted CNN-LSTM

(c) Testing adjusted CNN-LSTM

Figure 12. (**a**) Confusion matrix of the CNN-LSTM with adjusted hyperparameters training, (**b**) Confusion matrix of the CNN-LSTM with adjusted hyperparameters validation, (**c**) Testing of the CNN-LSTM with adjusted hyperparameters using a simulated voltage waveform.

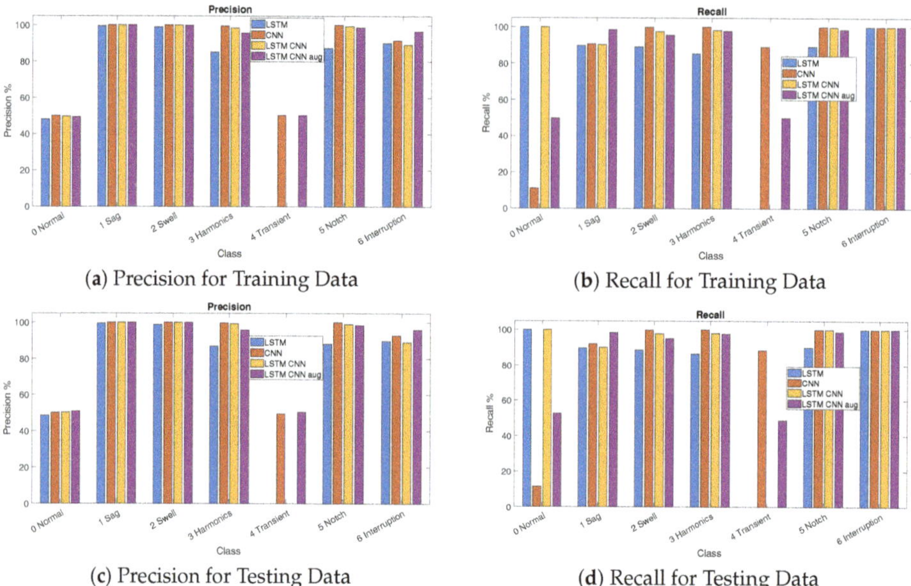

(a) Precision for Training Data

(b) Recall for Training Data

(c) Precision for Testing Data

(d) Recall for Testing Data

Figure 13. Bar chart of the comparison of the precision and recall of the LSTM (blue), CNN (red), LSTM-CNN(yellow) and LSTM-CNN with adjusted hyperparameters(purple). (**a**) Precision Training Data, (**b**) Recall Training Data, (**c**) Precision Testing Data, (**d**) Recall Testing Data.

6.2. Testing of the Detection Techniques Using Experimental Datasets

Other tests were conducted with experimental datasets using the test bench shown in Figure 8 in order to compare and prove the effectiveness of all the architectures previously mentioned. It has been possible to generate several experimental datasets of the interruption and of the sag disturbances. The experimental measurments are shown in Figures 14 and 15 along with the plots of the classification results below. Each classification point consist of 1000 samples of the measured signal. Once again, the CNN-LSTM with adjusted hyperparameters was the most consistent in classifying all the disturbances without misclassification. As mentioned before, the identification algorithms were tested with exerimental datasets containing interruption and sag disturbances. Concerning the sag disturbance, all of the four architectures performed correct identification. Some misclassifications occurred when testing the interruption disturbance with the CNN-LSTM and with the LSTM. Combining these results with the ones previously mentioned, the CNN-LSTM with adjusted hyperparameters is the is the one which performed best.

The event signal, shown in Figures 9c, 10c, 11c and 12c, is a voltage signal with a harmonic distortion, sag and an interruption. Each architecture had good results when tested using this signal. However, the LSTM classified the harmonics as a no fault and misclassified a section as a notch. The LSTM-CNN also showed the same problem. On the other hand the CNN misclassified the harmonic disturbance as a transient disturbance. While all architectures successfully classified only the Interruption and the Sag in the testing, the LSTM-CNN with adjusted hyperparameters was the one that had better results because it classified the harmonics, sag, interruption correctly without misclasification.

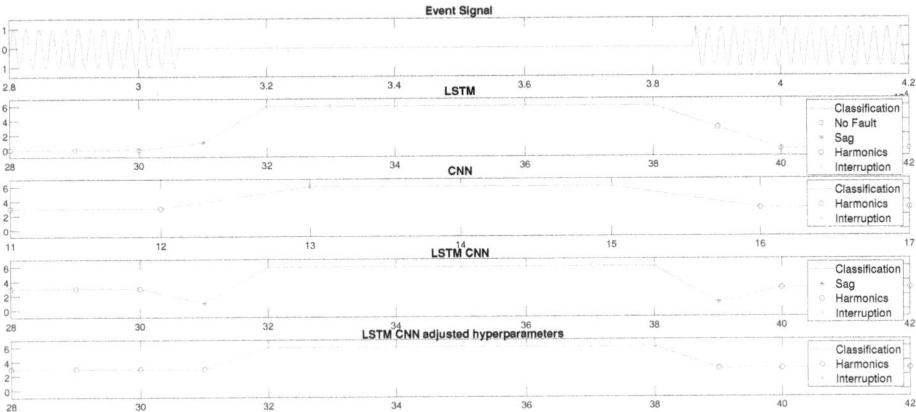

Figure 14. Voltage signal with an interruption measured on the test bench (top plot). From top to bottom, the classification performances of each architecture: LSTM, CNN, LSTM-CNN and the LSTM-CNN with adjusted hyperparameters.

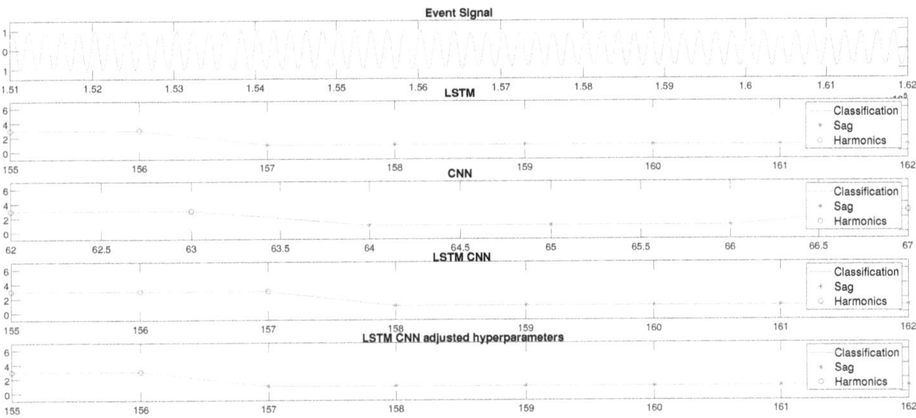

Figure 15. Voltage signal with an sag measured on the test bench (top plot). From top to bottom, the classification performances of each architecture: LSTM, CNN, LSTM-CNN and the LSTM-CNN with adjusted hyperparameters.

7. Conclusions

This work investigates the effectiveness of various deep learning architectures for Power Quality disturbances detection and classification. To do this, it is imperative to study the mechanism of these algorithms to extract the unique features of each disturbance and obtaining an efficient and accurate classification. The training and validation of deep learning architectures depend on a large number of data to better generalize the classification results. A Matlab/Simulink model has been designed and implemented in order to generate these disturbances. To improve the classification performances of the strategies under evaluation and converge to a generalized result, the data in the simulated dataset was augmented. Using the resulting datasets the authors have proposed a comparison among the LSTM, the CNN and a joint architecture that uses both the LSTM and CNN. All of the architectures were trained and validated using the augmented datasets and then tested using experimental data.

Concerning the experimental validation of the algorithms, it has been possible to generate an experimental dataset of the interruption and of the sag disturbances. The two datasets were processed by exploiting the four previously mentioned architectures. The first signal contained a train of interruptions and the second signal a train of sags. All of the four architectures successfully classified the sag signal. There were some discrepancies between the architectures while classifying the signal containing interruptions. Again, the LSTM-CNN with adjusted hyperparameters proved to be superior in classifying the disturbances.

These results show that it is possible to train deep learning architectures with simulated data and operate disturbance identification on experimental data. The transient disturbance appears to be hardly detectable for all of the architectures under evaluation, mainly due to the small duration of the disturbance. The architecture that best performed while classifying the transient disturbance was the LSTM-CNN with adjusted hyperparameters. Furthermore, concerning the classifications of other disturbances, the LSTM-CNN with adjusted hyperparameters was the most performing one, both considering the simulated and the experimental datasets.

Author Contributions: Data curation, C.I.G. and L.P.; Formal analysis, C.I.G.; Funding acquisition, F.G., L.P. and G.T.; Methodology, A.L., F.G. and G.T.; Software, C.I.G.; Supervision, A.L.; Validation, L.P. and G.T.; Writing—original draft, C.I.G. and M.C.P.; Writing—review and editing, M.C.P. All authors have read and agreed to the published version of the manuscript.

Funding: E-CUBE(E^3) Project, Energy Exchange and Efficiency, project funded by POR FESR 2014-2020 ASSE 1 AZIONE 1.1.5.

Conflicts of Interest: The authors declare no conflict of interest.

Appendix A

Appendix A.1. Simulink Schematic

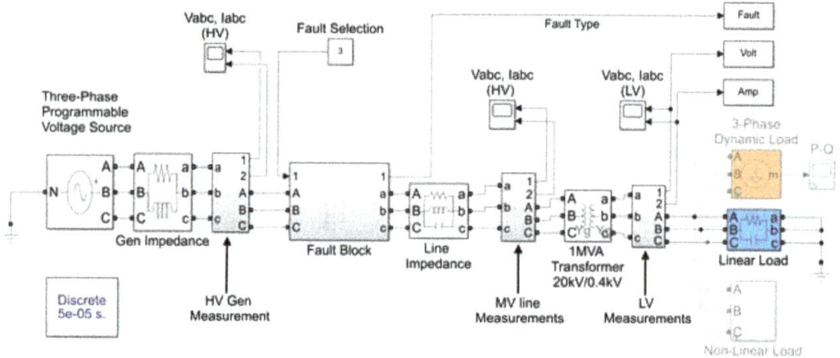

Figure A1. Simulink Model.

Appendix A.2. PQD and STFT

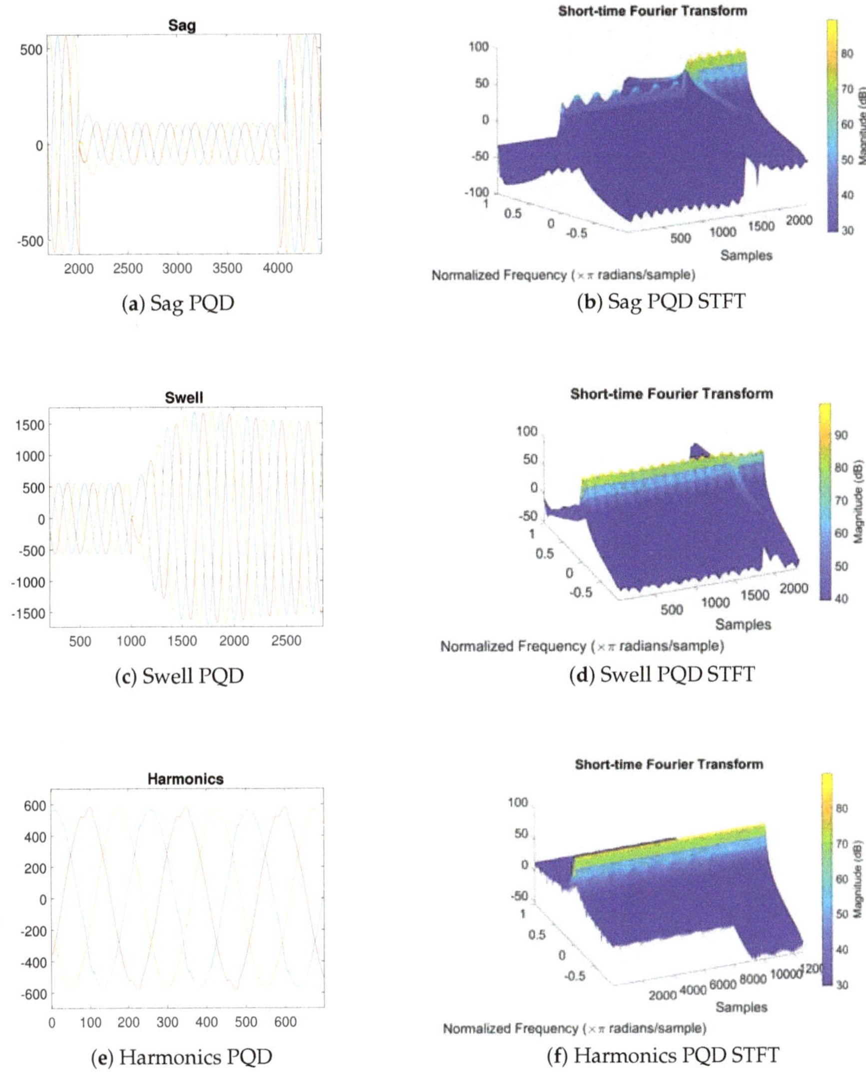

(**a**) Sag PQD (**b**) Sag PQD STFT

(**c**) Swell PQD (**d**) Swell PQD STFT

(**e**) Harmonics PQD (**f**) Harmonics PQD STFT

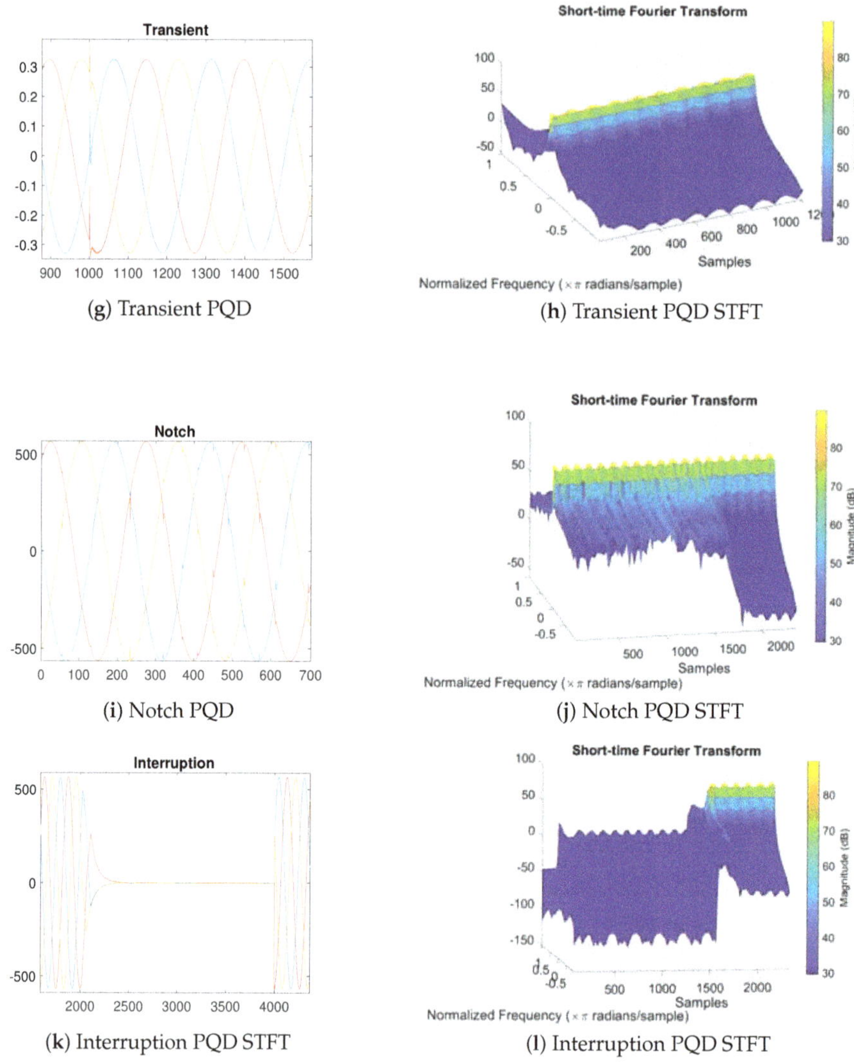

(g) Transient PQD (h) Transient PQD STFT

(i) Notch PQD (j) Notch PQD STFT

(k) Interruption PQD STFT (l) Interruption PQD STFT

Figure A2. Disturbance plots (Left) and the STFT (Right). (**a**,**b**) Sag, (**c**,**d**) Swell, (**e**,**f**) Harmonics, (**g**,**h**) Transient, (**i**,**j**) Notch, (**k**,**l**) Interruption.

References

1. Singh, B.; Al-Haddad, K.; Chandra, A. A review of active filters for power quality improvement. *IEEE Trans. Ind. Electron.* **1999**, *46*, 960–971. [CrossRef]
2. Chung, Y.; Kwon, G.H.; Park, T.; Kim, H.; Moon, J.I. Voltage sag, swell and flicker generator with series injected inverter. In Proceedings of the IEEE Power Engineering Society General Meeting, San Francisco, CA, USA, 16 June 2005; Volume 2, pp. 1308–1313.
3. Hafiz, F.; Swain, A.; Naik, C.; Abecrombie, S.; Eaton, A. Identification of power quality events selection of optimum base wavelet and machine learning algorithm. *IET Sci. Meas. Technol.* **2019**, *13*, 260–271. [CrossRef]
4. Mohan, N.; Soman, K.P.; Vinayakumar, R. Deep power: Deep learning architectures for power quality disturbances classification. In Proceedings of the International Conference on Technological Advancements in Power and Energy (TAP Energy), Kollam, India, 21–23 December 2017; pp. 1–6.

5. Alshahrani, S.; Abbod, M.; Alamri, B.; Taylor, G. Evaluation and classification of power quality disturbances based on discrete Wavelet Transform and artificial neural networks. In Proceedings of the 50th International Universities Power Engineering Conference (UPEC), Stoke on Trent, UK, 1–4 September 2015; pp. 1–5.
6. Bhavani, R.; Prabha, N.R. A hybrid classifier for power quality (PQ) problems using wavelets packet transform (WPT) and artificial neural networks (ANN). In Proceedings of the IEEE International Conference on Intelligent Techniques in Control, Optimization and Signal Processing (INCOS), Srivilliputhur, India, 23–25 March 2017; pp. 1–7.
7. Kwan, K.H.; So, P.L.; Chu, Y.C. An Output Regulation-Based Unified Power Quality Conditioner With Kalman Filters. *IEEE Trans. Ind. Electron.* **2012**, *59*, 4248–4262. [CrossRef]
8. Soo-Hwan, C.; Jeong-Uk, K.; Il-Yop, C.; Jong-Hoon, H. Determination of Power-Quality Disturbances Using Teager Energy Operator and Kalman Filter Algorithms. *Int. J. Fuzzy Log. Intell. Syst.* **2012**, *12*, 42–46.
9. Kumawat, P.N.; Verma, D.K.; Zaveri, N. Comparison between Wavelet Packet Transform and M-band Wavelet Packet Transform for Identification of Power Quality Disturbances. *Power Res.* **2018**, *14*, 37–45. [CrossRef]
10. Liquan, Z.; Meijiao, G.; Lin, W. Classification of multiple power quality disturbances based on the improved SVM. In Proceedings of the International Conference on Wireless Communications, Signal Processing and Networking (WiSPNET), Chennai, India, 22–24 March 2017; pp. 2625–2628.
11. Milchevski, A.; Taskovski, D. Classification of power quality disturbances using wavelet transform and SVM decision tree. In Proceedings of the 11th International Conference on Electrical Power Quality and Utilisation, Lisbon, Portugal, 17–19 October 2011; pp. 1–5.
12. Eristi, H.; Yıldırım, Ö.; Eristi, B.; Demir, Y. Optimal feature selection for classification of the power quality events using wavelet transform and least squares support vector machines. *Int. J. Electr. Power Energy Syst.* **2019**, *49*, 95–103. [CrossRef]
13. Cubuk, E.D.; Zoph, B.; Mane, D.; Vasudevan, V.; Le, Q.V. AutoAugment: Learning Augmentation Strategies From Data. In Proceedings of the IEEE/CVF Conference on Computer Vision and Pattern Recognition (CVPR), Long Beach, CA, USA, 16–20 June 2019.
14. Fawaz, H.I.; Forestier, G.; Weber, J.; Idoumghar, L.; Muller, P. Data augmentation using synthetic data for time series classification with deep residual networks. *arXiv* 2018, arXiv:1808.02455.
15. Xue, H.; Chen, A.; Zhang, D.; Zhang, C. A Novel Deep Convolution Neural Network and Spectrogram Based Microgrid Power Quality Disturbances Classification Method. In Proceedings of the 2020 IEEE Applied Power Electronics Conference and Exposition (APEC), New Orleans, LA, USA, 15–19 March 2020; pp. 2303–2307.
16. Aziz, S.; Khan, M.U.; Abdullah; Usman, A.; Mobeen, A. Pattern Analysis for Classification of Power Quality Disturbances. In Proceedings of the 2020 International Conference on Emerging Trends in Smart Technologies (ICETST), Karachi, Pakistan, 26–27 March 2020; pp. 1–5.
17. Benysek, G. *Improvement in the Quality of Delivery of Electrical Energy Using Power Electronics Systems*; Power Systems; Springer: Dordrecht, The Netherlands, 2007.
18. Grasso, F.; Paolucci, L.; Bacci, T.; Talluri, G.; Cenghialta, F.; D'Antuono, E.; Giorgis, S.D. Simulation Model and Experimental Setup for Power Quality Disturbances Methodologies Testing and Validation. In Proceedings of the 2019 IEEE 5th International Forum on Research and Technology for Society and Industry (RTSI), Florence, Italy, 9–12 September 2019; pp. 359–363.
19. Borges, F.A.S.; Fernandes, R.A.S.; Silva, I.N.; Silva, C.B.S. Feature Extraction and Power Quality Disturbances Classification Using Smart Meters Signals. *IEEE Trans. Ind. Inform.* **2016**, *12*, 824–833. [CrossRef]
20. Shen, Y.; Abubakar, M.; Liu, H.; Hussain, F. Power Quality Disturbance Monitoring and Classification Based on Improved PCA and Convolution Neural Network for Wind-Grid Distribution Systems. *Energies* **2019**, *12*, 1280. [CrossRef]
21. Wang, S.; Chen, H. A novel deep learning method for the classification of power quality disturbances using deep convolutional neural network. *Appl. Energy* **2019**, *235*, 1126–1140. [CrossRef]
22. Ma, J.; Zhang, J.; Xiao, L.; Chen, K.; Wu, J. Classification of Power Quality Disturbances via Deep Learning. *IETE Tech. Rev.* **2017**, *34*, 408–415. [CrossRef]

23. Ahajjam, M.A.; Licea, D.B.; Ghogho, M.; Kobbane, A. Electric Power Quality Disturbances Classification based on Temporal-Spectral Images and Deep Convolutional Neural Networks. In Proceedings of the 2020 International Wireless Communications and Mobile Computing (IWCMC), Limassol, Cyprus, 15–19 June 2020; pp. 1701–1706.
24. Pascanu, R.; Gulcehre, C.; Cho, K.; Bengio, Y. How to Construct Deep Recurrent Neural Networks. *arXiv* 2013, arXiv:1312.6026.
25. Hochreiter, S.; Schmidhuber, A.J. Long Short- Term Memory. *Neural Comput.* **1997**, *9*, 1735–1780. [CrossRef] [PubMed]
26. Goodfellow, I.; Bengio, Y.; Courville, A. *Deep Learning*; The MIT Press: Cambridge, MA, USA, 2016.
27. Han, D.; Liu, Q.; Fan, W. A new image classification method using CNN transfer learning and web data augmentation. *Expert Syst. Appl.* **2018**, *95*, 43–56. [CrossRef]
28. Huang, J.; Chen, B.; Yao, B.; He, W. ECG Arrhythmia Classification Using STFT-Based Spectrogram and Convolutional Neural Network. *IEEE Access* **2019**, *7*, 92871–92880. [CrossRef]
29. Yao, S.; Piao, A.; Jiang, W.; Zhao, Y.; Shao, H.; Liu, S.; Liu, D.; Li, J.; Wang, T.; Hu, S.; et al. STFNets: Learning Sensing Signals from the Time-Frequency Perspective with Short-Time Fourier Neural Networks. In Proceedings of the The World Wide Web Conference, San Francisco, CA, USA, 13–17 May 2019.
30. Tan, R.H.; Ramachandaramurthy, V.K. A Comprehensive Modeling and Simulation of Power Quality Disturbances Using MATLAB/SIMULINK. In *Power Quality Issues in Distributed Generation*; Luszcz, J., Ed.; IntechOpen: Rijeka, Croatia, 2015; Chapter 3.
31. Fayek, H.M.; Lech, M.; Cavedon, L. Evaluating deep learning architectures for Speech Emotion Recognition. *Neural Netw.* **2017**, *92*, 60–68. [CrossRef] [PubMed]
32. Buckland, M.; Gey, F. The relationship between Recall and Precision. *J. Am. Soc. Inf. Sci.* **1994**, *45*, 12–19. [CrossRef]

Article

Leaky Lamb Wave Radiation from a Waveguide Plate with Finite Width

Sang-Jin Park [1,2], Hoe-Woong Kim [1] and Young-Sang Joo [1,2,*]

1 Versatile System Technology Development Division, Korea Atomic Energy Research Institute, 111 Daedeok-daero 989beon-gil, Yuseong-gu, Daejeon 34057, Korea; sjpark@kaeri.re.kr (S.-J.P.); hwkim@kaeri.re.kr (H.-W.K.)

2 Department of Advanced Nuclear System Engineering, Korea University of Science and Technology, 217 Gajeong-ro, Yuseong-gu, Daejeon 34113, Korea

* Correspondence: ysjoo@kaeri.re.kr

Received: 8 October 2020; Accepted: 11 November 2020; Published: 16 November 2020

Abstract: In this paper, leaky Lamb wave radiation from a waveguide plate with finite width is investigated to gain a basic understanding of the radiation characteristics of the plate-type waveguide sensor. Although the leaky Lamb wave behavior has already been theoretically revealed, most studies have only dealt with two dimensional radiations of a single leaky Lamb wave mode in an infinitely wide plate, and the effect of the width modes (that are additionally formed by the lateral sides of the plate) on leaky Lamb wave radiation has not been fully addressed. This work aimed to explain the propagation behavior and characteristics of the Lamb waves induced by the existence of the width modes and to reveal their effects on leaky Lamb wave radiation for the performance improvement of the waveguide sensor. To investigate the effect of the width modes in a waveguide plate with finite width, propagation characteristics of the Lamb waves were analyzed by the semi-analytical finite element (SAFE) method. Then, the Lamb wave radiation was computationally modeled on the basis of the analyzed propagation characteristics and was also experimentally measured for comparison. From the modeled and measured results of the leaky radiation beam, it was found that the width modes could affect leaky Lamb wave radiation with the mode superposition and radiation characteristics were significantly changed depending on the wave phase of the superposed modes on the radiation surface.

Keywords: leaky Lamb wave; semi-analytical finite element (SAFE); waveguide sensor; finite-width plate; waveguide plate; width modes; spatial beating; Rayleigh–Sommerfeld integral (RSI)

1. Introduction

Elastic-guided waves can travel a long distance along the waveguide geometry from a single excitation location. This ability of guided waves not only makes it possible to inspect huge structures effectively, but can also allow remote inspection for hard-to-access structures in harsh environments and underground [1–8]. Therefore, guided waves have been widely used in non-destructive testing (NDT) and structural health monitoring (SHM) fields.

There are numerous industrial applications of guided waves; one that maximally uses the advantages of guided waves is waveguide sensor [9–13]. Waveguide sensors are excellent inspection alternatives for special NDT applications because they can perform remote inspection through a long waveguide without any damage to the main probe unit or the inspector under hazardous inspection environments. Hence, waveguide sensors have often been used in the field of power plants; one application example is under-sodium viewing (USV) in a sodium-cooled fast reactor (SFR), which uses liquid sodium as a core coolant. In an SFR, because of the optical opacity of the liquid sodium, USV for the in-vessel structures (including the reactor core) is conducted using the ultrasonic imaging technique. For these USV inspections, immersion sensors have been developed since the early stage of SFR development; however, there are remaining unresolved issues in their development, including thermal and radiation damage to the main actuating part submerged in the high temperature and radioactive liquid sodium [14]. On the other hand, waveguide sensors have been regarded as a promising USV alternative because there is no concern about any damage to the main actuating part.

The waveguide sensors under development can be classified based on the waveguide geometry of a rod or plate. A rod-type waveguide sensor uses a bundle rod and a rolled plate [14,15], whereas a plate-type waveguide sensor uses a plate strip with finite width [16–22]. Between these, the plate-type waveguide sensor has been established based on the long distance propagation ability and high radiation efficiency in a fluid of the lowest-order flexural mode of a Lamb wave and its concept for USV was proposed in the early 1980s [16,17]. In recent years, sensor development has resumed [18], with one advanced design concept newly adopting a beryllium coating layer [19]; most recently, USV performance of a 10 m full scale waveguide sensor has been demonstrated in a sodium environment (>200 °C) [20,21], and the underwater performance of ranging inspection for obstacle detection in refueling processes has also been validated [22]. Since its early development stage, however, the plate-type waveguide sensor has had a technical issue caused by the radiation characteristics of leaky Lamb waves. The immersion- and rod-type waveguide sensors generally use longitudinal wave radiation from the axisymmetric radiator; therefore, they have a single axisymmetric main beam and their USV resolutions are not constrained by the scanning direction. On the contrary, the plate-type waveguide sensor uses leaky Lamb wave radiation along the rectangular radiating face and it has a non-axisymmetric radiation beam, which has different radiation characteristics on the vertical and lateral planes (the vertical and lateral planes are on the median and transverse planes with respect to the waveguide plate). Here, the formation and characteristics of the vertical beam are understandable based on previous research on leaky Lamb wave radiation in an infinitely wide plate [23–26]. However, those of the lateral beam have not been completely explained, and moreover, there is little research that fully addresses the leaky Lamb wave radiation from a plate strip with finite width.

One effort has been made to understand leaky Lamb wave radiation from a plate strip with finite width [27]. In this previous research, characteristics of the leaky radiation beam radiated from the plate strip were acoustically analyzed on two median and lateral planes; however, this research assumed that the velocity distribution of leaky Lamb waves on the aperture was uniform in the width direction. In other words, the effect of the plate width on the leaky Lamb wave radiation was partially studied without full consideration of the propagation characteristics of the leaky Lamb waves in the plate strip.

This paper investigates leaky Lamb wave radiation from a waveguide plate with a finite width to gain a basic understanding of the radiation characteristics of the plate-type waveguide sensor. First, the propagation characteristics of the Lamb wave in the plate strip was evaluated using the semi-analytical finite element (SAFE) method. Then, leaky Lamb wave radiation was computationally modeled on the basis of the analyzed propagation characteristics, and was also measured experimentally for comparison. From this analysis and measurement, the formation and characteristics of the leaky radiation beam radiated from a waveguide plate are three-dimensionally revealed and the design direction for performance improvement of the waveguide sensor is briefly proposed.

2. Lamb Wave Propagation in a Plate Strip with Finite Width

2.1. Semi-Analytical Finite Element(SAFE) Method

To study the leaky Lamb wave radiation from a plate strip with finite width, it is necessary to sufficiently understand the propagation characteristics of the Lamb wave in the plate strip. Figure 1 illustrates the coordinate system of a plate strip with thickness h and width W; the plate is infinitely long to $\pm x$ direction and the plate material is assumed to be a homogeneous, isotropic, and lossless one. Wave propagation in the plate strip has been analytically studied by many researchers [28–36], but most of their analytical studies show limitations of effective frequency and thickness–width ratio by assumptions used in the solution derivation. Fortunately, the semi-analytical finite element (SAFE) method, a numerical method with no limitations of the analytical solutions, has been developed [35–37]. Since the SAFE method can calculate dispersion curves for arbitrary cross-section waveguide geometries, such as rails as well as plate strips [35–37], it is now prevalent in NDT and SHM fields.

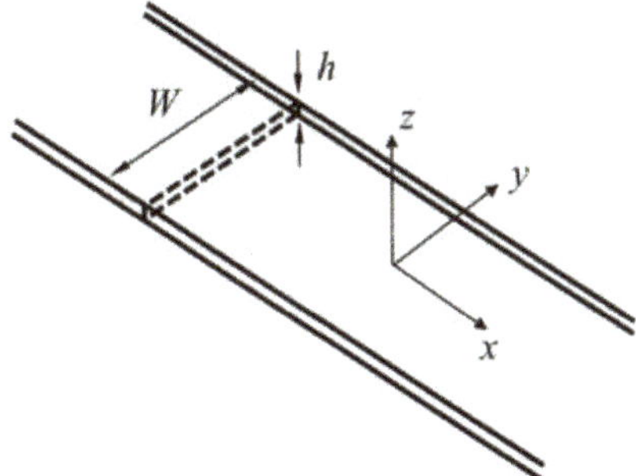

Figure 1. Coordinate system of a plate strip with finite width.

The SAFE method only constructs the two dimensional finite element (FE) model for the cross-section of the analysis object and the analytical solution of the wave propagation is then applied to the constructed FE model; the cross-section parallel to the *yz* plane is only modeled using FE and the equation of a traveling wave along the +*x* direction is applied. The governing equation of the constructed FE system without the external load can be written as follows:

$$\left\{\gamma^2\mathbf{K}_2 + i\gamma\mathbf{K}_1 + \mathbf{K}_0 + \omega^2\mathbf{M}\right\}u = 0 \tag{1}$$

where $\mathbf{K}_2$, $\mathbf{K}_1$ and $\mathbf{K}_0$ are stiffness matrices, $\mathbf{M}$ is the mass matrix of the cross section and the displacement vector u is given by:

$$u = \mathbf{U}(y,z)e^{i(\gamma x-\omega t)} \tag{2}$$

where $\mathbf{U}$ denotes displacement functions of the cross section. From this eigenvalue equation, the wavenumber γ can be solved at each frequency ω and the dispersion curves can be drawn within the interested frequency range. Recently, SAFE methods using commercial Finite Element Method (FEM) software have been introduced, and thereby SAFE modeling has become convenient [38,39]. In this study, to analyze the wave propagation in the plate strip, a modal analysis method under periodic boundary conditions [39] was employed in the commercial FEM software ANSYS (release 2017, ANSYS Inc., Canonsburg, PA, USA).

2.2. Dispersion Curves and Wave Structures

Figure 2 shows calculated dispersion curves of Lamb waves in a stainless steel (SS304) plate with 1.5 mm thickness and 15 mm width in a vacuum. Note that propagation characteristics in the plate coupled with liquid, such as the wave velocities and structures, are assumed to not be different from those in a vacuum [40]. Specifically, Figure 2a shows the phase velocity dispersion curves, whereas Figure 2b shows the group velocity dispersion curves. As shown in these dispersion curves, in contrast to those in an infinitely wide plate, Lamb waves in a finite width plate have numerous width modes. Therefore, a certain single mode of Lamb waves in the plate strip is named S(*m*,*n*) or A(*m*,*n*); the first index *m* is the order of the thickness mode; the second index *n* is that of the width mode. Furthermore, one can observe that the phase velocity increases and the group velocity decreases as the order of the width mode increases.

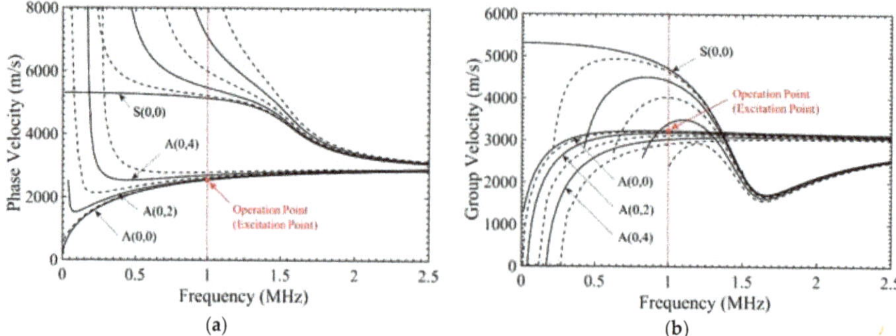

Figure 2. Dispersion curves of Lamb waves in a SS304 plate with 1.5 mm thickness and 15 mm width: (**a**) phase velocity, (**b**) group velocity; SS304 plate properties: density ρ = 7800 kg/m^3, longitudinal wave velocity C_L = 5800 m/s and shear wave velocity C_S = 3160 m/s; solid line: even order of width mode (symmetric width mode); dashed line: odd order of width mode (anti-symmetric width mode).

The most important observation in the dispersion curves of Figure 2 is that many Lamb wave modes are very close to each other at the operation point of the waveguide sensor, which indicates the difficulty of single mode excitation. In this high modal density area, neighbor width modes from the A(0,1) mode to the A(0,4) mode can have a high possibility of being excited with the lowest-order mode, the A(0,0) mode; the S(m,n) modes are not in our interest because the flexural modes, A(0,n) modes are only used for the high radiation efficiency in the waveguide sensor. In addition, it can be reasonably inferred that the multiple width modes cannot be separated during the short propagation distance because they are not significantly different in group velocities at the operation point.

Each higher-order width mode can be identified in the wave structure results, as shown in Figure 3. The out-of-plane velocity profiles in the width direction change appreciably according to the order of the width mode; they can be described as a combination of the trigonometric and hyperbolic functions [31,32]. Here, the non-flat velocity profile in case of n = 0 is estimated to be caused by the free–free boundary condition and high frequency range. From the dispersion curves and the wave structures, it can be assumed that leaky Lamb wave propagation in the plate strip coupled with liquid might be affected by superposition among the higher-order width modes.

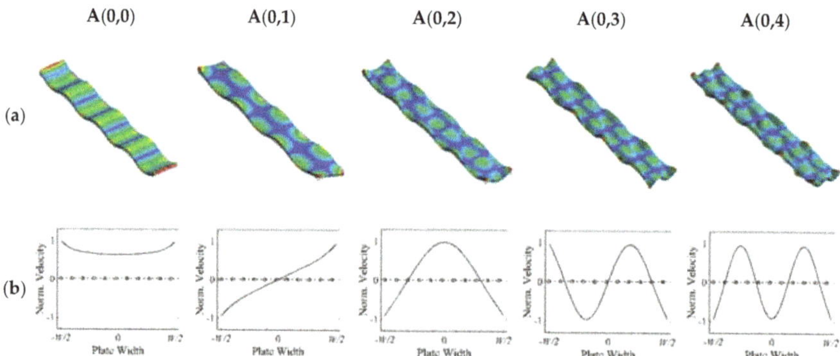

Figure 3. (**a**) Wave structures and (**b**) particle velocity profiles (width direction) of the A(0,n) modes in a SS304 plate with 1.5 mm thickness and 15 mm width at 1.0 MHz; the velocity profiles are extracted on the neutral plane (z = 0); solid line: out-of-plane direction (z direction); dashed line: in-plane horizontal direction (y direction); circle markers: in-plane extensional direction (x direction).

3. Leaky Lamb Wave Radiation from a Waveguide Plate with Finite Width

3.1. Rayleigh–Sommerfeld Integral Model

Figure 4 presents a radiation aperture area with length L and width W. A leaky Lamb wave is propagated along the $+x$ axis and is sequentially radiated into the surrounding liquid ($z > 0$) from the radiation surface S; the leaky Lamb wave radiation from the radiation surface is limited by $x_0 \in [-L/2, L/2]$ in an infinite plate in the $\pm x$ direction. In fact, leaky Lamb wave radiation from the radiation surface at the plate end is used in the practical waveguide sensor. However, the radiation of the backward leaky Lamb wave reflected from the aperture end is not considered in this study; its effect on the main beam generated by the forward wave is assumed to be negligible based on its different radiation angle and diminished energy. An acoustic pressure at a certain point p can be defined by the Rayleigh–Sommerfeld integral (RSI) [41]:

$$P(r,\theta,\varphi,t) = -\frac{i\rho\omega}{2\pi}\int_S v(x_0,y_0,t)\cdot\frac{e^{ik_l r'}}{r'}dS(x_0,y_0), \tag{3}$$

where ρ is the liquid density, k_l is the wavenumber in the liquid, and $v(x_0, y_0, t)$ is the velocity distribution on the aperture, given by:

$$v(x_0,y_0,t) = V(y_0)e^{-\alpha x_0}\cdot e^{i(k_P x_0-\omega t)},\ x_0 \in [-L/2, L/2] \tag{4}$$

where α and k_P are the attenuation coefficient and the wavenumber of the leaky Lamb wave, respectively, and $V(y_0)$ is the velocity profile of a certain width mode in the width direction. Based on this integral equation, the RSI model was constructed as a computational approach; this model is also called the Rayleigh–Sommerfeld numerical integration (RSNI) [42]. The RSI model computes the integral equation (Equation (3)) without any assumptions and approximations such as the far-field approximation; therefore, this model is known to have computation results of high accuracy [43]. In addition, RSI can provide the better computational speed compared with SAFE or FEM approaches, especially for the high-frequency model that requires both small wavelengths and an integration time step [44].

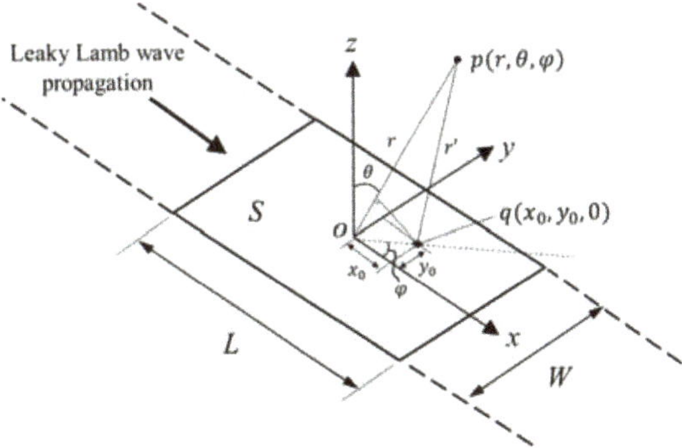

Figure 4. Coordinate system of radiation aperture area for leaky Lamb wave radiation.

The underwater beam profile of the leaky wave radiated from the unbaffled aperture shown in Figure 4 was calculated using the constructed RSI model. The water domain for calculating the beam profile was predetermined as 90 mm (x) × 50 mm (y) × 90 mm (z) and the total number of grid points was 300 × 167 × 300, with 0.3 mm grid spacing (about 1/5 times smaller than the wavelength of the leaky wave), and 0.5 mm source spacing (about 1/5 times smaller than the wavelength of the leaky Lamb wave). The out-of-plane velocity profiles obtained by the SAFE method were applied to the point sources with the curve-fitting technique. The 3D acoustic field by a certain point source on the aperture was calculated; then, the calculation process was repeated for all point sources. The acoustic field of the leaky wave radiated from the aperture was obtained by integrating the individual calculation results. From the calculated acoustic field, the beam profile was reconstructed by making an envelope for the maximum pressure peaks on the grid points; the beam profile was then normalized by the maximum value in the entire calculation domain. All parameters to calculate the radiation beam profiles of the leaky Lamb wave, such as the phase velocity, were determined on the basis of the dispersion curve results shown in Figure 2a. Other model information is presented in Table 1; the attenuation coefficient of the leaky Lamb wave was adopted from the attenuation dispersion curve in the infinite plate; the attenuation coefficient is strictly different depending on the width mode, but its value at 1.0 MHz is assumed to be equal to that of the fundamental flexural mode in the infinite plate.

Table 1. Rayleigh–Sommerfeld integral (RSI) model details.

Parameter	Value
Surrounding liquid (wave velocity)	Water (C_L = 1480 m/s)
Leaky Lamb wave modes	A(0,n)
Plate dimension	Thickness: 1.5 mm Width: 15mm
Plate material (wave velocity)	SS304 (C_L = 5800 m/s, C_S = 3160 m/s)
Aperture size	Length L: 18 mm Width W: 15 mm
Excitation frequency	1.0 MHz
Attenuation coefficient α	0.168 dB/mm at 1.0 MHz [40]

Finally, a 3D beam profile can be analyzed on two independent planes: the vertical and lateral planes. The vertical plane is the median plane (xz plane, $\varphi = 0$) with respect to the waveguide plate, and the lateral plane is a lateral cross-sectional plane across the main lobe at the radiation angle. The beam profile on the vertical plane is defined as the vertical beam profile; that on the lateral plane is defined as the lateral beam profile.

3.2. Leaky Radiation Beam Patterns and Beam Profiles

Beam profiles of the A(0,n) mode for n = 0, 1, 2, and 3 are shown in Figures 5–8, respectively. Figures 5a, 6a, 7a and 8a and Figures 5b, 6b, 7b and 8b represent the vertical and lateral beam profile results obtained from the RSI model, respectively.

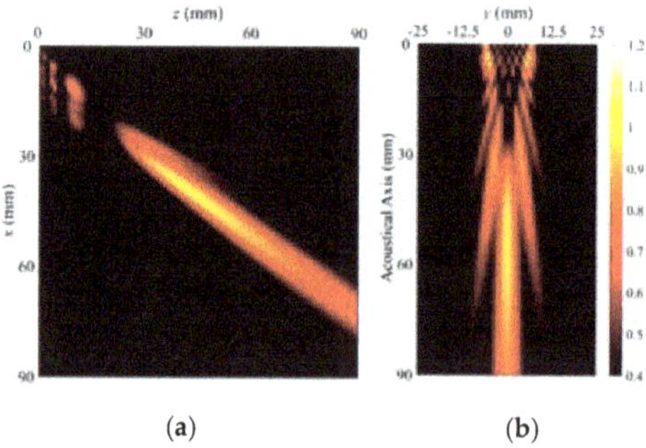

(a) (b)

Figure 5. Radiation patterns and beam profiles of leaky A(0,0) mode Lamb wave, (**a**) vertical beam profile, and (**b**) lateral beam profile.

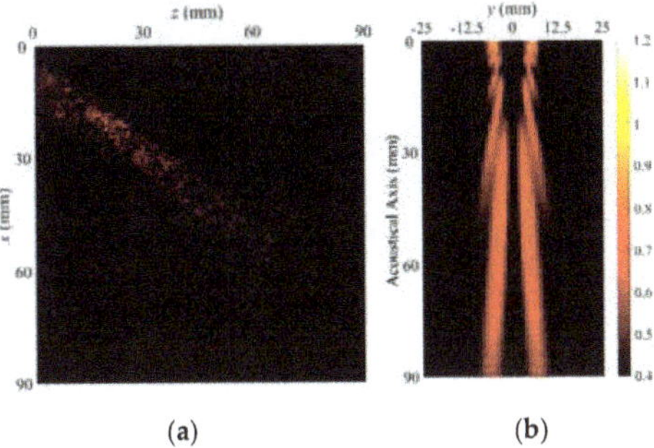

(**a**) (**b**)

Figure 6. Radiation patterns and beam profiles of leaky A(0,1) mode Lamb wave, (**a**) vertical beam profile, and (**b**) lateral beam profile.

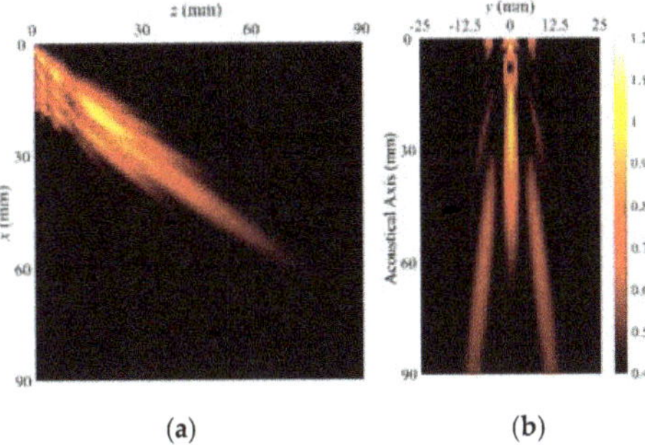

(**a**) (**b**)

Figure 7. Radiation patterns and beam profiles of leaky A(0,2) mode Lamb wave, (**a**) vertical beam profile, and (**b**) lateral beam profile.

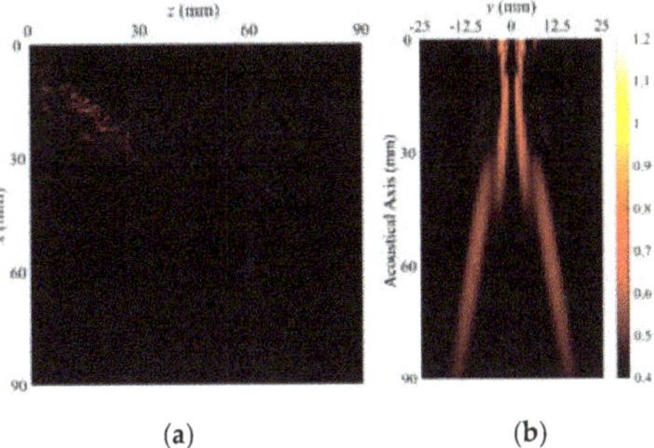

(**a**) (**b**)

Figure 8. Radiation patterns and beam profiles of leaky A(0,3) mode Lamb wave, (**a**) vertical beam profile, and (**b**) lateral beam profile.

The first trend identified in Figures 5a, 6a, 7a and 8a is that the radiation angles of the vertical beam show no big differences with changes in the width mode because the phase velocities at 1.0 MHz are close to each other; the radiation angles are calculated as 34.0°–35.5° with respect to the z axis.

The second trend is that the leaky radiation beam pattern of the A(0,0) mode has a single main lobe in the far-field as shown in Figure 5b, but the others have split ones as shown in Figures 6b, 7b and 8b, In other words, the A(0,n) mode Lamb waves above n = 1 induce beam splitting on the lateral plane. It is acoustically clear that the beam splitting is caused by the sinusoidal profiles in the width direction on the aperture. Also, the splitting angle increases as the order increases.

The third trend is that the odd order of the width mode has no vertical beam profile at $y = 0$ (see Figures 6b and 8b) due to the anti-symmetric profile in the width direction, which has a node point at the center of the plate width. As a result, the lowest-order width mode of a Lamb wave in a plate strip provides a single main beam similar to that of a conventional immersion probe, and thus it can be best to solely use the leaky radiation beam of the lowest-order width mode in the waveguide sensor for immersion inspection.

4. Beam Profile Measurements

4.1. Experimental Setup

Figure 9 shows the experimental setup for the measurement of the leaky radiation beam profile radiated from a plate strip in water. The underwater beam profile measurement system consists of an XYZ three-axis scanner, a hydrophone (ONDA HNR-0500, frequency range: 0.25–10 MHz) with a pre-amplifier (ONDA AH-1100, frequency range: 0.005–25 MHz), a waveform generator (Agilent 33521A), a gated amplifier (RITEC GA-2500A), a broadband receiver (RITEC BR-640), a noise suppressor with a 1.0 MHz center frequency (ORBISSYS NS-0017) and a computer with master control software (UTEX Winspect^{TM}). Material and dimensions of the 400 mm long plate strip used in the experiment are the same as those in the RSI model (1.5 mm thickness and 15 mm width); also, the aperture size is 18 mm × 15 mm, the same as that in the RSI model. The material of an ultrasonic wedge with 19 mm height and 20 mm width is Lucite (C_L = 2370 m/s); the incidence angle of the wedge is 70° for the generation of the A(0,0) mode Lamb wave at 1.0 MHz. The wedge was mounted with a PZT (lead zirconate titanate) transducer with a 0.5 inch diameter and 1.0 MHz center frequency (GE benchmark series) at three different excitation source positions (d = 300, 350, 400 mm), and the leaky radiation beam was measured for each excitation source position. A four-cycled tone burst signal generated from the waveform generator was input to the transducer with signal amplification by the gated amplifier. The leaky wave radiated from the plate strip was measured by the hydrophone; then, the measured wave signal was transferred to the computer after amplifying and band-pass filtering (from 25 kHz to 5 MHz) by the pre-amplifier, the noise suppressor, and the broadband receiver. Scanning volume size was 90 mm (x) × 50 mm (y) × 90 mm (z), the same as the analysis domain of the RSI model, and scan step was 1 mm. In the post-processing process, a 3D beam profile was reconstructed by mapping the measured 3D matrix data comprising the maximum peak (V_{Peak}) to the space coordinate; as shown in Figure 9c, the maximum peak (V_{Peak}) of the gated signal beyond the threshold value was extracted and saved. Finally, the reconstructed beam profile was normalized by the maximum value in the far-field domain.

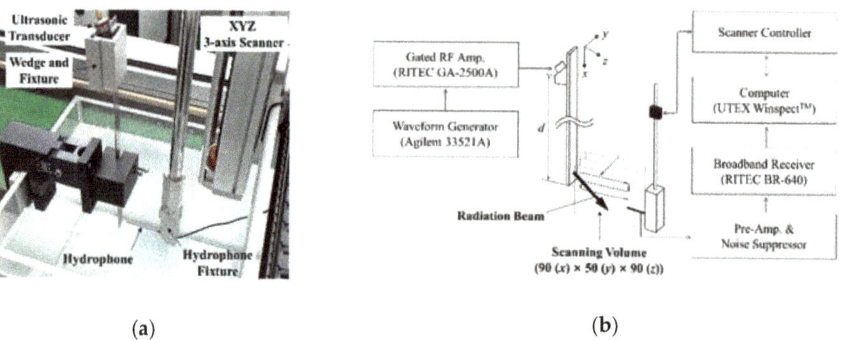

Figure 9. *Cont.*

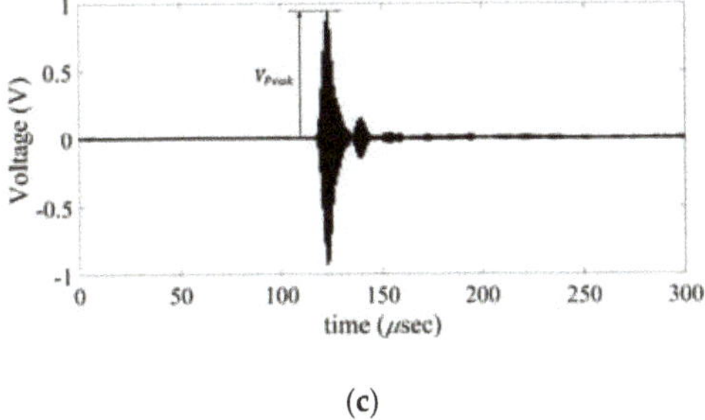

(c)

Figure 9. Experiment for the measurement of the 3D radiation beam profile of a leaky wave radiated from a SS304 plate with 1.5 mm thickness and 15 mm width in water, (**a**) experimental setup and (**b**) block diagram, and (**c**) A-scan signal of the leaky wave measured at the center of the radiation surface.

4.2. Measured Beam Profiles and Effects of the Mode Superposition

The measured vertical and lateral beam profiles radiated from the waveguide plate are shown in Figure 10. Without any quantitative evaluations on radiation characteristics, all three measured beam profiles seem to be visually different from those of the A(0,0) mode shown in Figure 5a,b, despite the attempt to excite only the A(0,0) mode, and they show different radiation characteristics depending on the excitation source position. In particular, drastic changes of the lateral beam are noticeable compared with those of the vertical beam; the lateral beams at d = 300 mm and d = 400 mm have a dual main beam, but one at d = 350 mm has a single main beam within the measurement domain. These characteristics cannot be explained by only pure width mode and demonstrate that the leaky radiation beams radiated from the waveguide plate with finite width are affected by the mode superposition of the width modes, as predicted from the dispersion curves of Figure 2.

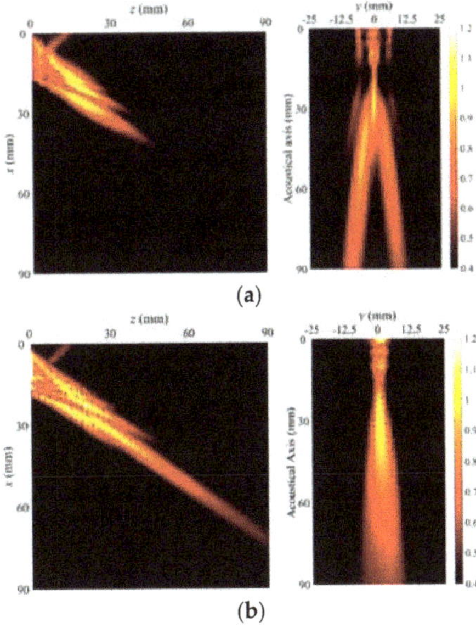

Figure 10. *Cont.*

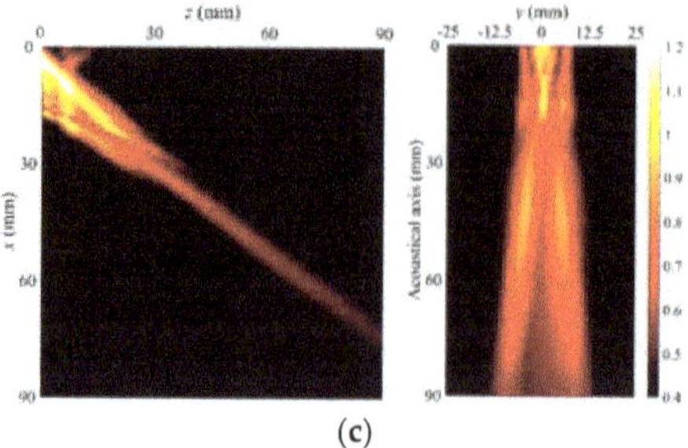

Figure 10. Measured leaky radiation beam profiles radiated from a SS304 plate (1.5 mm thickness, 15 mm width) with variation of the excitation source position, (a) d = 300 mm, (b) d = 350 mm, and (c) d = 400 mm.

Moreover, the different characteristics changed by the excitation source position are strongly estimated to be affected by the change of the wave phase by phase velocity differences between the superposed modes. Figure 11 shows the wave phase changes on the waveguide plate calculated from the following equation at time $t = 0$, weighting constants $a = b = 1$, and phase delay constants $\tau_0 = \tau_2 = 0$:

$$v(x_0, y_0, t) = a{\cdot}V_0(y_0)e^{i(k_{P0}x_0-\omega(t-\tau_0))} + b{\cdot}V_2(y_0)e^{i(k_{P2}x_0-\omega(t-\tau_2))} \quad (5)$$

where $V_0(y_0)$ and $V_2(y_0)$ are the normalized velocity distributions of the A(0,0) and A(0,2) modes identified in Figure 3, respectively, and k_{P0} and k_{P2} are the wavenumbers of the A(0,0) and A(0,2) modes, respectively. According to the obtained dispersion curves, the A(0,1) and A(0,2) modes are found to have a high possibility to be excited with the A(0,0) mode. However, only the mode superposition between the A(0,0) mode and the A(0,2) mode was investigated here; the reason why the A(0,1) mode is excluded from the mode superposition is that the anti-symmetric width modes (including the A(0,1) mode) cannot be easily generated by general Lamb wave excitation methods such as the angle–beam method because they have symmetric distribution of the input wave energy in the width direction. It is known that the profile of the input energy in the width direction should match that of the target width mode to be generated [45,46]. The two width modes excited by the high modal density are continually superposed in-phase or out-of-phase with each other during propagation along the waveguide plate. Therefore, from Figure 11, it can be recognized that the superposed modes make the spatial beating due to their differences in phase velocity and the beating on the radiation surface, varied by the excitation source position, results in the change in radiation characteristics.

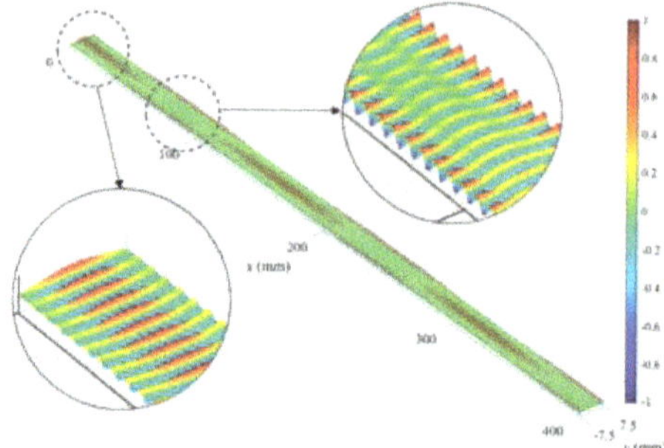

Figure 11. Change of wave phase on the waveguide plate by the mode superposition between the A(0,0) and A(0,2) modes.

To validate the effect of the spatial beating on the leaky radiation beam, using the constructed RSI model, the leaky radiation beam profiles were simulated with the phase variations of superposed modes on the radiation surface. Figures 12a, 13a and 14a and Figures 12b, 13b and 14b show the wave phase on the radiation surface used in the beam profile calculations of Figures 12c, 13c and 14c and Figures 12d, 13d and 14d; the wave phase on the radiation surface was simulated using Equation (5), including the leaky attenuation coefficient term with adjustment of the phase delay between the superposed modes, and they were presented in case of reasonable quantitative matching with the measurement results as shown in Figure 15 (a representative example of the comparison result). Although the constructed RSI model is a continuous wave model using a single frequency (only the center frequency component of the excitation input is considered), its simulation result shows the good agreement with the experimental one in far-field characteristics and beam pattern.

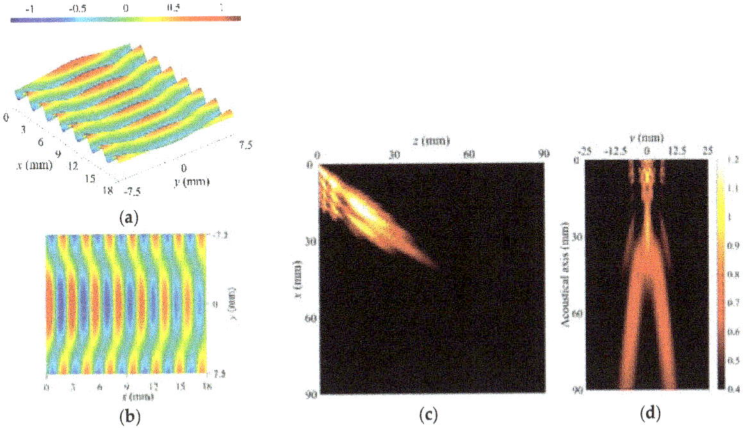

Figure 12. Simulation #1 of the leaky radiation beam profile based on the wave phase on the radiation surface. (**a**) Isotropic view of wave phase on radiation surface, (**b**) top view of wave phase on radiation surface, (**c**) vertical beam profile, and (**d**) lateral beam profile.

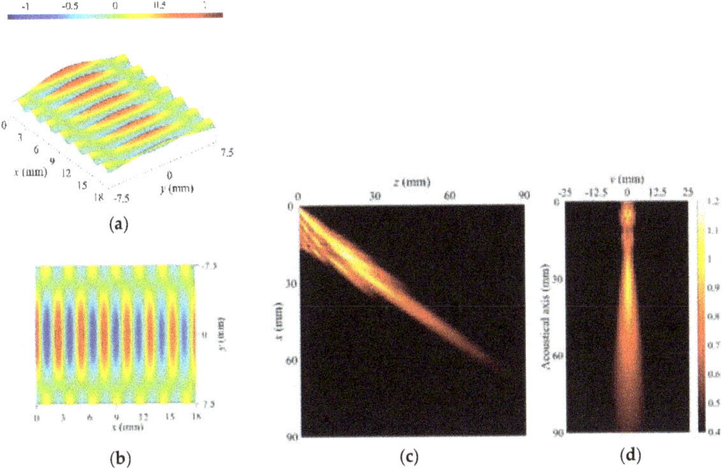

Figure 13. Simulation #2 of the leaky radiation beam profile based on the wave phase on the radiation surface. (**a**) Isotropic view of wave phase on radiation surface, (**b**) top view of wave phase on radiation surface, (**c**) vertical beam profile, and (**d**) lateral beam profile.

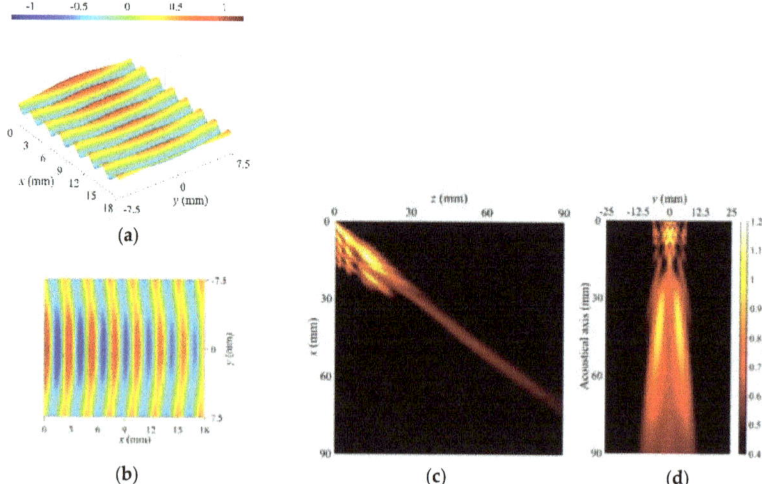

Figure 14. Simulation #3 of the leaky radiation beam profile based on the wave phase on the radiation surface. (**a**) Isotropic view of wave phase on radiation surface, (**b**) top view of wave phase on radiation surface, (**c**) vertical beam profile, and (**d**) lateral beam profile.

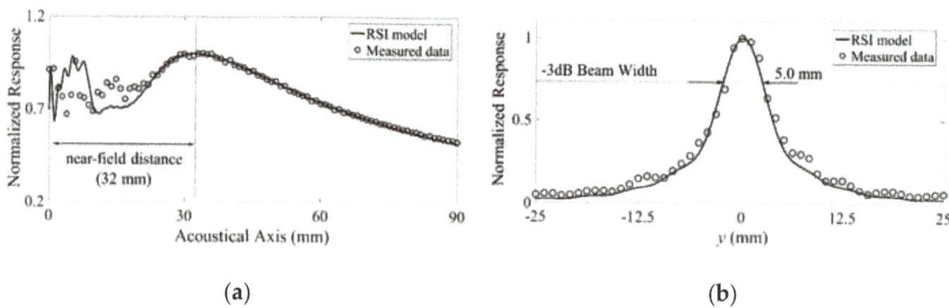

Figure 15. Comparison results between the measured beam profile of a leaky Lamb wave radiated at d = 350 mm (Figure 10b) and the simulation result of Figure 13. (**a**) Response on acoustical axis and (**b**) lateral beam width at maximum intensity of the main beam.

Consequentially, simulation results of the beam profiles describe the fact that the A(0,2) mode affects leaky Lamb wave radiation with the A(0,0) mode in the waveguide plate with finite width, and the change of the wave behavior on the radiation surface caused by the mode superposition and the wave velocity difference results in the change of the radiation characteristics of the leaky radiation beam profile.

5. Discussion and Design Direction for Performance Improvement of the Waveguide Sensor

It seems difficult to completely physically avoid the spatial beating phenomenon and the mode superposition within a short propagation distance because of the high modal density by numerous width modes of the Lamb wave in a waveguide plate with finite width. Therefore, if a single narrow main beam is needed for the application purpose, the position tuning for the excitation source can be recommended for the wave energy concentration at the center in the width direction. Figure 16 shows the measured beam profiles of the leaky Lamb wave excited at d = 330 mm and a clear single main beam can be seen compared with the beam profiles measured for other excitation source positions shown in Figure 10.

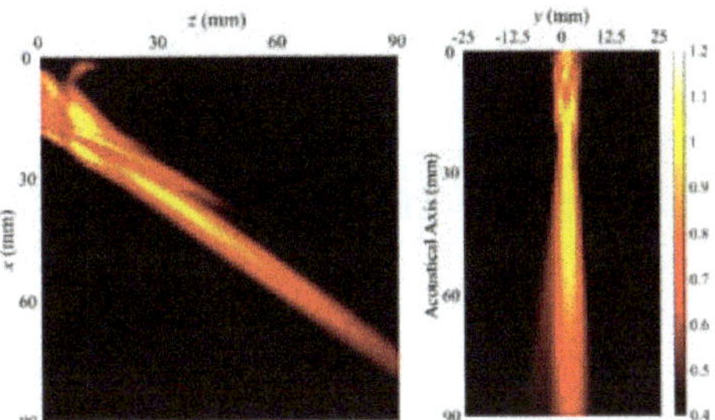

Figure 16. Measured beam profiles of the leaky Lamb wave excited at d = 330 mm (1.5 mm thickness and 15 mm width, 1.0 MHz center frequency).

In addition, as the plate width increases, the phase velocity curve of the A(0,2) mode converges to that of the A(0,0) mode as shown in Figure 17. This is because the plate strip gets close to the infinitely wide plate with extension of the plate width. Accordingly, unless the plate width is not constrained in the waveguide sensor design, the plate width increment can extend the beating length by a difference in reduction between the A(0,0) and A(0,2) modes in wave velocities, and thereby the effect of the spatial beating on the leaky radiation beam can be reduced.

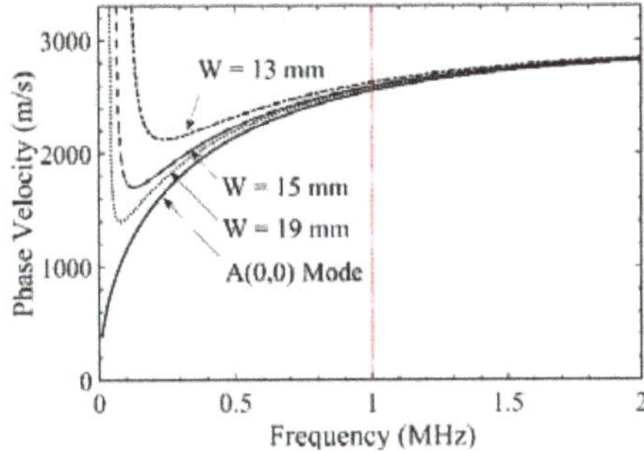

Figure 17. Phase velocity dispersion curves of the A(0,2) mode in a 1.5 mm thick SS304 plate with variation of the plate width.

Finally, a change in the frequency–thickness product (fh) can be also considered for performance improvement. Figure 18 shows the measured beam profile of the leaky Lamb wave excited at d = 350 mm with fh = 1.5 MHz·mm (1.5 MHz center frequency and 1.0 mm plate thickness), the same as the operational frequency–thickness product in the waveguide sensor (1.0 MHz center frequency and 1.5 mm plate thickness). As the excitation frequency increment makes the wavelength of the leaky Lamb wave short, the width–wavelength ratio is increased. Therefore, the effective plate width can be enlarged without an increment of the physical plate width, although the ultrasonic attenuation is increased and the radiation characteristics are also changed by the frequency increment of the leaky wave.

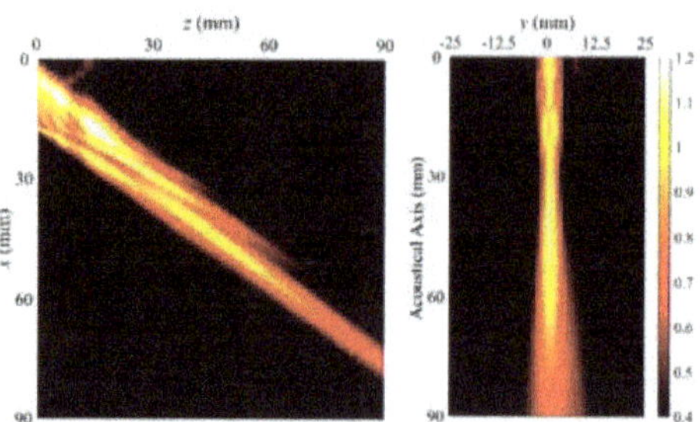

Figure 18. Measured beam profiles of the leaky Lamb wave excited at d = 350 mm (1.0 mm thickness and 15 mm width, 1.5 MHz center frequency).

As a result, leaky Lamb wave radiation from a waveguide plate with a finite width is affected by the mode superposition of the width modes. For this reason, radiation characteristics of the leaky radiation beam, in particular the lateral beam, are sensitive to the excitation conditions (excitation source position, excitation frequency, etc.). This can be an advantage or disadvantage for the plate-type waveguide sensor. Therefore, according to the application purposes and goals, it will be necessary to properly design and tune the excitation conditions in the plate-type waveguide sensor.

6. Conclusions

This work investigates leaky Lamb wave radiation from a waveguide plate with a finite width with consideration for the width modes of the Lamb wave and their superposition. Dispersion curves obtained using the SAFE method showed that Lamb waves in the plate strip have numerous width modes, in contrast to the case of an infinitely wide plate. These width modes were very close to each other in the dispersion curve, and thus it could be inferred from these results that the multiple width modes were bound to be excited and propagated together by the high modal density. In the beam profile measurement, one could observe that characteristics of the leaky radiation beam from the waveguide plate were noticeably changed with variation of the excitation source position. This characteristic was strongly estimated to be affected by the spatial beating induced by wave velocity differences of the superposed modes. Changes to the wave phase on the waveguide plate were validated by a simple computational result and then the leaky radiation beam profiles were simulated with variation of the wave phase of the superposed modes on the radiation surface. From this simulation, it was demonstrated that leaky Lamb wave radiation from the waveguide plate was affected by superposition of the width modes. In particular, the lateral beam was dominantly influenced by the wave phase on the radiation surface, changed by the excitation source position. Perfect avoidance of the spatial beating caused by the mode superposition of the width modes in the strip-like plate is expected to be physically difficult. Therefore, the excitation conditions, including the excitation frequency and the geometry of the waveguide plate, need to be properly designed for the waveguide sensor using leaky Lamb wave radiation from a waveguide plate with a finite width. Further study is necessary to analyze the radiation characteristics of practical plate-type waveguide sensors based on this preliminary work.

Author Contributions: Conceptualization and methodology: S.-J.P. and Y.-S.J.; investigation: S.-J.P. and Y.-S.J.; SAFE analysis: S.-J.P.; experimental validation: S.-J.P. and Y.-S.J.; writing—original draft preparation, S.-J.P.; writing—review and editing: H.-W.K., and Y.-S.J.; supervision: Y.-S.J.; All authors have read and agreed to the published version of the manuscript.

Funding: This work was supported by the Nuclear Research & Development Program of the National Research Foundation with a grant funded by the Korea Ministry of Science and ICT.

Conflicts of Interest: The authors declare no conflict of interest.

References

1. Ditri, J.; Rose, J.L.; Chen, G. Mode selection criteria for defect detection optimization using Lamb waves. In Proceedings of the 18th Annual Review of Progress in Quantitative NDE; Thompson, D.O., Chimenti, D.E., Eds.; Plenum Press: New York, NY, USA, 1992; Volume 11, pp. 2109–2115.
2. Cawley, P.; Alleyne, D.N. The use of Lamb waves for the long range inspection of large structure. *Ultrasonics* **1996**, *34*, 287–290. [CrossRef]
3. Alleyne, D.N.; Cawley, P. Long range propagation of Lamb waves in chemical plant pipework. *Mater. Eval.* **1997**, *55*, 504–508.
4. Na, W.-B.; Kundu, T. Underwater pipeline inspection using guided waves. *J. Press. Vessel Technol.* **2002**, *124*, 196–200. [CrossRef]
5. Gao, H.; Rose, J.L. Ice detection and classification on an aircraft wing with ultrasonic shear horizontal guided waves. *IEEE Trans. Ultrason. Ferroelectr. Freq. Control* **2009**, *56*, 334–344.
6. Chen, J.; Su, Z.; Cheng, L. Identification of corrosion damage in submerged structures using fundamental anti-symmetric Lamb waves. *Smart Mater. Struct.* **2010**, *19*, 015004. [CrossRef]
7. Gao, H.; Rose, J.L. Goodness dispersion curves for ultrasonic guided wave based SHM: A sample problem in corrosion monitoring. *Aeronaut. J. R. Aeronaut. Soc.* **2010**, *114*, 49–56. [CrossRef]
8. Leinov, E.; Lowe, M.-J.S.; Cawley, P. Investigation of guided wave propagation and attenuation in pipe buried in sand. *J. Sound Vib.* **2015**, *347*, 96–114. [CrossRef]
9. Lynnworth, L.C.; Liu, Y.; Umina, J.A. Extensional bundle waveguide techniques for measuring flow of hot fluids. *IEEE Trans. Ultrason. Ferroelectr. Freq. Control* **2005**, *52*, 538–544. [CrossRef]
10. Celga, F.B.; Cawley, P.; Allin, J.; Davies, J. High temperature(>500 °C) wall thickness monitoring using dry-coupled ultrasonic waveguide transducer. *IEEE Trans. Ultrason. Ferroelectr. Freq. Control* **2011**, *58*, 156–167.
11. Laws, M.; Ramadas, S.N.; Lynnworth, L.C.; Dixon, S. Parallel strip waveguide for ultrasonic flow measurement in harsh environments. *IEEE Trans. Ultrason. Ferroelectr. Freq. Control* **2015**, *62*, 697–708. [CrossRef]
12. Sun, F.; Sun, Z.; Chen, Q.; Murayama, R.; Nishino, H. Mode conversion behavior of guided wave in a pipe inspection system based on a long waveguide. *Sensors* **2016**, *16*, 1–14. [CrossRef] [PubMed]
13. Shah, H.; Balasubramaniam, K.; Rajagopal, R. In-situ process-and online structural health monitoring of composites using embedded acoustic waveguide sensors. *J. Phys. Commun.* **2017**, *1*, 1–11. [CrossRef]
14. Griffin, J.W.; Bond, L.J.; Peters, T.J.; Denslow, K.M.; Posakony, G.J.; Sheen, S.H.; Chien, H.T.; Raptis, A.C. *Under-Sodium Viewing: A Review of Ultrasonic Imaging Technology for Liquid Metal Fast Reactors*; Pacific Northwest National Laboratory: Richland, WA, USA, 2009; pp. 1–64.
15. Wang, K.; Chien, H.; Elmer, T.W.; Lawrence, W.P.; Engel, D.M.; Sheen, S. Development of ultrasonic waveguide techniques for under-sodium viewing. *NDT E Int.* **2012**, *49*, 71–76. [CrossRef]
16. Watkins, R.D.; Deighton, M.O.; Gillespie, A.B.; Pike, R.B. A proposed method for generating and receiving narrow beams of ultrasound in the fast reactor liquid sodium environment. *Ultrasonics* **1982**, *20*, 7–12. [CrossRef]
17. Watkins, R.D.; Barrett, L.M.; McKnight, J.A. Ultrasonic waveguide for use in the sodium coolant of fast reactors. *Nucl. Energy* **1988**, *27*, 85–89.
18. Joo, Y.-S.; Park, C.-G.; Lee, J.-H.; Kim, J.-B. Development of ultrasonic waveguide sensor for under-sodium inspection in a sodium-cooled fast reactor. *NDT E Int.* **2011**, *44*, 239–246. [CrossRef]
19. Joo, Y.-S.; Bae, J.-H.; Kim, J.-B.; Kim, J.-Y. Effects of beryllium coating layer on performance of the ultrasonic waveguide sensor. *Ultrasonics* **2013**, *53*, 387–395. [CrossRef]
20. Kim, H.-W.; Joo, Y.-S.; Park, C.-G.; Kim, J.-B.; Bae, J.-H. Ultrasonic imaging in hot liquid sodium using a plate-type ultrasonic waveguide sensor. *J. Nondestruct. Eval.* **2014**, *33*, 676–683. [CrossRef]
21. Joo, Y.-S. Under-sodium viewing techniques in sodium-cooled fast reactors. *J. Korean Soc. Nondestruct. Test.* **2012**, *32*, 439–447.
22. Kim, H.-W.; Joo, Y.-S.; Park, S.-J.; Kim, S.-K. Ultrasonic ranging technique for obstacle monitoring above reactor core in prototype generation IV sodium-cooled fast reactor. *Nucl. Eng. Technol.* **2020**, *52*, 776–783. [CrossRef]
23. Banerjee, S.; Kundu, T. Ultrasonic field modeling in plates immersed in fluid. *Int. J. Sol. Struct.* **2007**, *44*, 6013–6029. [CrossRef]

24. Hayashi, T.; Inoue, D. Calculation of leaky Lamb waves with a semi-analytical finite element. *Ultrasonics* **2014**, *54*, 1460–1469. [CrossRef] [PubMed]
25. Inoue, D.; Hayashi, T. Transient analysis of leaky Lamb waves with a semi-analytical finite element. *Ultrasonics* **2015**, *62*, 80–88. [CrossRef] [PubMed]
26. An, Z.; Hu, J.; Mao, J.; Lian, G.; Wang, X. Optical visualization of leaky Lamb wave and its application in nondestructive testing. In Proceedings of the 2017 ICU Honolulu Sixth International Congress on Ultrasonics, Honolulu, HI, USA, 18–22 December 2017.
27. Deighton, M.O.; Gillespie, A.-B.; Pike, R.-B.; Watkins, R.-D. Mode conversion of Rayleigh and Lamb waves to compression waves at a metal-liquid interface. *Ultrasonics* **1981**, *19*, 249–258. [CrossRef]
28. Morse, R.W. Dispersion of compressional waves in isotropic rods of rectangular cross section. *J. Acoust. Soc. Am.* **1948**, *20*, 833–838. [CrossRef]
29. Morse, R.W. The velocity of compressional waves in rods of rectangular cross section. *J. Acoust. Soc. Am.* **1950**, *22*, 219–223. [CrossRef]
30. Mindlin, R.D.; Fox, E.A. Vibrations and waves in elastic bars of rectangular cross section. *Trans. ASME J. Appl. Mech.* **1960**, *27*, 152–158. [CrossRef]
31. Konenkov, Y.K. Plate waves and flexural oscillations of a plate. *Sov. Phys. Acoust.* **1960**, *6*, 52–59.
32. Konenkov, Y.K.; Namukina, N.I.; Tartakovskii, B.D. Investigation of forced flexural vibrations of an elastic strip. *Sov. Phys. Acoust.* **1966**, *11*, 288–295.
33. Fraser, W.B. Stress wave propagation in rectangular bars. *Int. J. Solids Struct.* **1969**, *5*, 379–397. [CrossRef]
34. Krushynska, A.A.; Meleshko, V.V. Normal waves in elastic bars of rectangular cross section. *J. Acoust. Soc. Am.* **2011**, *129*, 1324–1335. [CrossRef]
35. Hayashi, T.; Song, W.-J.; Rose, J.L. Guided wave dispersion curves for a bar with an arbitrary cross-section, a rod and rail example. *Ultrasonics* **2003**, *41*, 175–183. [CrossRef]
36. Hayashi, T.; Tanaka, T. Lamb wave propagation in a plate with a finite width. *Trans. Jpn. Soc. Mech. Eng. A* **2006**, *72*, 149–154. [CrossRef]
37. Mukdadi, O.M.; Datta, S.K.; Dunn, M.L. Elastic guided waves in a layered plate with a rectangular cross section. *J. Press. Vessel Technol.* **2002**, *124*, 319–325. [CrossRef]
38. Hakoda, C.; Rose, J.; Shokouhi, P.; Lissenden, C. Using Floquet periodicity to easily calculate dispersion curves and wave structures of homogeneous waveguides. In *44th Annual Review of Progress in Quantitative NDE*; Chimenti, D.E., Bond, L.J., Eds.; Springer: New York, NY, USA, 2017; Volume 37, pp. 020016:1–020016:10.
39. Groth, E.B.; Iturrioz, I.; Clarke, T.-G.R. The dispersion curve applied in guided wave propagation in prismatic rods. *Lat. Am. J. Solids Struct.* **2018**, *15*, 1–27. [CrossRef]
40. Pavlakovic, B.; Lowe, M. *DISPERSE: User's Manual Version 2.0.16B*; Imperial College London: London, UK, 2003; pp. 1–209.
41. Rayleigh, J.W.S. *The Theory of Sound*; Dover Publications: New York, NY, USA, 1945; Volume 2, pp. 1–520.
42. Nakahata, K.; Kono, N. 3-D modelings of an ultrasonic phased array transducer and its radiation properties in solid. In *Ultrasonic Waves*; Santos, J.A., Ed.; IntechOpen: London, UK, 2012; Available online: http://www.intechopen.com/books/ultrasonic-waves/3-d-modelings-of-an-ultrasonic-phasedarray-transducer-and-its-radiation-properties-in-solid (accessed on 10 August 2019).
43. Song, S.-J.; Kim, H.-J. Modeling of radiation beams from ultrasonic transducers in a single medium. *J. Korean Soc. Nondestruct. Test.* **2000**, *20*, 91–101.
44. Kundu, T.; Placko, D.; Rahani, E.K.; Yanagita, T.; Dao, C.M. Ultrasonic field modeling: A comparison of analytical, semi-analytical, and numerical techniques. *IEEE Trans. Ultrason. Ferroelectr. Freq. Control* **2010**, *57*, 2795–2807. [CrossRef]
45. Duan, W.; Niu, X.; Gan, T.-H.; Kanfoud, J.; Chen, H.-P. A numerical study on the excitation of guided waves in rectangular plates using multiple point sources. *Metals* **2017**, *7*, 1–21. [CrossRef]
46. Serey, V.; Quaegebeur, N.; Micheau, P.; Masson, P.; Castaings, M.; Renier, M. Selective generation of ultrasonic guided waves in a bi-dimensional waveguide. *SHM* **2018**, *18*, 1–13. [CrossRef]

Article

Evaluation of Cracks on the Welding of Austenitic Stainless Steel Using Experimental and Numerical Techniques

Azouaou Berkache [1,*], Jinyi Lee [2,3,*] and Eunho Choe [1]

1 IT-Based Real-Time NDT Center, Chosun University, Gwangju 61452, Korea; 1989.choe@gmail.com
2 Department of Electronics Engineering, Chosun University, Gwangju 61452, Korea
3 Interdisciplinary Program in IT-Bio Convergence System, Chosun University, Gwangju 61452, Korea
* Correspondence: azouaoubrk@yahoo.fr (A.B.); jinyilee@chosun.ac.kr (J.L.)

Abstract: This paper deals with investigation and characterization of weld circumferential thin cracks in austenitic stainless steel (AISI 304) pipe with eddy current nondestructive testing technique (EC-NDT). During welding process, the heat source applied to the AISI 304 was not uniform, accompanied by a change of the physical property. To take into consideration this change, the relative magnetic permeability was considered as a gradiently changed variable in the weld and the heat affected zone (HAZ), which was generated by the Monte Carlo Method based on pseudo random number generation (PRNG). Numerical simulations were performed by means of MATLAB software using 2D finite element method to solve the problem. To verify, results from the modeling works were conducted and contrasted with findings from experimental ones. Indeed, the results of comparison agreed well. In addition, they show that considering this changing of this magnetic property allows distinguishing the thin cracks in the weld area.

Keywords: weld cracks; eddy current nondestructive testing; gradiently relative magnetic permeability; heat affected zone; austenitic stainless steel

Citation: Berkache, A.; Lee, J.; Choe, E. Evaluation of Cracks on the Welding of Austenitic Stainless Steel Using Experimental and Numerical Techniques. *Appl. Sci.* **2021**, *11*, 2182. https://doi.org/10.3390/app11052182

Academic Editor: Alberto Campagnolo

Received: 22 January 2021
Accepted: 25 February 2021
Published: 2 March 2021

1. Introduction

The austenitic stainless steel 304 is suitable for large field applications, such as heat exchangers, power plants, oil and gas industry, chemical engineering and especially in nuclear power plants, because of its useful characteristics, such as high temperature service and environment, corrosion resistance, weldability, formability and mechanical properties [1,2]. Pipes are exposed to a variety of environmental influences, and high temperature and high pressure that cause severe corrosive and environmental deterioration as results in fatigue cracks and flaws in pipes that can appear and grow. A pipe failure can lead to serious ecological disasters, human casualties and financial loss. To predict and avoid such threats and maintain the safety and integrity of pipes, periodic nondestructive testing inspections are necessary [3–5].

Eddy-current testing (ECT) is adapted to solve such problems. Applied to inspect conductive and ferromagnetic devices to examine their structural integrity [6–9], it has certain advantages in terms of safety as a testing tool—rapid inspection, high sensitivity, and minimizing contact with the specimen. It is efficiently associated with automatic detection in various work [10,11]. The ECT technique is frequently used for nondestructive defect inspection of tubes welders and circumferential welds [5,12,13].

Many welding processes are presented in several papers; they require a lot of effort and experiments to better understand the parameters [14–16]. The finite element method associated with experimental investigations is a powerful tool to identify and analyze welding parameters and to obtain an optimal solution in a short time [17]. Usually, in simulation work, the mathematical model of the problem to be treated contains all the necessary information.

In this work, a study of weld circumferential thin cracks in AISI 304 pipe was carried out. The used AISI 304 pipe was joined by gas tungsten arc welding (GTAW).

Several previous studies focused on crack detection and characterization of cracks in austenitic stainless steel, for instance, detection and evaluation of weld defects using 3-dimension tunnel, and show excellent inspection results for a weld in stainless steel [18]. Saito et al. [19] described weld defects and evaluation of weld quality and how to achieve weld quality improvement. Park et al. [20] investigated cracking behavior of AISI 304 exposed to high temperature and revealed that strain-induced martensitic transformation in stainless steel has a negative effect on stress corrosion effect. Hu et al. [21] studied microstructure residual stress and corrosion cracking of repair welding on 304 AISI by experiment and simulation, and found that repair welding in 304 stainless steel is recommended no more than two times. All these studies are related to the evaluation of defects in using different methods.

However, there are few considerations of delta-ferrite structures, which show ferromagnetic properties, inspite of that the austenitic stainless steel AISI 304 around the welding for analyzing numerical and experimental studies. AISI 304 is classified as paramagnetic material with $\mu_r = 1 + \mathcal{X}_m$, where relative magnetic permeability μ_r and the magnetic susceptibility $\mathcal{X}_m$, are 1 and 0, respectively. However, after heat treatment (heat welding), the property of AISI 304 as paramagnetic material will disappear and it will become a partial ferromagnetic material named delta-ferrite structure, which results in the change of the relative magnetic permeability at each region $\mu_r \neq 1$,with high permeability in the weld area, low permeability in the raw material $\mu_r = 1$ and decreasing permeability in the HAZ $\mu_r = 3 \sim 1$ [22–24]. Therefore, it can be considered that the assembled pipe has three regions with three magnetic permeabilities. Modeling this change on this input parameter and characterizing its consequence on the output response under MATLAB, in order to be able to both reproduce the distribution of this parameter and to compare this response with experimental, is an important step of this work. For this purpose, the relative magnetic permeability is gradiently distributed in the weld and the HAZ regions according to the experimental measurement. The Monte Carlo method based on pseudo random number generation (PRNG) is used and then coupled with the finite element method. The numerical analysis using the Stochastic Finite Element Method (SFEM) that models the eddy-current testing of the problem is presented. A comparison against constant relative permeability and experimental ones was done, and the results show that it is important to consider this change in magnetic property of the material.

2. Materials

Figure 1 shows the experiment system. It consists mainly of two joined AISI 304 pipe test samples, a rotating motorized stage platform, an ECT system, a data acquisition instrument (DAQ) and a laptop computer for data control and storage. The ECT system is an Olympus Nortek 500 eddy-current flaw detector, which controls AC power supply and frequency. The output signal obtained by the ECT equipment is transmitted to a laptop computer via an analog-to-digital converter. Absolute probes with different frequency ranges were fixed and the pipe was mounted and precisely positioned on the rotary stage platform, which is controlled by software based on the LabVIEW program and rotated at a speed of 21 mm/s and scan interval of 1mm. The experiments were performed by an ASNT Level II qualified examiner by following ISO 7912 instructions [25].

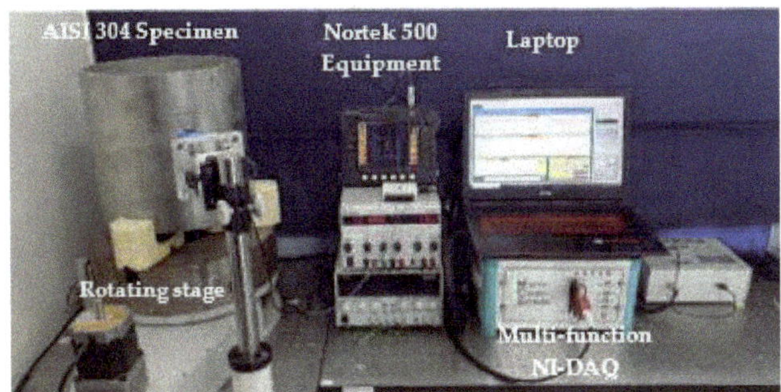

Figure 1. Experimental setup.

Artificial cracks of the same length (5 mm) and width (0.2 mm) with different depths (0.3 mm, 0.5 mm and 1.0 mm) referred as (d0.3, d0.5 and d1.0), respectively, were manufactured by a Sinker type ZNC electrical discharge machining in the weld area; they are spaced 7 mm apart, as given by Figure 2. In order not to influence the measurements, an appropriately sized ECT probe was used. The cracks position was intended to simulate the most frequent ones, which can occur during the welding process or during the pipe daily service. The geometrical and physical parameters are summarized in Table 1. Moreover, microscopic analyses of the width and length of these cracks are checked, and the depth is measured using digital calipers and presented in Figures 3 and 4, respectively.

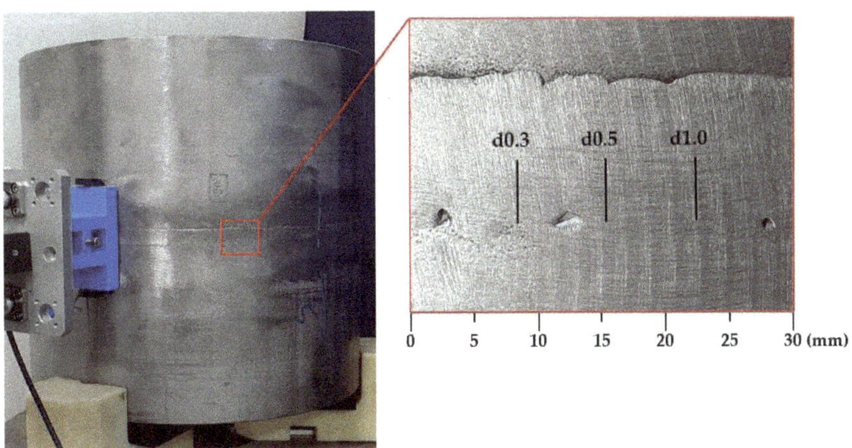

Figure 2. The structure of joined pipe and the configuration type eddy current nondestructive testing technique (EC-NDT).

Table 1. Test setup parameters.

Probe		Test Specimen	
Inner diameter	3 mm	Thickness	9 mm
Outer diameter	4 mm	Conductivity	1.38 Ms/m
Height	4 mm	Permeability1	1
Numbers of turns	N/A	Permeability2	Random
Lift-off	0.1 mm	Permeability3	Random
		Weld width	12 mm

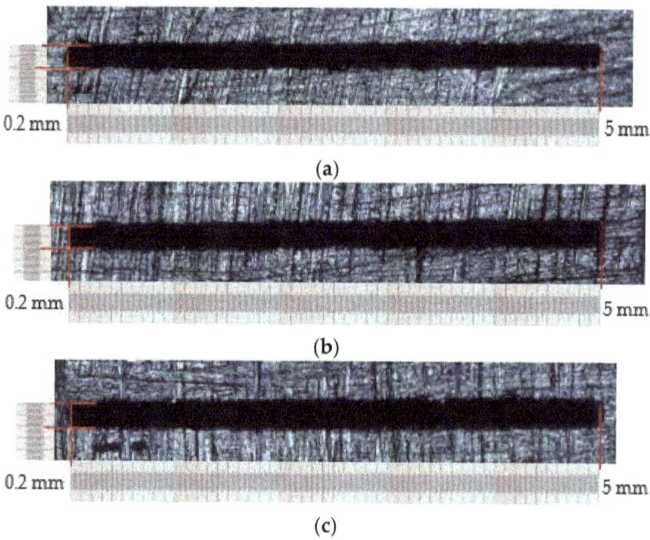

Figure 3. Microscopic analysis: measurement of length and width of the cracks with (**a**) depth = 0.3 mm, (**b**) depth = 0.5 mm, and (**c**) depth = 1 mm.

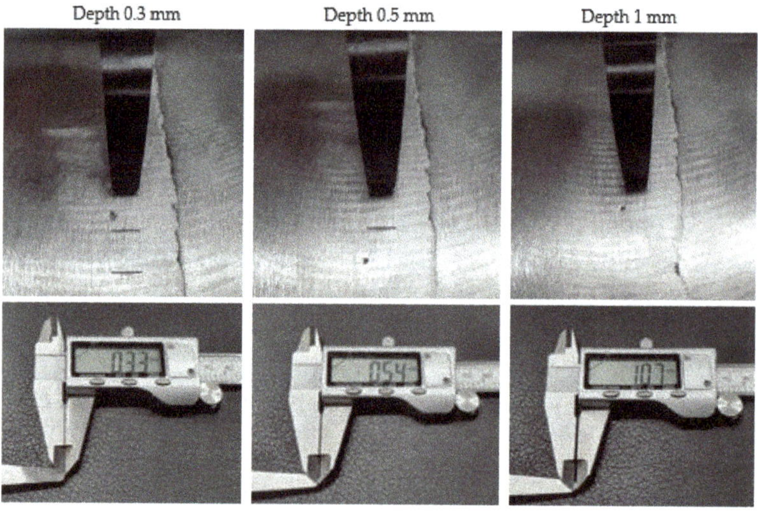

Figure 4. Cracks depth measurement.

3. Global Equations and Parameters

3.1. Random Numbers Generation

Random numbers are useful in several different kinds of applications, such as simulation, statistics, machine learning, sampling and in other areas [26,27]—in this section according to the experimental measurement of the relative magnetic permeability, which is gradiently distributed in the heat affected zone (HAZ) and weld area due to the heat welding. To simulate this stochastic model, a source of randomness is required to reproduce the real distribution of the magnetic permeability in these concerned zones. A pseudo-random number is a best way to solve this problem, by generating a sequence of independent uniform variable real between 0 and 1 or integer. Various pseudo-random number generators (PRNGs) exist, and the most popular random number generation technique is the linear congruential generator (LCG) for several reasons. LCGs are the most popular generators, implemented in the MATLAB programming software used, and have many properties; for instance, ease of use, reproducibility, uniformity, independence, large period and efficiency [28,29]. The main advantages of these properties are that they generate easily and directly pseudo-random numbers without storing them, memory savings, time savings and simulation control. A quick overview of the LCGs mainly used in computer programming is given; for more details, see references [30–32]. The LCGs are based on linear recursions in modular arithmetic. Their general form, represented by:

$$x_{n+1} = (ax_n + c) \bmod m, n \geq 0 \tag{1}$$

here $m > 0$ is the modulus, a is the multiplier, c is the increment and x_0 is the seed or the starting value; $0 \leq a < m, 0 \leq c < m, 0 \leq x_0 < m$. Selection of the numbers m, a, c and x_0 is crucial for getting a random sequence of numbers.

After the step of generating pseudo-random numbers, apply to these numbers an appropriate transformation according to the number of elements contained in the surfaces of the studied areas obtained by finite element meshing. Then the algorithms are combined with the finite element code.

3.2. Electromagnetic Equation

The governing equations of the numerical model used in this paper are obtained with the consideration of assumptions that the conduction current is dominated, to describe electromagnetic eddy-current problems extracted from Maxwell's equations, which describe the basics of electromagnetic theory given as follows [33,34]:

$$\vec{\nabla} \times \vec{H} = \vec{J} \tag{2}$$

$$\vec{\nabla} \times \vec{E} = -\frac{\partial \vec{B}}{\partial t} \tag{3}$$

$$\vec{J} = \vec{J}_s - \sigma \frac{\partial \vec{A}}{\partial t} \tag{4}$$

Using the relation $\vec{B} = \mu \vec{H}$ and the magnetic flux density potential $\vec{B} = \vec{\nabla} \wedge \vec{A}$. After replacing we obtain the 2D electromagnetic harmonic equation in terms of the Magnetic Vector Potential (MVP) with only the z direction component $\vec{A}(0, 0, A_Z)$:

$$\vec{\nabla} \times \left(\frac{1}{\mu} \left(\vec{\nabla} \times \vec{A}_Z \right) \right) - j\omega\sigma \vec{A}_Z = -\vec{J}_{SZ} \tag{5}$$

where: $\vec{J}$—the total current density, $\vec{J}_{SZ}$—the source current density, μ—is magnetic permeability in the specimen, HAZ and in the weld zone respectively, σ—is the electrical conductivity, ω—the angular frequency.

In the Heat affected zone and the weld zone, the magnetic permeability is noted with indices (1), (2) and given as, $\mu_1 = [\mu\prime_a \ldots\ldots\ldots\ldots \mu_{THAZ}]$ and $\mu_2 = [\mu_a \ldots\ldots\ldots\ldots \mu_{Tw}]$. $THAZ$ and T_W denote the total number of triangular elements obtained from the finite element meshing in the HAZ zone and weld area, respectively.

$$[M] + j\omega[N][A] = [F] \tag{6}$$

with: $[M]$—stiffness matrix, $[N]$—dynamic matrix, $[A]$—unknowns vector and $[F]$—source vector.

3.3. Impedance Computation

The presence of possible defects in the weld zone lead to a change in the physical characteristics, which results in the variation of the coil impedance. Several methods exist for impedance calculation; the difference lies in the choice of the state variable which has a direct relationship with the solution resulting from the numerical model and the configuration of the device to be studied. In this application the impedance Z is calculated from the MVP (the real and imaginary parts) as follows [35,36]:

$$Re(Z) = -\frac{N^2}{J\,S^2}\omega \iint_S 2\pi r\, Im(A)\, ds \quad (7)$$

$$Im(Z) = -\frac{N^2}{J\,S^2}\omega \iint_S 2\pi r\, Re(A)\, ds \quad (8)$$

with: N—coils number, S—surface of inductor coil, r—inductor radius.

4. Results and Discussion

In the current application, the numerical and experimental investigation of weld thin cracks in joined AISI 304 pipe were carried out, using SFEM code analysis implemented under MATLAB.

The pipe radius is 160 mm, far greater than the probe size, so the pipe wall can be considered as a conducting plate [37,38]. According to this, the studied problem is simplified and becomes two-dimensional (2D) in the (x, y) plane as shown in Figure 5.

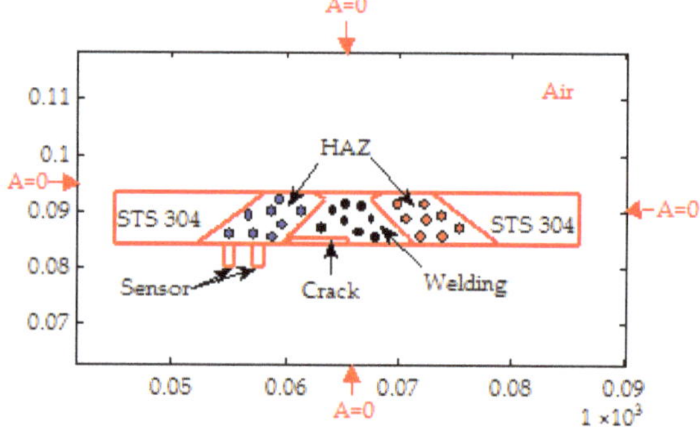

Figure 5. Solving domain and boundary conditions.

The solving domain resolution concerns the studied electromagnetic device and the air; it is divided into six regions with different physical properties, with physical boundary conditions of homogeneous Dirichlet type applied at the fields of study. The scheme is illustrated in Figure 5, which shows that the mesh air domain that we have taken into consideration is large enough to contain the zone of influence of the probe, so that the emitted field is negligible at the border of the field of study.

The field of study is covered by finite element mesh as illustrated by Figure 6a. To reproduce the real geometry of the studied device and to approach the measurement results, it is discretized by subdividing it into subdomains, with 38,400 triangular elements and 19,269 nodes generated automatically. Eddy current distributes locally near the coil [39], to consider this fact in the simulation work. A remeshing is done at each probe displacement. This technique allows obtaining dense and fine mesh around the probe with good quality elements. The triangular mesh quality as a function of the probe displacement is shown in Figure 6b. The mesh quality is over than 0.75 and consequently more accurate simulation results.

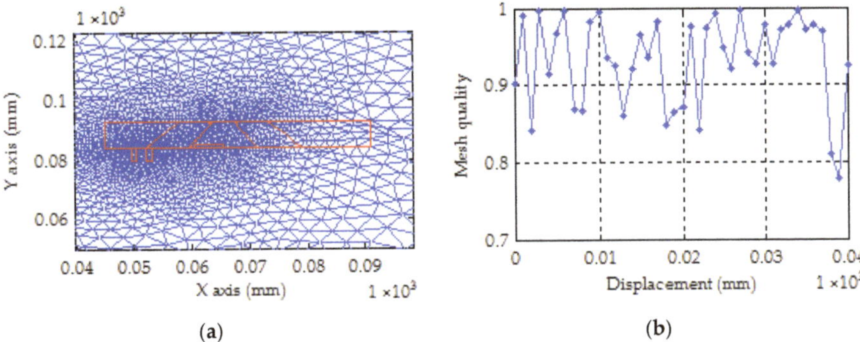

Figure 6. (**a**) Mesh of the solving domain and (**b**) Mesh quality according to the displacement.

As explained above, after a step of random-numbers generation, a computational technique was introduced by which was obtained the relative magnetic permeability. Depending on the experimental measurement for each pitch measurement, this magnetic property distribution is not the same and it varies from 1 to 20, hence the assimilation to a delta-ferrite structure.

The simulated relative magnetic permeability distribution (to reproduce the behavior as the experimental) is illustrated by Figure 7, covering a displacement of 20 mm from the middle length of the inspected devise, by cause of symmetry distribution.

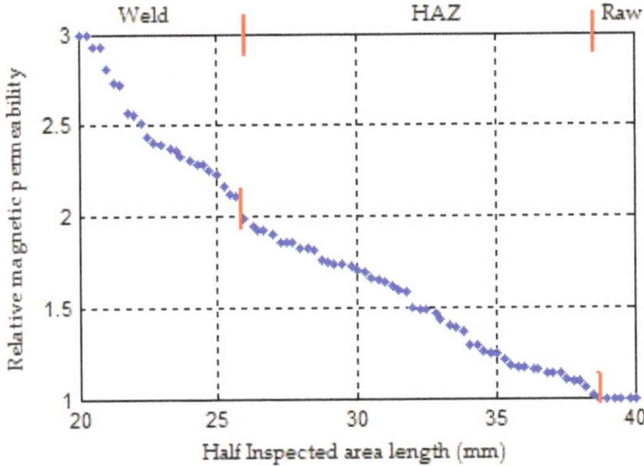

Figure 7. The magnetic relative permeability distribution.

The ECT probe moves along the direction of the cracks, from the position x = 0 to x = 40 mm, in step of 1 mm. The objective here is to conduct a qualitative study relating to the presence or absence of the most frequent thin cracks in circumferential girth weld, considering the influence of the heat welding on the HAZ and the weld area. The HAZ length is to be assumed 12.7 mm ($\frac{1}{2}$ inches), refer to the KEPIC MI Technical Standard [40]. To achieve both larger skin depth and to control the surface of inspected pipe, the measurements have been realized using six frequencies, operating frequency 20 kHz, detection frequency 40 kHz, optimum frequency 50 kHz and resonant frequency 300 kHz. In addition, 10 kHz and 100 kHz were added for comparison. The experiments were performed by an ASNT Level II qualified examiner by following ISO 7912 instructions [25].

The measurement of the eddy currents resulting from experimentation and simulation was exploited by the measurement of the related quantity, which results in the impedance measurement. The results of the comparisons were normalized and given in Figures 8–13 for different crack depths. The impedance variation was analyzed for weld cracks and stored with the corresponding displacement coordinate of the ECT probe. The normalized impedance was computed by $Delta\ Z(\%) = (Z - Z_0)/(Z - Z_0)_{max}$. With Z_0 and Z are the impedances of the raw material without crack and the impedance from the HAZ and the weld with crack, respectively.

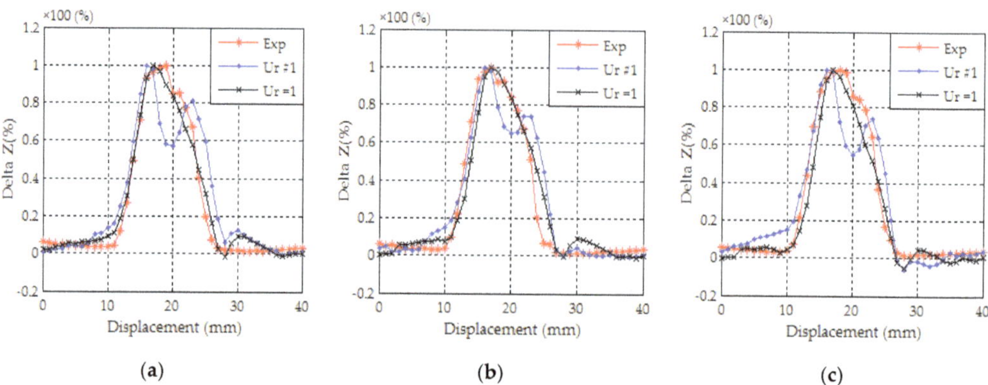

Figure 8. Impedance variation in (%) at 10 kHz with (**a**) depth = 0.3 mm, (**b**) depth = 0.5 mm, and (**c**) depth = 1 mm.

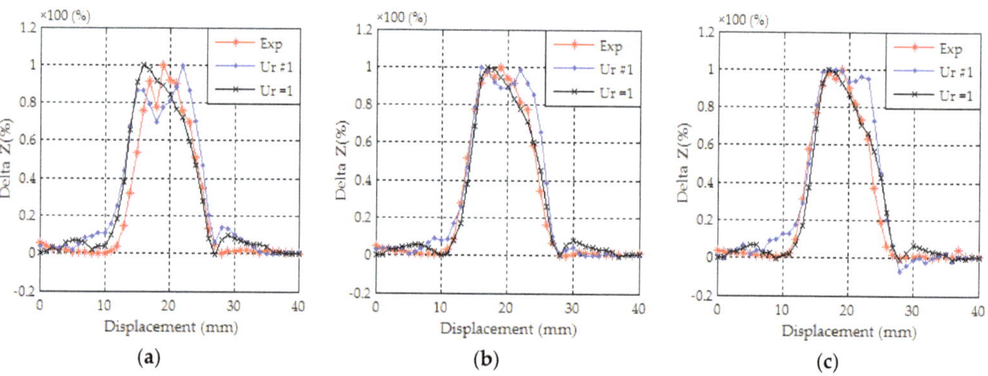

Figure 9. Impedance variation in (%) at 20 kHz with (**a**) depth = 0.3 mm, (**b**) depth = 0.5 mm, and (**c**) depth = 1 mm.

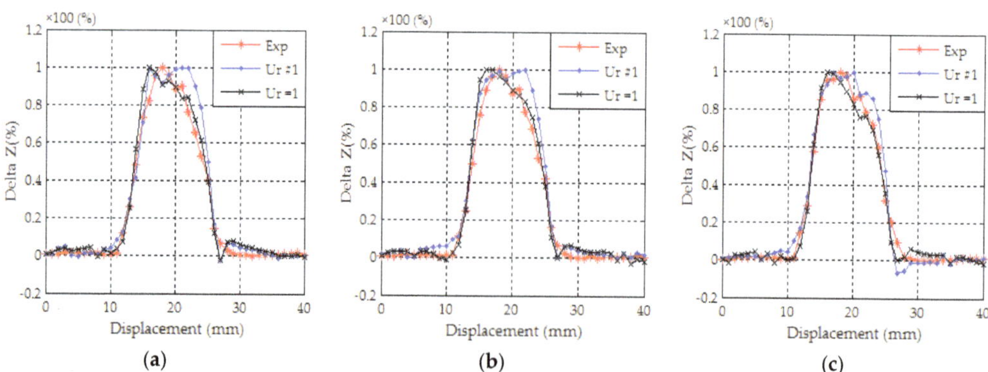

Figure 10. Impedance variation in (%) at 40 kHz with (**a**) depth = 0.3 mm, (**b**) depth = 0.5 mm, and (**c**) depth = 1 mm.

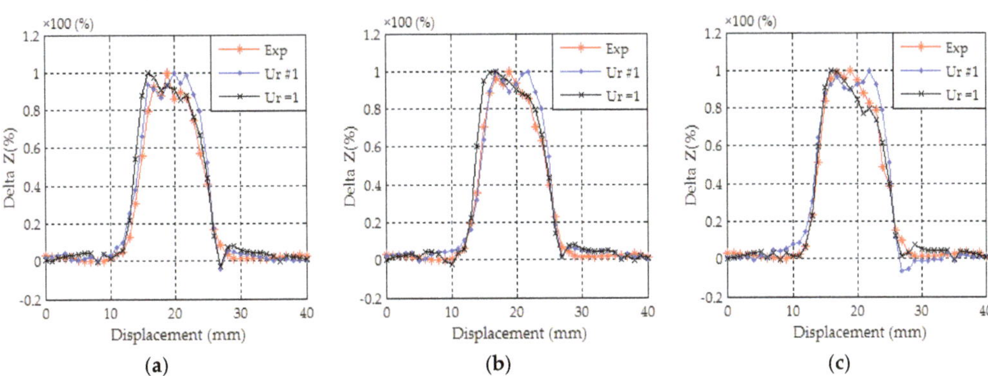

Figure 11. Impedance variation in (%) at 50 kHz with (**a**) depth = 0.3 mm, (**b**) depth = 0.5 mm, and (**c**) depth = 1 mm.

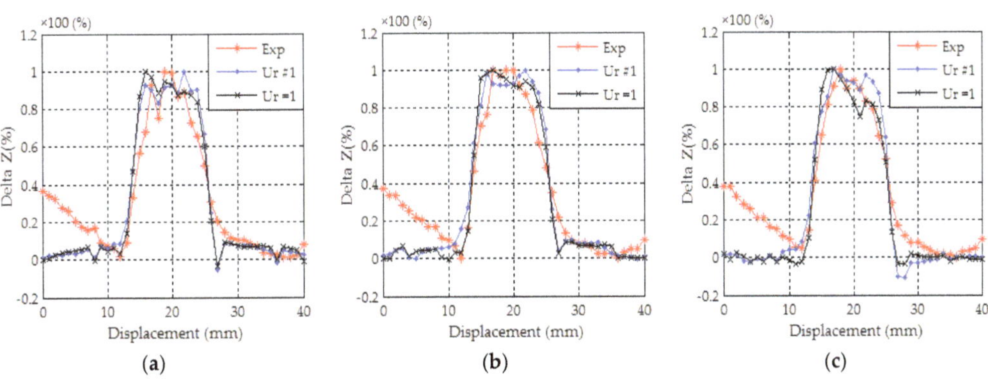

Figure 12. Impedance variation in (%) at 100 kHz with (**a**) depth = 0.3 mm, (**b**) depth = 0.5 mm, and (**c**) depth = 1 mm.

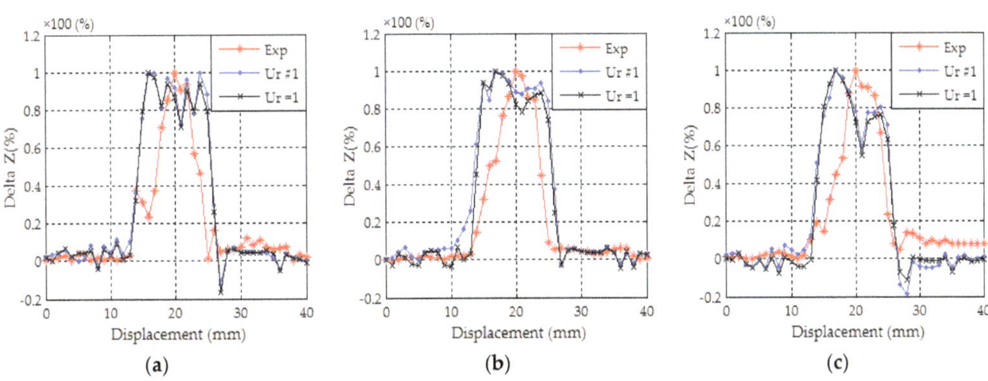

Figure 13. Impedance variation in (%) at 300 kHz with (**a**) depth = 0.3 mm, (**b**) depth = 0.5 mm, and (**c**) depth = 1 mm.

The method applied here to characterize the thin cracks relies on the fact that the eddy-current response depends on the probe excitation frequency.

By considering the operating frequency, detection and the optimum frequency, the penetration depth of eddy currents was significant. Therefore, the eddy-current response would be sensitive to the crack surface in-depth direction. When $\mu_r = 1$ the impedance signal is smooth for each frequency. Fine cracks do not appear, or are almost indistinguishable, as shown by the results of Figures 8–11. Therefore, the results obtained for constant relative magnetic permeability cannot provide right information regarding the presence or absence of thin cracks in the weld area.

In contrast, with $\mu_r \neq 1$ this model is able to reproduce the experimental results, while giving a better prediction of cracks. The shape of the impedance signal is not the same for each depth and each frequency. The signal presents fluctuations at the peak, which corresponds to the change of environment (weld and crack). Thus, this is comprehensible as it corresponds to a presence of the thin cracks in the weld area. Figure 14 shows three shapes of shallow mountains that correspond to the three cracks referred as d0.3, d0.5 and d1.0. A small peak at 180 mm on the X axis indicates a small defect as shown in Figure 2. The output signal of the Nortec 500 equipment is entered into the DAQ and expressed as a 3D surface graph using LabVIEW as given in Figure 14.

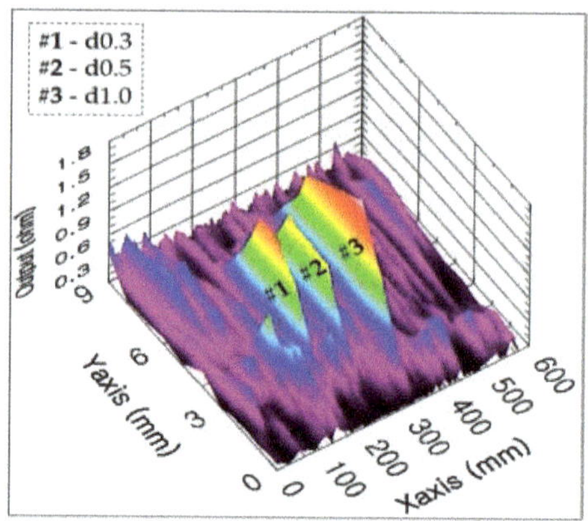

Figure 14. Cracks imaging obtained with 300 kHz.

Figures 12 and 13 correspond respectively to 100 and 300 kHz. At these frequencies, the impedance of the probe is maximum, and the current flow is stronger at the surface and decreases rapidly in-depth direction. Thus, the current response would be particularly sensitive to surface cracking. The same signal was obtained with the two relative magnetic permeabilities. An insignificant signal difference was observed with 100 kHz.

The simulation results reproduce the trend and the shape of the experimental signals; however, a small deviation can be observed that is probably due to several parameters; the universe of the experimental is in 3D while the simulations are in 2D, the machining of the cracks, the lift-off and another factor come from the impedance measurement carried out.

The impedance variation as a function of the probe displacement reflects the change in the distribution of physical properties over the part of the pipe being inspected. The proposed approach is validated by a comparison and shows a satisfactory concordance with the experimental ones in all cases for all the frequencies used, which proves:

- Eddy currents are well adapted to the detection of thin surface cracks under the stress of the heat welding which affects the relative magnetic permeability locally.
- The validity of the modeling and the analysis approach.

5. Conclusions

The stochastic finite element method was applied to study weld cracks in AISI 304 pipe used in nuclear power plants by nondestructive testing. The relative magnetic permeability was gradiently generated using the Monte Carlo Method based on pseudo-random number generation. It is considered as an essential property in this study to characterize weld cracks areas.

A qualitative interpretation of the eddy-current probe output and comparison of both experiment and simulation were carried out. In all cases the comparisons show a good agreement between the two results. Compared to the constant relative permeability, $\mu_r \neq 1$ showed a greater sensitivity with respect to the change caused by the presence of thin cracks in the weld. On the other hand, considering $\mu_r \neq 1$ is more sensitive than $\mu_r = 1$ to distinguish the thin cracks with 10, 20, 40 and 50 kHz. Furthermore, the results confirm that taking into account the influence of the heat treatment induced by the welding process is more effective for this purpose. Thus, in the framework of future research, it will be interesting to use artificial intelligence based on deep learning exploiting big data applied in the field of nondestructive testing techniques for surface and subsurface scanning.

Author Contributions: Conceptualization, J.L.; methodology, J.L., A.B. and E.C.; software, A.B.; validation, A.B. and J.L.; formal analysis, A.B.; investigation, J.L. and A.B.; resources, J.L. and E.C.; data curation, J.L., A.B. and E.C.; writing—original draft preparation, A.B.; writing—review and editing, A.B. and J.L.; visualization, E.C. and J.L.; supervision, J.L.; project administration, J.L.; funding acquisition, J.L. This paper was prepared with the contributions of all authors. All authors have read and agreed to the published version of the manuscript.

Funding: This work was supported by KOREA HYDRO & NUCLEAR POWER Co., LTD. (No.2018-T-14).

Institutional Review Board Statement: Not applicable.

Informed Consent Statement: Not applicable.

Data Availability Statement: Not applicable.

Conflicts of Interest: The authors declare no conflict of interest.

References

1. Nam, T.H.; An, E.; Kim, B.J.; Shin, S.; Ko, W.S.; Park, N.; Kang, N.; Jong, B.J. Effect of post weld heat treatment on the microstructure and mechanical properties of a submerged-arc-welded 304 stainless steel. *Metals* **2018**, *8*, 26. [CrossRef]
2. Jang, D.; Kim, K.; Kim, H.C.; Jeon, J.B.; Nam, D.G.; Sohn, K.Y.; Kim, B.J. Evaluation of Mechanical property for welded austenitic stainless steel 304 by following post weld heat treatment. *Korean J. Met. Mater.* **2017**, *55*, 664–670. [CrossRef]
3. Kawahara, Y. Application of high temperature corrosion-resistant materials and coatings under severe corrosive environment in waste-to-energy boilers. *J. Therm. Spray Technol.* **2007**, *16*, 202–213. [CrossRef]
4. Adegboye, M.A.; Fung, W.K.; Karnik, A. Recent advances in pipelines monitoring and oil leakage detection Technologies: Principles and approaches. *Sensors* **2019**, *19*, 2548. [CrossRef]
5. Dai, L.; Feng, H.; Wang, T.; Xuan, W.; Liang, Z.; Yang, X. Pipe crack recognition based on eddy current NDT and 2D impedance characteristics. *Appl. Sci.* **2019**, *9*, 689. [CrossRef]
6. Mehaddene, H.; Mohellebi, H.; Berkache, A.; Berthiau, G. Experimental and numerical multi-defects analysis in ferromagnetic medium. *Przegląd Elektrotechniczny* **2020**, *96*, 57–61. [CrossRef]
7. AbdAlla, A.N.; Faraj, M.A.; Samsuri, F.; Rifai, D.; Ali, K.; Al-Douri, Y. Challenges in improving the performance of eddy current testing: Review. *Meas. Control* **2019**, *52*, 46–64. [CrossRef]
8. Zhang, H.; Wei, Z.; Xie, F.; Sun, B. Assessment of the Properties of AISI 410 Martensitic Stainless Steel by an Eddy Current Method. *Materials* **2019**, *12*, 1290. [CrossRef] [PubMed]
9. ISO 12718. *Non-Destructive Testing—Eddy Current Testing—Vocabulary*, 2nd ed.; ISO: Geneva, Switzerland, 2019.
10. Berkache, A.; Oudni, Z.; Mehaddene, H.; Mohellebi, H.; Lee, J. Inspection and characterization of random physical property defects by stochastic finite element method. *PrzeglądElektrotechniczny* **2019**, *95*, 96–101. [CrossRef]
11. Kim, J.; Le, M.; Lee, J.; Hwang, Y.H. Eddy current testing and evaluation of far-side corrosion around rivet in jet-engine of aging supersonic aircraft. *J. Nondestruct. Eval.* **2014**, *33*, 471–480. [CrossRef]
12. Tsuboi, H.; Ikeda, K.; Kurata, M.; Kainuma, K.; Nakamura, K. Eddy current analysis for the pipe welding. *IEEE Trans. Magn.* **1998**, *34*, 1234–1236. [CrossRef]
13. Sim, S.; Le, M.; Kim, J.; Lee, J.; Kim, H.; Do, H.S. Nondestructive inspection of control rods in nuclear power plants using an encircling-type magnetic camera. *Int. J. Appl. Electromagn. Mech.* **2016**, *52*, 1561–1567. [CrossRef]
14. Costanza, G.; Sili, A.; Tata, M.E. Weldability of austenitic stainless steel by metal arc welding with different shielding gas. *Procedia Struct. Integr.* **2016**, *2*, 3508–3514. [CrossRef]
15. Cho, D.W.; Cho, W.I.; Na, S.J. Modeling and simulation of arc: Laser and hybrid welding process. *J. Manuf. Process.* **2013**, *16*, 26–55. [CrossRef]
16. Kah, P.; Layus, P.; Hiltunen, E.; Martikainen, J. Real-time weld process monitoring. *Adv. Mater. Res.* **2014**, *933*, 117–124. [CrossRef]

17. Wu, C.; Kim, J.W. Review on mitigation of welding-induced distortion based on FEM analysis. *J. Weld. Join.* **2020**, *38*, 56–66. [CrossRef]
18. Li, W.; Ma, W.; Qi, P.; Wen-jiao, D.; Yuan, X.; Yin, X. Detection and evaluation of weld defects in stainless steel using alternating current field measurement. In Proceedings of the 44th Annual Review of Progress in QNDE, Provo, Utah, USA, 16–21 July 2017; American Institute of Physics: College Park, MD, USA, 2018; Volume 1949, p. 230019. [CrossRef]
19. Saito, T.; Ichiyama, Y. Weld defects and evaluation of weld quality: Welding phenomena and process control in flash welding of steel sheets. *Weld. Int.* **1996**, *10*, 117–123. [CrossRef]
20. Park, I.; Kim, E.-Y.; Yang, W.-J. Microstructural Investigation of Stress Corrosion Cracking in Cold-Formed AISI 304 Reactor. *Metals* **2021**, *11*, 7. [CrossRef]
21. Hu, X.; Jiang, H.-Y.; Luo, Y.; Jin, Q.; Peng, W.; Yi, C.-M. A Study on Microstructure, Residual Stresses and Stress Corrosion Cracking of Repair Welding on 304 Stainless Steel: Part II-Effects of Reinforcement Height. *Materials* **2020**, *13*, 2434. [CrossRef] [PubMed]
22. Oršulová, T.; Palček, P.; Kúdelčík, J. Effect of plastic deformation on the magnetic properties of selected austenitic stainless steels. *Prod. Eng. Arch.* **2017**, *14*, 15–18. [CrossRef]
23. American Society of Mechanical Engineers. *ASME Boiler and Pressure Vessel Code, 2010 ed.*; ASME: New York, NY, USA, 2010; p. 108.
24. UGITECH. Magnetism and Stainless Steel. Available online: https://www.swisssteelinternational.us/fileadmin/user_upload/Ugitec/Documents_publics/AC/MARCHES/Magnetism/7512UGITECHMagnetGB.pdf (accessed on 1 January 2009).
25. ISO 9712. *Non-Destructive Testing—Qualification and Certification of NDT Personnel*, 4th ed.; ISO: Geneva, Switzerland, 2012.
26. Pasqualini, L.; Parton, M. Pseudo random number generation through reinforcement learning and recurrent neural networks. *Algorithms* **2020**, *13*, 307. [CrossRef]
27. Irfan, M.; Ali, A.; Khan, M.A.; ul-Haq, M.E.; Shah, S.N.M.; Saboor, A.; Ahmad, W. Pseudorandom number generator (PRNG) design using hyper-chaotic modified robust logistic map (HC-MRLM). *Electronics* **2020**, *9*, 104. [CrossRef]
28. L'Ecuyer, P. Pseudorandom number generators. In *Encyclopedia of Quantitative Finance*; Simulation Methods in Financial Engineering; Cont, R., Ed.; John Wiley: Chichester, UK, 2010; pp. 1431–1437.
29. Bhattacharjee, K.; Maity, K.; Das, S. Search for good pseudo-random number generators: Survey and empirical studies. *arXiv* **2018**, arXiv:1811.04035.
30. Greenberger, M. Notes on a new pseudo-random number generator. *JACM* **1961**, *8*, 163–167. [CrossRef]
31. Salmon, J.K.; Moraes, M.A.; Dror, R.O.; Shaw, D.E. Parallel Random Numbers: As Easy as 1,2,3. In Proceedings of the 2011 International Conference for High Performance Computing, Networking, Storage and Analysis, Seattle, WA, USA, 12–18 November 2011. [CrossRef]
32. Datcu, O.; Macovei, C.; Hobincu, R. Chaos based cryptographic pseudo-random number generator template with dynamic state change. *Appl. Sci.* **2020**, *10*, 451. [CrossRef]
33. Silvester, P.; Chari, M.V.K. Finite element solution of saturable magnetic field problems. *IEEE Trans. Power Appar. Syst.* **1970**, *PAS-89*, 1642–1651. [CrossRef]
34. Ida, N.; Betzold, K.; Lord, W. Finite element modeling of absolute eddy current probe signals. *J. Nondestruct. Eval.* **1982**, *3*, 147–154. [CrossRef]
35. Thomas, J.L. *Simplified Modeling of Eddy Current Control of Steam Generator Tubes Report of Internship ESA IGELEC*; University of Nantes: Nantes, France, 1998.
36. Oudni, Z.; Feliachi, M.; Mohellebi, H. Assessment of the probability of failure for EC nondestructive testing based on intrusive spectral stochastic finite element method. *Eur. Phys. J. Appl. Phys.* **2014**, *66*, 30904. [CrossRef]
37. Chen, D.; Shao, K.R.; Lavers, J.D. Very fast numerical analysis of benchmark models of eddy-current testing for steam generator tube. *IEEE Trans. Magn.* **2002**, *38*, 2355–2357. [CrossRef]
38. Takagi, T.; Hashimoto, M.; Fukutomi, H.; Kurohwa, M.; Miya, K.; Tsuboi, H.; Tanaka, M.; Tani, J.; Serizawa, T.; Harada, Y.; et al. Benchmark models of eddy current testing for steam generator tube: Experiment and numerical analysis. *Int. J. Appl. Electromagn. Mater.* **1994**, *5*, 149–162. [CrossRef]
39. Takagi, T.; Hashimoto, M.; Sugiura, T.; Norimatsu, S.; Arita, S.; Miya, K. 3D Numerical Simulation of Eddy Current Testing of a Block with a Crack. *Rev. Prog. Nondestr. Eval.* **1990**, *9*, 327–334. [CrossRef]
40. Korea Electric Association. *MI in Service Inspection of N.P.P*; MIA Special ed.; 2005 ed∼2009 add; Korea Electric Association: Seoul, Korea, 2009; p. 32.

Article

Measurement of Thinned Water-Cooled Wall in a Circulating Fluidized Bed Boiler Using Ultrasonic and Magnetic Methods

Jinyi Lee [1,2,*], Eunho Choe [3], Cong-Thuong Pham [4] and Minhhuy Le [4,5,*]

1 Department of Electronic Engineering, Chosun University, Gwangju 61452, Korea
2 Interdisciplinary Program in IT-Bio Convergence System, Chosun University, Gwangju 61452, Korea
3 Department of Control and Instrumentation, Graduate School of Chosun University, Gwangju 61452, Korea; fordo2@nate.com
4 Faculty of Electrical and Electronic Engineering, Phenikaa University, Hanoi 12116, Vietnam; thuong.phamcong@phenikaa-uni.edu.vn
5 A&A Green Phoenix Group JSC, Phenikaa Research and Technology Institute (PRATI), No.167 Hoang Ngan, Trung Hoa, Cau Giay, Hanoi 11313, Vietnam
* Correspondence: jinyilee@chosun.ac.kr (J.L.); huy.leminh@phenikaa-uni.edu.vn (M.L.)

Abstract: In this paper, a nondestructive inspection system is proposed to detect and quantitatively evaluate the size of the near- and far-side damages on the tube, membrane, and weld of the water-cooled wall in the fluidized bed boiler. The shape and size of the surface damages can be evaluated from the magnetic flux density distribution measured by the magnetic sensor array on one side from the center of the magnetizer. The magnetic sensors were arrayed on a curved shape probe according to the tube's cross-sectional shape, membrane, and weld. On the other hand, the couplant was doped to the water-cooled wall, and a thin film was formed thereon by polyethylene terephthalate. Then, the measured signal of the flexible ultrasonic probe was used to detect and evaluate the depth of the damages. The combination of the magnetic and ultrasonic methods helps to detect and evaluate both near and far-side damages. Near-side damages with a minimum depth of 0.3 mm were detected, and the depth from the surface of the far-side damage was evaluated with a standard deviation of 0.089 mm.

Keywords: circulating fluidized bed combustion boiler; water-cooled wall tube; magnetic sensor array; magnetic flux density; flexible ultrasonic probe

Citation: Lee, J.; Choe, E.; Pham, C.-T.; Le, M. Measurement of Thinned Water-Cooled Wall in a Circulating Fluidized Bed Boiler Using Ultrasonic and Magnetic Methods. *Appl. Sci.* **2021**, *11*, 2498. https://doi.org/10.3390/app11062498

Academic Editor: Giuseppe Lacidogna

Received: 26 January 2021
Accepted: 8 March 2021
Published: 11 March 2021

1. Introduction

Circulating fluidized bed combustion boilers burn various fuels such as wood, coal, and combustible waste together with solid fluidized media such as sand and ash [1]. In addition, combustion air is injected at high speed through a distribution plate at the bottom of the furnace to burn coal in a gas-solid flow condition inside the furnace. The high temperature of the heated fluid medium particles scatter and circulate in a suspended state to transfer heat to the heat transfer tube. Since heat is transferred through collisional contact with the fluid particles, the heat transfer coefficient is very superior compared to the convection heat exchange method of the existing boiler. However, due to the collision contact between the surface of the water-cooled wall and the fluid particles, which is repeated as the operation time elapses, erosion due to direct exposure to combustion flames, corrosion due to high-temperature combustion and formation of potassium chloride, and acceleration of corrosion due to adhesion could appear. Thus, the life cycle of the water-cooled wall is shorter than that of the existing boiler systems. In addition, the lower part where the concentration of the fluid medium is high is a splash area where the fluid medium violently behaves, and the water-cooled wall is severely damaged. These damages intensify in the kick-out area located at the boundary between the lower fireproof part and the water-cooled wall [2]. On the other hand, when abrasion and corrosion occur on the water-cooled wall tube, leakage and secondary damage due to the leakage may occur. It is also very important to periodically monitor and maintain the thickness of the water-cooled wall since damage to the tube, membrane, and welding portion of the water-cooled wall can cause a decrease in power generation efficiency.

Recently, numerous NDT methods have been developed for the inspection of damage on the water-cooled wall. Ultrasonic testing (UT) applies an acoustic medium to the inspection area of the water-cooled wall and measures the change in the reflection time of the ultrasonic wave according to the thickness change when the ultrasonic wave is incident [3]. Although it is possible to perform a precise inspection of the inner surface of the tube, it is difficult to measure the surface corrosion, and the incident angle of ultrasonic waves may vary according to the manual inspection of the operator, resulting in an error in thickness measurement. Phased array ultrasonic testing (PAUT) has been developed to reduce mechanical errors while scanning the probe on the specimens and signal enhancement. It provides excellent results of damage detection and quantitative evaluation of the damage size, such as depth and length [4,5]. However, the UT and PAUT methods require continuous supplement of the coupling materials such as water for the propagation of ultrasonic wave between the probes and the tube, and the surface of the tube should be cleaned before the inspection. Therefore, it is difficult to operate in the inspection of the water-cooled wall tubes in the power generation, and it also requires high technical trained operators to use the UT and PAUT systems. Electromagnetic testing methods, including eddy current testing (ECT), remote field eddy current testing (RFECT), and magnetic flux leakage testing (MFLT), are the fast, reliable, and easy operation methods for the inspection of damages in the tubes. These are non-contact inspection methods that do not require the coupling material during the inspection. ECT is a widely used method for the inspection of heat exchanger tubes and boilers of nuclear power plants [6–11]. This method is highly sensitive to the surface cracks, but it is limited to detecting deep defects due to the high concentration of eddy current on the specimen surface in the skin depth effect. Especially, the eddy current has more difficulty penetrating the wall thickness of the water-cooled tube because it has high magnetic permeability. RFECT [12–16] uses a probe consisting of an excitation coil and a measuring coil that can be inserted into a ferromagnetic heat pipe tube such as the water-cooled tubes. The magnetic energy generated by the interpolation type excitation coil goes from the excitation coil to the outside of the tube and flows in the axial direction, and then back to the inside from the remote field area at a certain distance. The measuring coil can sense the energy delivered without receiving it from the excitation coil. In order to increase the ratio of the signal to noise, it is necessary to increase the cross-sectional area and the number of turns of the excitation coil and the measurement coil so that the spatial resolution of the probe is low. Therefore, there is a limitation in quantitatively evaluating where the damage is occurring on the water-cooled wall tube, the weld, and the membrane. For further improvement of the sensitivity, a giant magnetoresistance (GMR) and Hall sensors were used to measure the low magnetic leakage signal in the MFLT systems [17,18]. This method makes it possible to detect a defect on the surface and near the surface of the water-cooled wall tubes. However, it is still difficult to measure the thickness changes of the tube due to the damages. The combination of the ultrasonic and electromagnetic field has been developed in the electromagnetic acoustic transducer (EMAT) system for the inspection of the water-cooled tubes [19,20]. The magneto-elastic phenomenon and Lorentz force help the EMAT inspect a deeper damage without the need for coupling material. However, the signal is weak and requires advanced signal processing circuits and algorithms. In addition, the EMAT probe has a big size, and thus, it is not efficient to build an array EMAT probe with a high spatial resolution for quantitative evaluation of damage sizes.

This study proposed a combination of the magnetic flux leakage testing and ultrasonic testing methods for the efficient detection and quantitative evaluation of the depth and residual thickness distribution of the near-side and far-side corrosion of the water-cooled wall. A Hall sensor array probe with 48 elements arrayed in an interval of 2.5 mm was developed to detect the near-side damages and thus make it possible to evaluate the damage size. A flexible ultrasonic probe (FUP) was developed to detect the far-side damages on the tube, membrane, and welding lines of the water-cooled wall. The FUP was incorporated with a flexible membrane that allows the transmission of the ultrasonic wave from the probe to the water-cool tube surface efficiently. Thus, it is not required to largely supply coupling material during the inspection. In addition, the FUPs could be arrayed according to the water-cool plates for fast inspection. For the verification of the proposed method, artificial tapper-type wears and slit-type damages with different sizes were produced on the tube, membrane, and welding lines of the water-cooled wall. Both the detection and size/depth evaluation of the damages will be discussed.

2. Materials and Methods

2.1. Measurement of Magnetic Flux Density

Figure 1 shows the simplified 2D dipole model of the thickness changing on the boiler water-cooled wall tube due to corrosion [21,22]. A U-type magnetizer is placed on the surface of the membrane. The width, distance between poles, and height of the magnetizer are expressed as W, D, and H, respectively. The corrosion depth and length are d and $D/2 + W$, respectively. The distance between the magnetizer and the specimen, i.e., lift-off, is expressed as h. Then, the lift-off at the corrosion is $h + d$. In the dipole model, magnetic charges $\pm m$ per unit area are assumed to be distributed along the length of the magnetizer poles, membrane length, and corrosion length according to the assumption in the dipole model [21,22]. The magnetic flux density in the y-axis direction at the position of $P(x_p,y_p)$ is the summary of the magnetic field produced from the magnetic charges, as expressed in Equation (1). The vertical magnetic field from the left magnetizer pole, right magnetizer pole, no-corrosion specimen length, and corrosion specimen length are expressed as H_{LU}, H_{RU}, H_{LD}, H_{RD} in Equations (2)–(5), respectively.

$$H_y = H_{LU} + H_{LD} + H_{RU} + H_{RD} \tag{1}$$

$$H_{LU} = \frac{+m}{4\pi\mu}\int_{-\frac{D}{2}-W}^{\frac{D}{2}} \frac{y_p}{\left\{(x_p-u)^2+(y_p)^2\right\}^{\frac{3}{2}}}du \tag{2}$$

$$H_{RU} = \frac{-m}{4\pi\mu}\int_{\frac{D}{2}}^{\frac{D}{2}+W} \frac{y_p}{\left\{(x_p-u)^2+(y_p)^2\right\}^{\frac{3}{2}}}du \tag{3}$$

$$H_{LD} = \frac{-m}{4\pi\mu}\int_{-\frac{D}{2}-W}^{0} \frac{(y_p+h)}{\left\{(x_p-u)^2+(y_p+h)^2\right\}^{\frac{3}{2}}}du \tag{4}$$

$$H_{RD} = \frac{+m}{4\pi\mu}\int_{0}^{\frac{D}{2}+W} \frac{(y_p+h+d)}{\left\{(x_p-u)^2+(y_p+h+d)^2\right\}^{\frac{3}{2}}}du \tag{5}$$

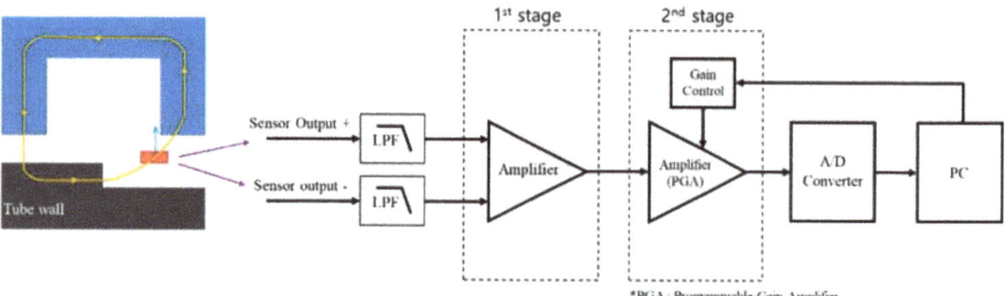

Figure 4. Signal processing block diagram of the magnetic flux leakage testing system. The block diagram is for a single Hall sensor.

Figure 5 shows the proposed magnetic leakage testing system to inspect the corrosion in the water-cooled tube wall. A magnetic sensor array and magnetizer were manufactured to fit with the water-cooled tube's surface, as shown in the left and middle drawing. The magnetic sensor array was placed at the middle of magnetizer poles for measuring the distribution of vertical magnetic flux, as discussed in the previous paragraphs. There were three wheels (a front and two rears wheels) used to maintain the lift-off between the sensor and the tube and help scan the tube easily.

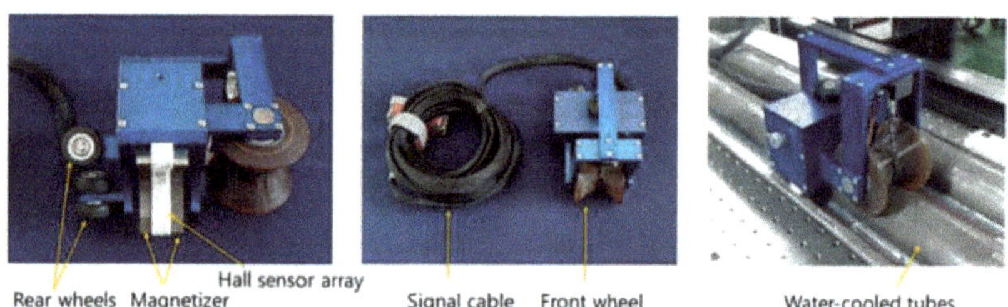

Figure 5. Configurations of the magnetic flux leakage testing system.

2.2. Flexible Ultrasonic Testing

Ultrasonic probe usually requires a supplement of a coupling material for transmitting ultrasonic wave from the transducer to the test specimen. It complicates the inspection system and is waste of coupling material. In addition, the test specimen surface should be flat enough to maintain a positive lift-off (non-contact) for protecting the collision of the transducer with the test specimen. It is hard for the inspection of near-surface defects in the water-cooled tube because the changes of the tube wall could make an unpredictable lift-off that could lead to the collision and break the transducer. Therefore, we propose using a flexible transducer that the lift-off could be varied and not require using the coupling material [23]. At the head of a normal transducer, we attached a flexible membrane that was water-filled. The membrane has a sphere shape after filing the water and maintains contact with the tube even though the lift-off can vary. Also, the ultrasound wave can still propagate from the transducer to the water membrane and come to the test specimen.

A sample flexible transducer is shown in Figure 6a. The transducer has a spring that keeps the contact between the membrane with the test specimen during the scan. The received time-domain signal of the transducer, which is A-scan signal ($u(t)$), is shown in Figure 6b. For a better signal-to-noise ratio, the spectrogram of the A-scan signal was processed ($S(\tau,f)$) and extracted only the signal ($S_A(\tau)$) at the center frequency of the transducer (f_c), as shown in Figure 6c,d. The spectrogram ($S(\tau,f)$) and extracted signals ($S_A(\tau)$) are calculated as expressed in Equations (10) and (11), respectively; where, h is a sliding Gaussian window. The extracted signal ($S_A(\tau)$) was then stacked to form the B-scan signal while scanning the transducer along with the test specimen.

$$S(\tau, f) = \left| \int_{-\infty}^{\infty} u(t)h(t-\tau)e^{-j2\pi ft}dt \right|^2 \tag{10}$$

$$S_A(\tau) = S(\tau, f)|_{f=f_c} \tag{11}$$

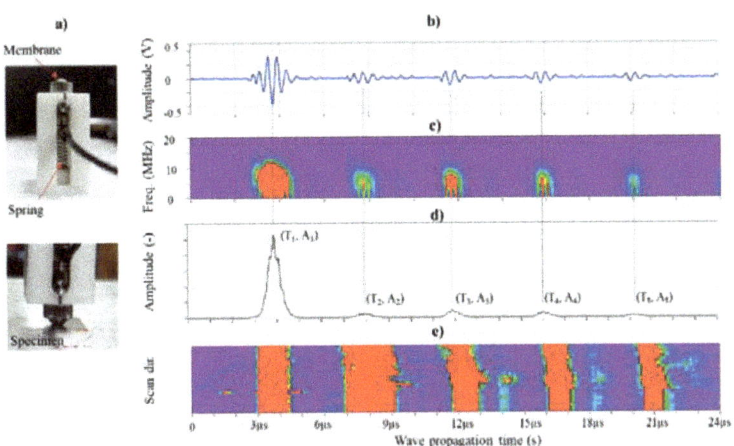

Figure 6. (**a**) A single flexible ultrasonic transducer measuring a thickness of a specimen, (**b**) its time-domain signal, (**c**) spectrogram signal, (**d**) the cross-section view signal of the spectrogram, and (**e**) the stacked cross-section view signal (B-scan).

It is observed from the ultrasonic transducer signal that there are multiple peaks. The first peaks are the reflected wave from the specimen surface. It has a delay time of about 4 μs (t_1, u_1), which is the propagation time from within the probe membrane. This delay time could be varied due to the flexibility of the membrane (lift-off). Thus, it is necessary to eliminate this delay time by shifting the signal with an amount of time $-t_1$. In addition, there are four peaks (t_2, u_2), (t_3, u_3), (t_4, u_4), (t_5, u_5) next to the specimen surface peak (t_1, u_1), which correspond to the repetitions from the bottom surface of the specimen. The time intervals of these four peaks are the same and can be used to calculate the specimen thickness, as shown in Equation (12); where v is the speed of the ultrasound in the specimen.

$$d = (t_2 - t_1) \times \frac{v}{2} = (t_3 - t_2) \times \frac{v}{2} = \cdots = (t_5 - t_4) \times \frac{v}{2} \tag{12}$$

Figure 7 is a schematic of the flexible ultrasonic probe (FUP) for quantitatively measuring the specimen thickness. The FUP is an array of multiple transducers (i.e., 6) arranged for covering the tube wall, welding lines, and specimen membrane area. The FUP could be alternated the magnetizer and magnetic sensor array modules in Figure 5.

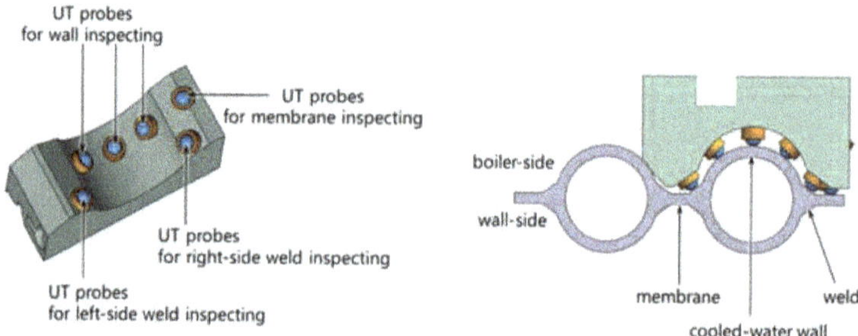

Figure 7. Schematic of the flexible ultrasonic probe (FUP) for measuring the water-cooled wall thicknesses.

3. Experiment and Results

3.1. Specimen

Figure 8 shows the shape and location of damages on a specimen. A total of four water-cooled tubes (SA210C) with inner and outer diameters of 47.3 to 51.3 mm and 63.5 mm, respectively, one-sided (t11–t14) and double-sided artificial damages (t41–44) simulated for wears were produced on Tube-1 and Tube-4, respectively. In Tube-2, slit-type artificial damages (t21–t28) with the same width of 7.0 mm, depth of 0.9 mm, and lengths from 20 to 100 mm were produced. In Tube-3, slit-type artificial damages (t31–t38) having the same width of 7 mm and length of 60 mm and different depths from 0.3 to 3.1 mm were produced. The detailed location and size are as shown in Tables 1 and 2.

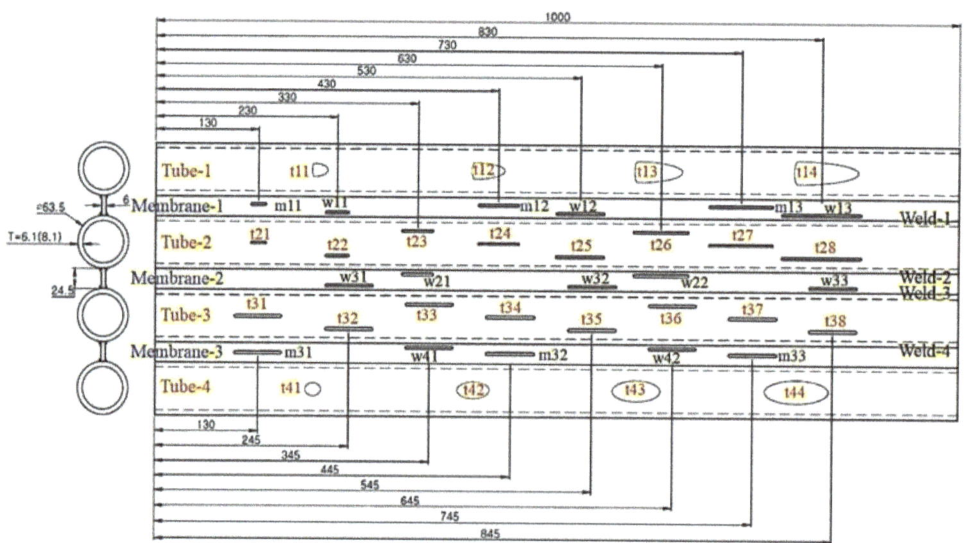

Figure 8. Specimen with different shape and size of artificial damages.

Table 1. Specification of artificial taper-type wear on the Tube-1 and Tube-4.

#	Width [mm]	Length [mm]	Depth (mm)	Position (mm)	#	Width [mm]	Length [mm]	Depth (mm)	Position (mm)
Tube-1 (One-Side Taper-Type Wear)					Tube-4 (Two-Side Taper-Type Wear)				
t11	16	20	0.9	200	t41	16	20	0.9	200
t12	20	40	1.5	400	t42	20	40	1.5	400
t13	25	60	2.5	600	t43	25	60	2.5	600
t14	28	80	3.1	800	t44	28	80	3.1	800

Table 2. Specification of artificial slit-type damages on the Tube-2 and Tube-3.

#	Width [mm]	Length [mm]	Depth (mm)	Position (mm)	#	Width [mm]	Length [mm]	Depth (mm)	Position (mm)
Tube-2 (Same Width)					Tube-3 (Same Length)				
t21	7	20	0.9	130	t31	7	60	0.23	130
t22	7	30	0.9	230	t32	7	60	0.50	245
t23	7	40	0.9	330	t33	7	60	0.96	345
t24	7	50	0.9	430	t34	7	60	1.08	445
t25	7	60	0.9	530	t35	7	60	1.5	545
t26	7	70	0.9	630	t36	7	60	1.86	645
t27	7	80	0.9	730	t37	7	60	2.34	745
t28	7	100	0.9	830	t38	7	60	2.64	845

The tubes were welded with a 6.0 mm thick membrane. There six slit-type artificial damages (w11, w12, w13, w31, w31, w33) on the two Membrane −1 and −2. The damages have the same width of 7.0 mm, different lengths from 20 to 80 mm, and different depths from 0.3 to 2.4 mm, as shown in Table 3. On the four welding lines (Welds 1, 2, 3, and 4), there are ten slit-type artificial damages (w11–w42) with the same width of 7.0 mm, different lengths from 30 to 100 mm, and different depths from 0.6 to 3.0 mm, as shown in Table 4. Totally, there are 40 artificial damages produced on the tubes, membranes, and welding lines. The picture of the specimen with damages is shown in Figure 9.

Table 3. Specification of artificial slit-type damages on the Membrane-1 and Membrane-3.

#	Width [mm]	Length [mm]	Depth (mm)	Position (mm)	#	Width [mm]	Length [mm]	Depth (mm)	Position (mm)
Membrane-1					Membrane-3				
m11	7	20	0.9	130	m31	7	60	0.3	130
m12	7	50	0.9	430	m32	7	60	1.2	445
m13	7	80	0.9	730	m33	7	60	2.4	745

Table 4. Specification of slit-type damages on the Weld-1~4.

#	Width [mm]	Length [mm]	Depth (mm)	Position (mm)	#	Width [mm]	Length [mm]	Depth (mm)	Position (mm)
		Weld-1					**Weld-3**		
w11	7	30	0.9	230	w31	7	60	0.6	245
w12	7	60	0.9	530	w32	7	60	1.5	545
w13	7	100	0.9	830	w33	7	60	3.0	845
		Weld-2					**Weld-4**		
w21	7	40	0.9	330	w41	7	60	0.9	345
w22	7	70	0.9	630	w42	7	60	1.8	645

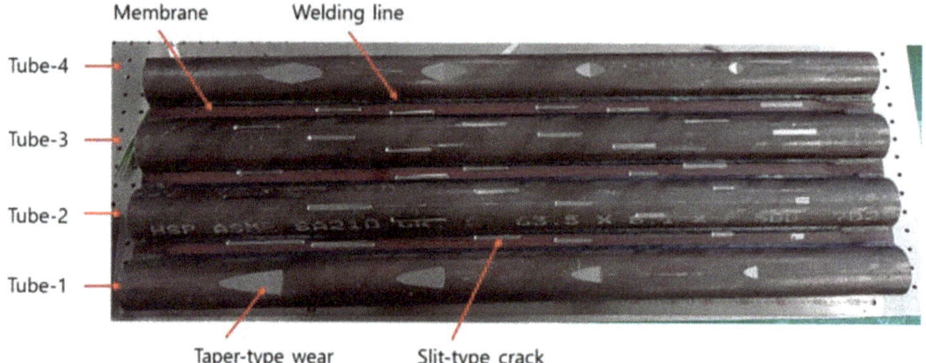

Figure 9. Sample specimens with four water-cooled tubes and artificial damages.

3.2. Inspection System

Figure 10 shows the prototype of the inspection system. In the magnetic leakage testing (MFLT) module, the magnetizer has a pole distance of 15 mm and has manufactured the profile following the tube and membrane surfaces. It maintains about 1.0 mm of distance above the specimen surface by the support of the three wheels. The magnetizer has 100 turns of copper wire and supplied by a current of about 200 mA to produce a magnetic field into the specimen. There are 48 Hall sensors arrayed at an interval of 2.5 mm on a curve following the tube and membrane surfaces. The MFLT probe scanned the specimen with steps of 4.0 mm. In the FUP, there are 6 flexible ultrasound transducers having a center frequency of 5 MHz. The MFLT module, including the magnetizer and magnetic sensor array, can be exchanged with the FUP module. The measured signal can be processed and displayed in real-time in a LabVIEW software on a notebook.

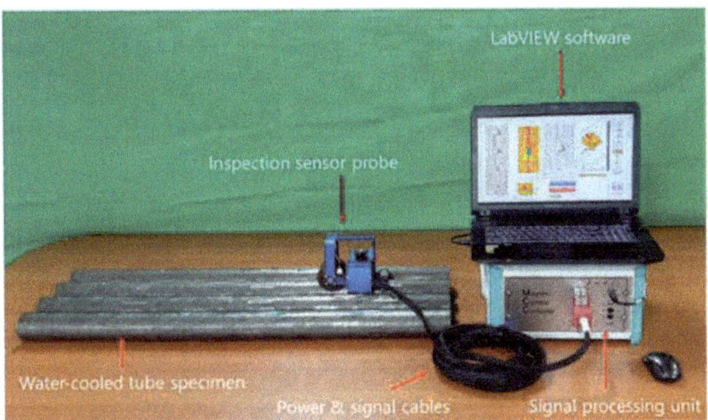

Figure 10. The prototype of the inspection system for water-cooled tubes.

3.3. Experiment Results

Figure 11 shows the scan results of the MFLT module on the Tube-1. Magnetic field distribution around the damages could help to recognize the presence of the damages. All the taper-type wears (t11, t12, t13, and t14) could be detected, and the magnetic field intensity increases as the size of the wear increases. In addition, the slit-type damages on the membrane (m11, m12, and m13) could also be detected, but the damages on the welding line (w11, w12, and w13) were out of the sensing area. The smallest slit-type damage (m11) has a length of 20 mm, depth of 0.9 mm, and 7 mm width that could be detected. Similarly, the taper-type wears on Tube-4 (t41, t42, t43, and t44) could be detected, as shown in Figure 12. However, the smallest size of slit-type damage on Membrane-3 (m31) having a length of 60 mm, depth of 0.3 mm, and width of 7 mm could not be detected; this is because the damage has a smallest depth of 0.3 mm. The damages (m32 and m33) which have depths of 1.2 mm and 2.4 mm, could be detected. The slit-type damages on the Weld-4 (w41 and w42) were out of the sensing area, but a part of the w42 signal could be measured because the damage has the deepest depth of 1.8 mm.

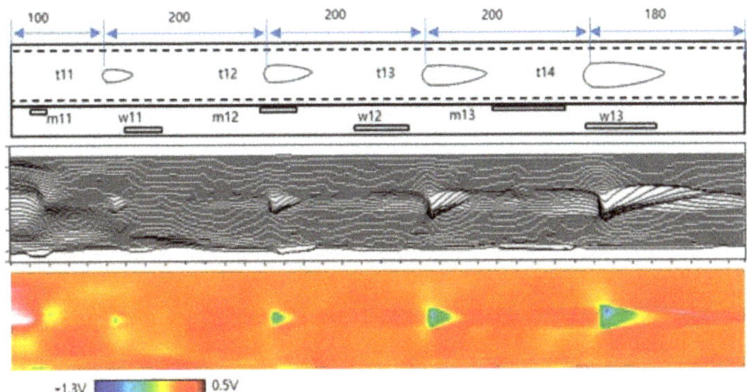

Figure 11. Distribution of magnetic field on the Tube-1, Membrane-1, and Weld-1 having artificial taper-type wear (single side) and slit-type damages.

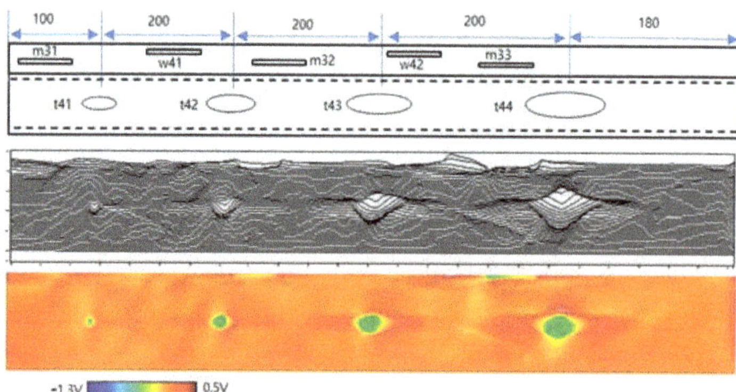

Figure 12. Distribution of magnetic field on the Tube-4, Membrane-3, and Weld-4 having artificial taper-type wear (double side) and slit-type damages.

Figures 13 and 14 show the detection result of the slit-type damages. All defects (t21~t28, t31~t38) with a depth of 0.3 mm or more located in Tube-2 and Tube-3 could be detected. In addition, damages of 0.9 mm in depth and 30 mm in length (w11) or more were detected in Weld-1 could also be detected. However, damages (w21, w22, w31, w32, w33) located in Weld-2 and Weld-3 were difficult to detect. Nevertheless, damages with a depth of 0.9 mm or more in Weld-4 (w41, w42) and damages with a depth of 1.2 mm or more in Membrane-3 (m32, m33) could be detected. This is because that when the sensor for magnetic flux density measurement scans Tube-4, it is skewed toward Membrane-3, and the lift-offs of Membrane-2 and Membrane-3 are not the same.

From the above results, the depths of damages detectable in the tube and membrane are 0.3 mm and 0.9 mm, respectively. In addition, some damages having a depth of 0.9 mm or more could be detected in the welded part due to the influence of the welding beads.

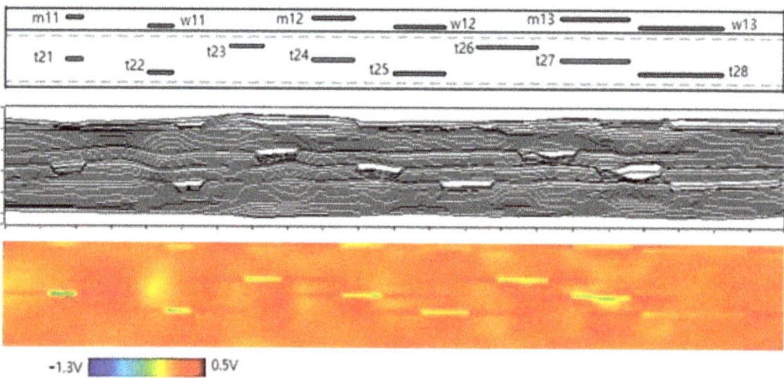

Figure 13. Distribution of magnetic field on the Tube-2, Membrane-1 (Weld-1, 2) having slit-type damages.

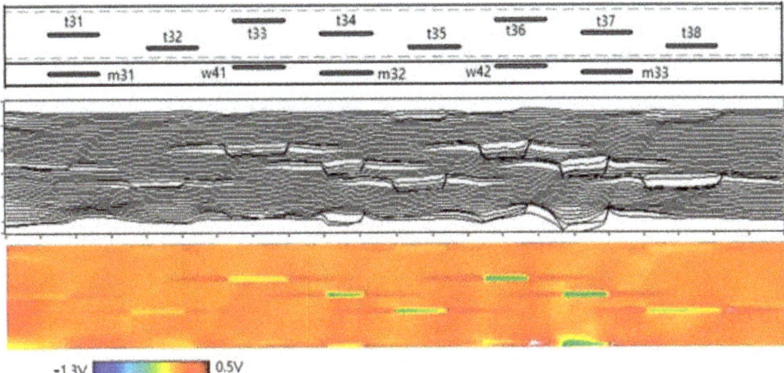

Figure 14. Distribution of magnetic field on the Tube-3, Membrane-3, and Weld-4 having slit-type damages.

Figure 15 shows the graph showing the relationship measured data with the depth of the defect on Tube-3. The damages have the same width of 7 mm and length of 60 mm. The measured data is the minimum data points selected from the numbers of Hall sensor array that are on the damages during the scan. There are 30 sensors, and 15 sensors data plotted on Figure 15a,b. The data in Figure 15a has more noise than in Figure 15b because some sensors are located far from the damages. Then, data of 15 sensors is used for further evaluation of the damages' depth. The average data of the sensors are used to reduce noise that may occur rather than a single sensor. Also, the relationship between the measured data with the damages' depth is expressed in Equation (13). This form of the experimental equation is same as the theoretical analysis by the dipole model of the previous section (Equation (9)). The factors c_1, c_2, and a_2 are 121.08, 6.65, and 2.52, respectively.

$$d = \sqrt{121.08\, V_H + 6.65} + 2.52 \tag{13}$$

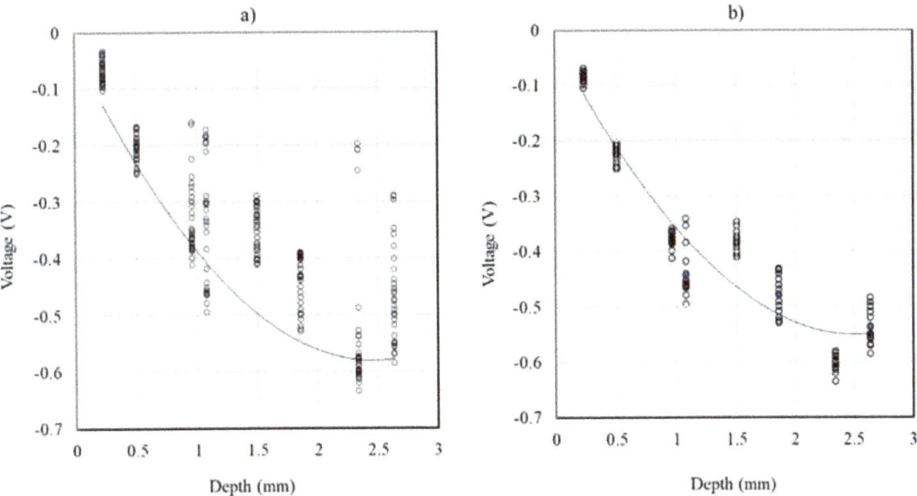

Figure 15. Relationship between the depth of flaw and the measured data with magnetic flux density method: (**a**) data of 30 sensors and (**b**) data of 15 sensors.

From Equation (13), the depth of damages on the tube (○), membrane (□), and weld (Δ) were estimated, as shown in Figure 16. The standard deviations of the depth estimation are 0.329, 0.269, and 0.523 mm for the damages on the tube, membrane and weld, respectively. The best estimation result is for damages on the membrane because the surface specimen is flat. The worst case for the damages on weld were due to the roughness of the weld surface, the sensor lift-off variation due to welding bead, and the edge effect at the terminal of the magnetizer.

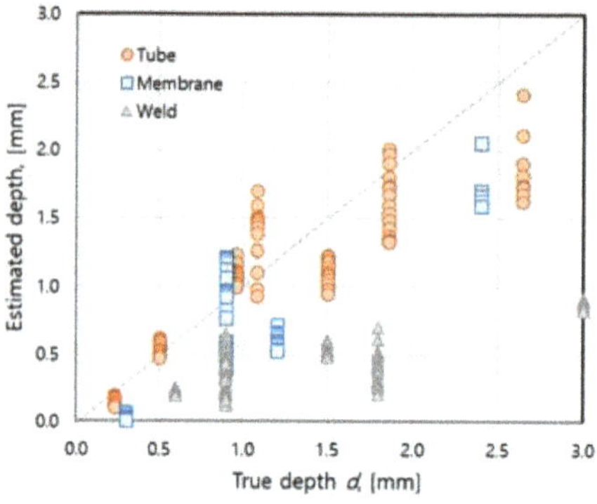

Figure 16. Estimation of depth of damages on the tube, membrane, and weld.

Figure 17 shows the B-scan result measured by the FUP after filling the acoustic medium and wrapping it with PET to the damage of Tube-1 (t11~t14). The horizontal axis represents time (T_i), and the vertical axis represents the moving distance of the FUP. The position of T_1 for each movement distance was about 20 μs before the start of the scan, but after 550 mm, it appeared at 17–18 μs. This is because of the variation of the inclined angle of the FUP and deformation of the membrane due to variation of the FUP lift-off. Therefore, it is necessary to shift the flying time on specimen surface (T_1) for each scan position, as expressed in Equation (14). Furthermore, it can be seen that in the vicinity of 100, 300, 525, and 725 mm, the delay of the FUP signal is longer than that of other locations, and near-side damage occurs in the corresponding region. It can also be determined from the delay of the signal that the shape of the damage is inclined to one side, and the depth can be estimated.

$$[\vec{T}_1] = [\vec{0}] \tag{14}$$

Figure 18 shows the B-scan results of the FUP measured from the back surface of the damages (m31–33) of Membrane-3 and (w21, w22) of Weld-2 using FUP. Similar to the previous experimental results. The position of T_1 for each movement distance was about 22 μs before the start of the scan, but after 450 mm past m32, it is back to 21 μs. Unlike the case of the near-side damage in Figure 17, it is possible to recognize that there is no near-side damage because the FUP signal appears continuously. On the other hand, it is observed that m31, m32, and m33 damages occur around 130 mm, 460 mm, and 750 mm, respectively. In addition, signal attenuation appears in the range of 625–700 mm. This is because the ultrasonic wave attenuates at the edge of the weld defect w42 located in Membrane-3. A similar phenomenon occurred near the weld defect (w41) at 325–380 mm. The depth of the damages was estimated, as shown in Figure 19. The standard deviation of the depth estimation is about 0.089 mm, which is much more accurate than using the magnetic flux leakage testing method.

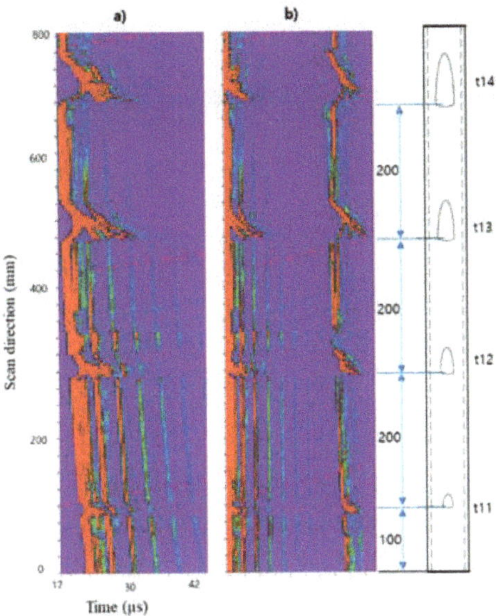

Figure 17. B-Scan results with FUP on the near-side damages of Tube-1 (**a**) before and (**b**) after T_1 adjustment.

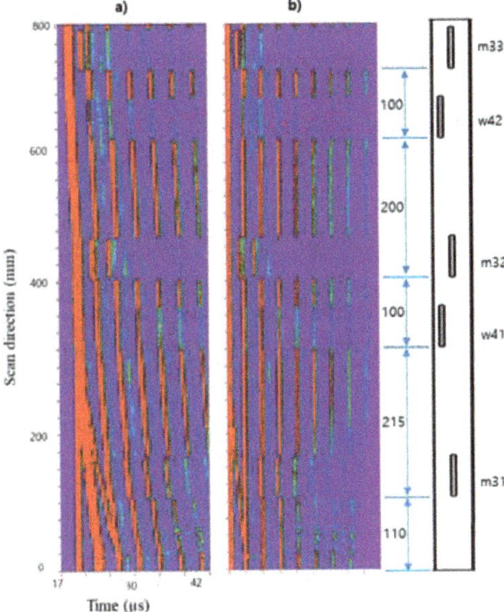

Figure 18. B-Scan results with FUP on the far-side damages of Membrane-3 and Weld-2 (**a**) before and (**b**) after T_1 adjustment.

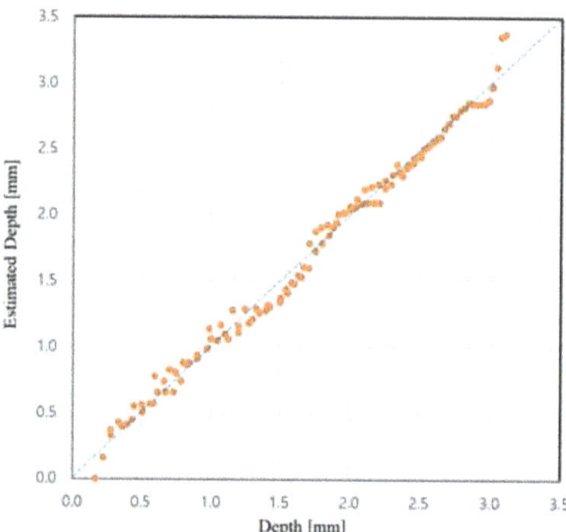

Figure 19. Estimation of damages' depth using the FUP.

4. Conclusions

In this paper, a nondestructive inspection system was proposed to detect defects on the near-side and far-side of the boiler water-cooled wall tube, membrane, and weld and to quantitatively evaluate the size of the defects. A magnetizer manufactured in a curved shape according to the cross-sectional shape of the tube, membrane, and welding part magnetizes a portion of the water-cooled wall in the axial direction. In addition, the shape of the surface defect can be qualitatively determined from the magnetic flux density distribution measured by the magnetic sensor array deflected from the center of the magnetizer to one side. The minimum depths of surface defects that can be measured are 0.3 mm, 0.9 mm, and 1.2 mm in each case of the tube, membrane, and weld. The depth of defects located in the tube, membrane, and weld can be quantitatively evaluated with a standard deviation of 0.329, 0.269, and 0.523 mm. A method of scanning with a flexible ultrasonic probe (FUP) after applying an acoustic medium to the defect surface of the water-cooled wall, covering a thin film of PET (polyethylene terephthalate), and applying a separate acoustic medium was proposed. According to the FUP arranged in a direction perpendicular to each cross-section of the tube, membrane, and weld, the location and shape of the surface defect and the back defect can be distinguished. Furthermore, the depth of the defect can be quantitatively evaluated with a standard deviation of 0.089 mm.

By combination of the magnetic flux leakage testing and ultrasonic testing, both the near-side and far-side defects could be detected and a quantitative evaluation of the depth could be made. Furthermore, the system is also expected to detect and evaluate the internal surface defects. For instance, if the defect is shallow in the near-surface, then the magnetic flux leakage testing is efficient for detection; otherwise, if the defect is deep to near the far-side, then the ultrasonic is more efficient. The further development of the proposed system should quantitatively evaluate different sizes of the defect such as length and width, or recognize the shape of the defects.

Author Contributions: Conceptualization, M.L. and J.L.; methodology, M.L.; software, M.L. and C.-T.P.; validation, J.L., M.L. and E.C.; formal analysis, E.C.; investigation, J.L. and M.L.; resources, J.L. and M.L.; data curation, E.C. and M.L.; writing—original draft preparation, J.L.; writing—review and editing, M.L.; visualization, C.-T.P. and E.C.; supervision, J.L. and M.L.; project administration, M.L. and J.L.; funding acquisition, M.L. and J.L. All authors have read and agreed to the published version of the manuscript.

Funding: This work was supported by the Vietnam National Foundation for Science and Technology Development (NAFOSTED) under grant number 103.02-2019.342, and the National Research Foundation of Korea (NRF) grant funded by the Korea government (MSIT) (No.NRF-2019R1A2C2006064).

Conflicts of Interest: The authors declare no conflict of interest.

References

1. Basu, P. Combustion of coal in circulating fluidized-bed boilers: A review. *Chem. Eng. Sci.* **1999**, *54*, 701–764. [CrossRef]
2. Wang, B. Erosion-corrosion of thermal sprayed coatings in FBC boilers. *Wear* **1996**, *199*, 24–32. [CrossRef]
3. Kim, T.-W.; Choi, J.-H.; Shun, D.-W.; Son, J.-E.; Jung, B.; Kim, S.-S.; Kim, S.-D. A Tube Thickness Map of Water Wall in a Commercial Circulating Fluidized Bed Combustor. *Korean Chem. Eng. Res.* **2005**, *43*, 412–418.
4. Jung, M.-J.; Park, B.-C.; Bae, J.-H.; Shin, S.-C. PAUT-based defect detection method for submarine pressure hulls. *Int. J. Nav. Archit. Ocean Eng.* **2018**, *10*, 153–169. [CrossRef]
5. Kim, Y.; Cho, S.; Park, I.K. Analysis of Flaw Detection Sensitivity of Phased Array Ultrasonics in Austenitic Steel Welds According to Inspection Conditions. *Sensors* **2021**, *21*, 242. [CrossRef] [PubMed]
6. Yusa, N.; Chen, Z.; Miya, K. Quantitative profile evaluation of natural defects in a steam generator tube from eddy current signals. *Int. J. Appl. Electromagn. Mech.* **2000**, *12*, 139–150. [CrossRef]
7. Xie, S.; Chen, Z.; Takagi, T.; Uchimoto, T. Evaluation of Wall Thinning in Carbon Steel Piping Based on Magnetic Saturation Pulsed Eddy Current Testing Method. *Stud. Appl. Electromagn. Mech.* **2013**, *39*, 296–303.
8. Le, M.; Kim, J.; Kim, J.; Do, H.S.; Lee, J. "Electromagnetic testing of moisture separator reheater tubes using a bobbin-type integrated Hall sensor array. *Int. J. Appl. Electromagn. Mech.* **2017**, *55*, S203–S209. [CrossRef]
9. Le, M.; Kim, J.; Kim, J.; Do, H.S.; Lee, J. Nondestructive testing of moisture separator reheater tubing system using Hall sensor array. *Nondestruct. Test. Eval.* **2018**, *33*, 35–44. [CrossRef]
10. Lafontaine, G.; Hardy, F.; Renaud, J. X-Probe®ECT array: A high-Speed Replacement for Rotating Probes. In Proceedings of the Third International Conference on NDE in Relation to Structural Integrity for Nuclear and Pressurized Components, Seville, Spain, 14–16 November 2001.
11. Joubert, P.Y.; Bihan, Y.L.; Placko, D. Localization of defects in steam generator tubes using a multi-coil eddy current probe dedicated to high speed inspection. *NDT E Int.* **2002**, *35*, 53–59. [CrossRef]
12. Choi, C.D.; Lim, I.S. Comparative Reliability of Nondestructive Testing for Weld: Water Wall Tube in Thermal Power Plant Boiler Case Study. *J. Appl. Reliab.* **2018**, *18*, 240–249. [CrossRef]
13. Vajpayee, A.; Russell, D. Automated Condition Assessment of Boiler Water Wall Tubes Using Remote Field Technology. A Revolution over Traditional and Existing Techniques. In Proceedings of the 10th International Conference of the Slovenian Society for Nondestructive Testing, Ljubljana, Slovenia, 1–3 September 2009; pp. 523–530.
14. Gil, D.-S.; Jung, G.-J.; Seo, J.-S.; Kim, H.-J.; Kwon, C.-W. Simulation of Remote Field Scanner for Defect Evaluation of Water Wall Tube within the Fluidized Bed Boiler. *KEPCO J. Electr. Power Energy* **2020**, *6*, 145–150.
15. Atherton, D.L.; Mackintosh, D.D.; Sullivan, S.P.; Dubois, J.M.S.; Schmidt, T.R. Remote-Field Eddy Current Signal Representation. *Mater. Eval.* **1993**, *51*, 782–789.
16. Park, J.-H.; Yoo, H.-R.; Kim, D.-K.; Kim, H.-J.; Cho, S.-H.; Song, S.-J.; Kim, H.-M.; Park, G.-S.; Rho, Y.-W. Development of RFECT System for In-Line Inspection Robot Considering Extendibility of Receiving Senosors based on Parallel Lock-in Amplifer. *Int. J. Precis. Eng. Manuf.* **2017**, *18*, 145–153. [CrossRef]
17. Park, J.W.; Park, J.H.; Song, S.J.; Kishore, M.B.; Kwon, S.G.; Kim, H.J. Enhanced Detection of Defects Using GMR Sensor Based Remote Field Eddy Current Techique. *J. Magn.* **2017**, *22*, 531–538. [CrossRef]
18. Lee, H.; Choe, E.; Lee, J.; Jung, G. Magnetic Flux Leakage Measurement System for Nondestructive Testing of Water-Cooled Wall. In Proceedings of the 2019 IEEE International Instrumentation and Measurement Technology Conference (I2MTC), Auckland, New Zealand, 20–23 May 2019; pp. 1–5.
19. Vakhguelt, A.; Kapayeva, S.D.; Bergander, M.J. Combination Non-Destructive Test (NDT) Method for Early Damage Detection and Condition Assessment of Boiler Tubes. *Procedia Eng.* **2017**, *188*, 125–132. [CrossRef]
20. Kapayeva, S.D.; Bergander, M.J.; Vakhguelt, A.; Khairaliyev, S.I. Remaining life assessment for boiler tubes affected by combined effect of wall thinning and overheating. *Int. J. Vibroeng.* **2017**, *19*, 5892–5907. [CrossRef]
21. Hwang, J.; Lee, J. Modeling of a Scan Type Magnetic Camera Image Using the Improved Dipole Model. *J. Mech. Sci. Technol.* **2006**, *20*, 1691–1701. [CrossRef]
22. Lee, J.; Jun, J.W.; Hwang, J.S.; Lee, S.H. Development of Numerical Analysis Software for the NDE by using Dipole Model. *Key Eng. Mater.* **2007**, *353–358*, 2383–2386. [CrossRef]
23. Le, M.; Kim, J.; Kim, S.; Lee, J. Nondestructive Testing of Pitting Corrosion Cracks in Rivet of Multilayer Structures. *IJPEM* **2016**, *17*, 1433–1442. [CrossRef]

Article

Micromagnetic Characterization of Operation-Induced Damage in Charpy Specimens of RPV Steels

Madalina Rabung [1,*], Melanie Kopp [1], Antal Gasparics [2], Gábor Vértesy [2], Ildikó Szenthe [2], Inge Uytdenhouwen [3] and Klaus Szielasko [1]

[1] Fraunhofer Institute for Nondestructive Testing (IZFP), 66123 Saarbrücken, Germany; melanie.kopp@izfp.fraunhofer.de (M.K.); klaus.szielasko@izfp.fraunhofer.de (K.S.)
[2] Centre for Energy Research, 1121 Budapest, Hungary; gasparics.antal@energia.mta.hu (A.G.); vertesy.gabor@energia.mta.hu (G.V.); szenthe.ildiko@energia.mta.hu (I.S.)
[3] SCK CEN Belgian Nuclear Research Centre, 2400 Mol, Belgium; inge.uytdenhouwen@sckcen.be
* Correspondence: madalina.rabung@izfp.fraunhofer.de; Tel.: +49-(0)681-9302-3882

Featured Application: In this study, the influence of neutron irradiation on the mechanical properties of nuclear pressure vessel materials is investigated using two independent methods of nondestructive magnetic testing. A correlation was found between magnetic characteristics and neutron irradiation-induced damage, regardless of the applied measurement technique. Additionally, by merging the outcome of both testing methods and applying a calibration/training procedure, the damage to reactor steel was successfully predicted. The results are helpful for the potential future practical application of these techniques to the regular inspection of nuclear reactors.

Citation: Rabung, M.; Kopp, M.; Gasparics, A.; Vértesy, G.; Szenthe, I.; Uytdenhouwen, I.; Szielasko, K. Micromagnetic Characterization of Operation-Induced Damage in Charpy Specimens of RPV Steels. *Appl. Sci.* **2021**, *11*, 2917. https://doi.org/10.3390/app11072917

Academic Editors: Jinyi Lee and Hoyong Lee

Received: 29 January 2021
Accepted: 20 March 2021
Published: 24 March 2021

Abstract: The embrittlement of two types of nuclear pressure vessel steel, 15Kh2NMFA and A508 Cl.2, was studied using two different methods of magnetic nondestructive testing: micromagnetic multiparameter microstructure and stress analysis (3MA-X8) and magnetic adaptive testing (MAT). The microstructure and mechanical properties of reactor pressure vessel (RPV) materials are modified due to neutron irradiation; this material degradation can be characterized using magnetic methods. For the first time, the progressive change in material properties due to neutron irradiation was investigated on the same specimens, before and after neutron irradiation. A correlation was found between magnetic characteristics and neutron-irradiation-induced damage, regardless of the type of material or the applied measurement technique. The results of the individual micromagnetic measurements proved their suitability for characterizing the degradation of RPV steel caused by simulated operating conditions. A calibration/training procedure was applied on the merged outcome of both testing methods, producing excellent results in predicting transition temperature, yield strength, and mechanical hardness for both materials.

Keywords: neutron irradiation embrittlement; reactor pressure vessel; magnetic nondestructive evaluation; micromagnetic multiparameter microstructure and stress analysis 3MA; magnetic adaptive testing

1. Introduction

The safe operational lifetime of reactor pressure vessels depends on a number of factors, including design, chemical composition, microstructure, and mechanical characteristics of the reactor pressure vessel (RPV) steels and their in-service-induced change in properties, defect occurrence, and tolerance, as well as operating conditions. Regarding defects, their nature, location, size, density, and growth rate also need to be considered. Operation conditions that affect operational lifetime are neutron exposure (fluence), operation temperature, and the number and magnitude of temperature/pressure cycles in normal conditions and hypothetical accidental conditions. During their operation, RPVs are prone to neutron-irradiation-induced embrittlement.

Currently, progressive material degradation is assessed using destructive tests performed on surveillance specimens in the frame of periodic safety reviews (PSRs). These are standard tensile specimens and ISO-V Charpy specimens of exactly the same RPV steels and their welds. Charpy impact tests are systematically used to assess structural materials over a long period [1]. The impact energy is recorded as a function of the temperature, where the temperature corresponding to an impact energy index value of 41 J, representing the ductile to brittle transition temperature (DBTT), and upper shelf energy (USE), representing fully ductile behavior, are determined. To obtain a single DBTT value, several specimens must be tested. In parallel, tensile specimens tested under a quasistatic loading rate are used to determine yield strength, tensile strength, uniform elongation, total elongation, and reduction in diameter. The disadvantage of destructive methods is that they do not allow for the characterization of the progress of material properties of the same specimen when successively damaged, and they are not applicable to the actual component.

In this context, the development of nondestructive evaluation (NDE) technologies can significantly contribute to the characterization of embrittlement in reactor pressure vessel (RPV) materials by providing complementary information about the progress of material properties. Tests performed nondestructively, in general, do not directly determine the material properties, which are determined by destructive tests. To quantify the material properties nondestructively, nondestructive methods must be correlated with the standardized data measured destructively.

Numerous nondestructive methods are suitable for the characterization of operation-induced damage to RPVs: measurement of the Seebeck coefficient [2,3], ultrasonic techniques [4–6], magnetic testing methods [7–12], and magnetoacoustic emission [13].

In ferromagnetic materials, the correlation between mechanical and magnetic hardness is well-known and understood [14,15]. Magnetic methods are advantageous because they are technically simple, inexpensive, and can be applied easily, even on active materials in hot cells.

A micromagnetic nondestructive method that is basically suitable for the characterization of damage to ferromagnetic materials such as RPV steel and for monitoring the progress of material properties is the micromagnetic multiparameter microstructure and stress analysis (3MA) approach, which uses several methods [16–18]. In this study, we applied a new version of the 3MA technique, 3MA-X8 [19]. This technique is presented in more detail in Section 2.2. This novel method offers improved sensitivity and reliability compared with the previous implementation of the 3MA method.

Another promising candidate for magnetic nondestructive testing methods is based on the detection of minor magnetic hysteresis loops. The philosophy behind this technique is that by measuring minor loops instead of major hysteresis loops, the sensitivity and reliability of the measurement can be significantly improved [20,21]. Similar to 3MA-X8, magnetic adaptive testing (MAT) is also a multiparametric, powerful, and sensitive procedure for magnetic inspection [22,23].

The purpose of this work is to analyze parameters derived from the 3MA-X8 and MAT methods separately, as well as in combination, to predict material properties. The reason for combining several measured parameters for material characterization is the increased robustness against disturbing influences such as material variations and surface treatment.

In previous work, a similar attempt was made: power scale laws (PSLs), magnetic Barkhausen noise (MBN), and MAT results were compared with each other on the same series of neutron-irradiated nuclear reactor pressure vessel steel material [24]. However, DBTT values were taken from the literature; they were not directly measured in the investigated specimens.

In the present work, specimens of two different steels of eastern and western RPV design were investigated. The results obtained from measurements carried out using 3MA-

X8 and MAT techniques on Charpy specimens were compared, before and after neutron irradiation at different neutron fluences, and were correlated with different measured mechanical properties. 3MA-X8 and MAT data, measured on the same specimen series, were collected into a common database and normalized so that they could be quantitatively compared with each other to study their reliability and sensitivity. Finally, a regression analysis was performed to predict neutron-irradiation-induced damage, and conclusions were drawn about the potential applicability of the method for nondestructive evaluation of RPV steel degradation. These results provide information complementary to that obtained from destructive tests of surveillance specimens, which are currently assumed to represent the whole component and cannot account for possible local material variations.

2. Materials and Methods

2.1. Materials and Mechanical Tests

To characterize the damage caused by neutron irradiation, two types of RPV materials were considered: western RPV material A508 Cl.2 and eastern RPV material 15Kh2NMFA. A large part of the Lemoniz reactor vessel, a Spanish reactor of the western type that was never in operation, was chosen to manufacture Charpy specimens at the Belgian Nuclear Research Centre (SCK CEN) [25,26]. ISO-V Charpy specimens were cut out from 3/4 depth. Figure 1 shows the geometry of the Charpy specimen and the dimensions as well as the definition of the T-L orientation. The T-L specimen orientation was selected according to ASTM E23-16b (Standard Test Methods for Notched Bar Impact Testing of Metallic Materials) except if otherwise stated.

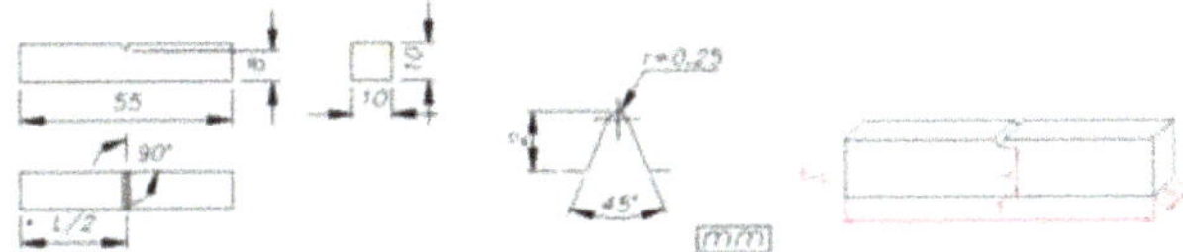

Figure 1. Schematic representation of ISO-V Charpy specimens.

The chemical composition was measured with a spark atomic emission spectrometer (Spectromaxx LMX06) (Table 1). The working method and manipulations were conducted according to the ASTM E415 standard. The typical heat treatment of RPV forgings consists of quenching, tempering, and postweld heat treatment; methods and conditions are described in ASME and ASTM specifications.

Table 1. Chemical composition (wt %) of A508 Cl.2 base metal, as measured by optical emission spectroscopy at the Belgian Nuclear Research Centre (SCK CEN) on a Charpy specimen.

C	Mn	Si	S	P	Cr	Ni	Mo	Cu
0.201	0.578	0.27	0.0085	0.0091	0.372	0.668	0.599	0.0472

A part from original eastern 1000 MW RPV steel was provided. ISO-V Charpy specimens were cut out from the $\frac{1}{4}$ depth. The 15Kh2NMFA (CrNiMoV) forging steel was manufactured by the IZHORA company (Russia) for a 1000 MW WWER (Water Water Energy Reactor). The original heat number was 181,358, and the forging steel was produced according to Russian specification TU 108.765-78. The chemical composition of 15Kh2NMFA steel is provided in Table 2.

Table 2. Chemical composition (wt %) of the 15Kh2NMFA material.

C	Mn	Si	S	P	Cr	Ni	Mo	V	Cu
0.16	0.42	0.29	0.008	0.012	1.97	1.29	0.52	0.12	0.12

The microstructure of the as-received specimens was a mixed-tempered ferrite–bainite structure. After manufacturing Charpy specimens from western RPV material A508 Cl.2 and from eastern RPV material 15Kh2NMFA, one part of them was mechanically tested and the other parts were nondestructively investigated. After the nondestructive examination (Sections 2.2 and 3.2), this set of specimens was divided into three sets of specimens for neutron irradiation. They were neutron-irradiated in the BR2 reactor at three different irradiation fluences (E > 1 MeV) in the primary water pool at an irradiation temperature of ~100–120 °C (Tables 3 and 4). The neutron irradiation was performed in a specially designed rig called NOMAD_3 [27], where 24 Charpy specimens were directly irradiated. This ensured that the damage created was large enough to be detected by nondestructive analysis. The fluence achieved was between 1.55 and 7.90 $\times$ 10^{19} n/cm^2 (E > 1 MeV). Four Charpy specimens for each irradiation level (low, middle, and high fluence) were available. Subsequently, the irradiated specimens were nondestructively investigated. To correlate the results of the micromagnetic testing measurements with the mechanical properties of the same specimens, several mechanical tests were conducted on Charpy specimens after they were nondestructively characterized. They were tested using an instrumented pendulum according to ISO 148-1 and ASTM E23 for the as-received nonirradiated and neutron-irradiated materials. Uniaxial tensile tests were performed with a crosshead rate of 0.2 mm/min, according to ASTM standards E8M and E21 on a conventional static universal tensile test machine. Vickers hardness HV10 tests were performed according to ASTM 92-17 on each Charpy specimen after neutron irradiation.

2.2. Micromagnetic Methods

Nondestructive methods for materials characterization are based on physical principles that are correlated to macroscopic physical properties and microscopic effects.

Micromagnetic techniques are widely used for the nondestructive characterization of the material properties of ferromagnetic steels and are based on the correlation between the magnetic properties of ferromagnetic materials and their mechanical–technological characteristics, which are dependent on the microstructure. This correlation is related to microstructure interaction with both the magnetic structure (Bloch walls) as well as the dislocations [7,14,15].

The requirements for the procedure for measurement of magnetic hysteresis behavior are strict: the test specimen to be measured must be long and rod-shaped and must be magnetized as homogeneously as possible at low frequency (mHz range). The magnetic flux induced in the test specimen must be measured by a coil surrounding the test specimen. For these reasons, direct measurement of the hysteresis curve is unsuitable for practical application to components. Several magnetic methods are suitable for the characterization of ferromagnetic materials (such as RPV base and weld materials). The following effects are often used in micromagnetic nondestructive testing: minor hysteresis loops, magnetic Barkhausen noise, harmonics analysis in the time domain signal of the magnetic tangential field strength, eddy currents, and incremental permeability.

Generally, micromagnetic measuring devices contain a magnetization unit, a probe, and a unit for measurement control and data processing (usually a PC). Depending on the design of the magnetization unit as well as the measurement parameters, different material depths and areas can be investigated. Micromagnetic methods can, therefore, be used to analyze a controllable fraction of the specimen volume. Commercial micromagnetic devices use different effects to describe the material's condition.

Fraunhofer IZFP developed the 3MA technique, which indirectly and nondestructively determines mechanical material properties using a one-sided access micromagnetic sensor. 3MA is based on a combination of several magnetic methods and has been described in previous studies in detail [16–18]. The latest implementation of 3MA is 3MA-X8, which was applied in this work [19]. 3MA-X8 is a variant of 3MA, defined around a minimalistic, rugged sensor design, using the magnetization coil on a U-shaped core as the only sensing element. 3MA-X8 uses low-frequency excitation (f < 20 kHz) and offers high-speed multichannel measurement (>>100 measurements/s on 3, 8, or more channels synchronously, depending on the exact device variant). Compared with previous implementations of 3MA, 3MA-X8 does not contain Barkhausen noise analysis. Harmonics analysis, eddy current incremental permeability analysis, and eddy current impedance analysis are accomplished by supplying a voltage signal of two superimposed frequencies to the electromagnet. The drive current depends on the material contacted by the probe and is analyzed along with the voltage to extract characteristic parameters (Figures 2–4) that describe the magnetization behavior, which is then correlated to the mechanical material properties. Due to the lower frequency range compared with previous 3MA implementations, a higher penetration depth of the magnetic field is reached. Within the present application, this is an important advantage, given the correlation of the magnetic properties to the material properties, such as DBTT and yield strength, as these are also integral values of the examined specimens.

The following section describes the 3MA-X8 analysis in more detail:

I. Eddy currents (ECs) arise in a material if it is exposed to an alternating magnetic field and depend on the electrical conductivity σ and the magnetic permeability μ of the material [16,17]. The higher-frequency EC-affected impedance (Z) is not constant throughout the rather low-frequency hysteresis loop. Therefore, Z describes a loop in the impedance plane throughout one hysteresis cycle [19]. Figure 2 illustrates the parameters derived from this loop.

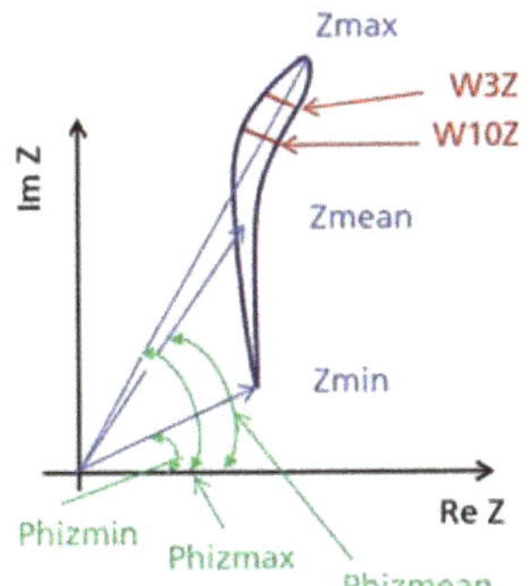

Parameter	Explanation
Zmean	Average of absolute impedance of impedance loop
Zmin	Minimum of absolute impedance of impedance loop
Zmax	Maximum of absolute impedance of impedance loop
Phizmean	Phase of the average impedance
Phizmin	Phase of the minimum impedance
Phizmax	Phase of the maximum impedance
W3Z	Width of the impedance loop at 3% of maximum
W10Z	Width of the impedance loop at 10% of maximum

Figure 2. Schematic illustration of the impedance loop and derived parameters.

II. Incremental permeability (IP) analysis is a method of separating magnetic permeability information from electrical conductivity information in EC analysis [16,17]. Plotting the change in EC coil impedance (DZ) against drive voltage U lecads to an incremental permeability plot (Figure 3, left) [19]. Several parameters are derived from the incremental permeability curve (Figure 3).

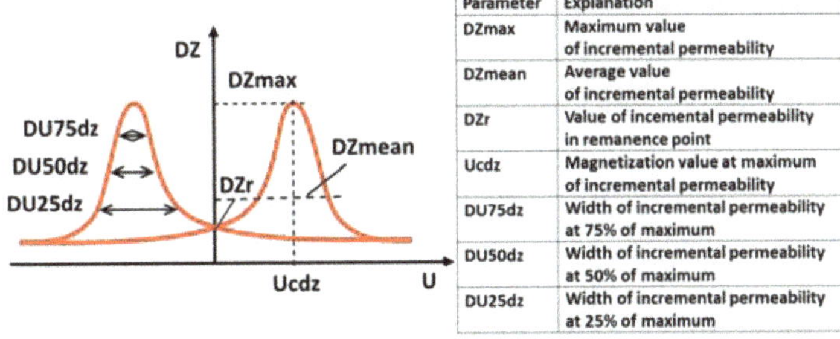

Parameter	Explanation
DZmax	Maximum value of incremental permeability
DZmean	Average value of incremental permeability
DZr	Value of incemental permeability in remanence point
Ucdz	Magnetization value at maximum of incremental permeability
DU75dz	Width of incremental permeability at 75% of maximum
DU50dz	Width of incremental permeability at 50% of maximum
DU25dz	Width of incremental permeability at 25% of maximum

Figure 3. Schematic illustration of the incremental permeability curve and the parameters derived.

III. The measured magnetizing current exhibits a low-frequency distortion due to hysteresis in the magnetic circuit. The fundamental and harmonic components can be determined numerically using a fast Fourier transform; thus, the distortions of the magnetizing current are quantified. The harmonic components determined through this procedure allow for the determination of material properties (Figure 4) [19].

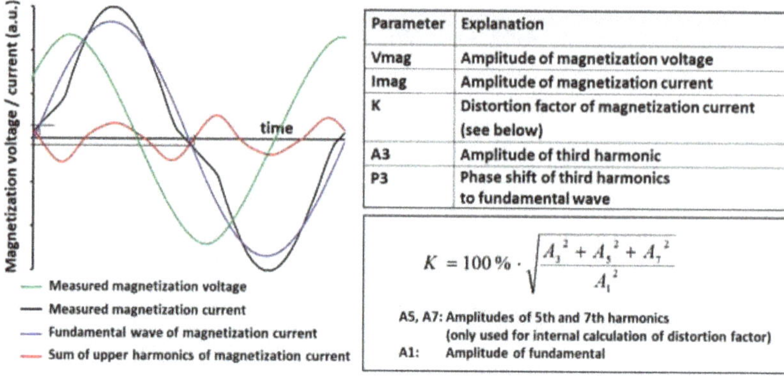

Parameter	Explanation
Vmag	Amplitude of magnetization voltage
Imag	Amplitude of magnetization current
K	Distortion factor of magnetization current (see below)
A3	Amplitude of third harmonic
P3	Phase shift of third harmonics to fundamental wave

$$K = 100\% \cdot \sqrt{\frac{A_3^2 + A_5^2 + A_7^2}{A_1^2}}$$

A5, A7: Amplitudes of 5th and 7th harmonics (only used for internal calculation of distortion factor)
A1: Amplitude of fundamental

Figure 4. Schematic illustration of the harmonic analysis in the time domain signal of the magnetization current.

Figure 5 shows the 3MA-X8 device, including a probe and a PC. The 3MA-X8 measurements were performed at a magnetization frequency of 50 Hz, a magnetization current voltage of 2 V, and a superimposed EC frequency of 2000 Hz. The 3MA-X8 device is controlled by modular measuring system (MMS) software. The output of the data acquisition and evaluation module was a set of 21 magnetic parameters derived from the 3MA-X8 measurements, as listed on the right-hand side of Figures 2–4. A specimen holder and a probe holder improved reproducibility by minimizing positioning variations.

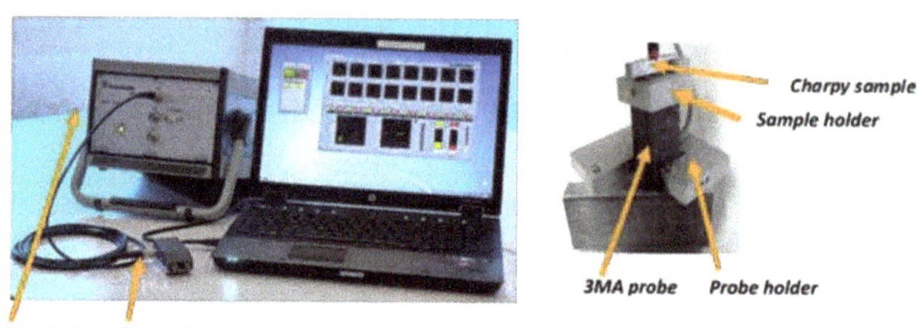

Figure 5. The micromagnetic multiparameter microstructure and stress analysis (3MA)-X8 system, including 3MA-X8 device, probe, and PC (**left**); (**right**) specimen holder and probe holder.

The Centre for Energy Research developed and applied the MAT method. The MAT method uses the systematic measurement of large families of minor hysteresis loops from minimum amplitudes up to, possibly, the maximum (major) on degraded ferromagnetic samples/objects. From the large volume of recorded data, the data that reflect material degradation in the most sensitive, or otherwise most convenient, manner are applied for evaluation of the degradation. These data, best-adapted for the investigated case, were used as the MAT parameter(s), and their dependence on an independent variable accompanying the inspected degradation is referred to as the MAT degradation function(s).

The magnetic induction method appears to be the easiest method of systematic measurement for MAT. A specially designed permeameter [28], with a magnetizing yoke, was applied for the measurement of families of minor loops of a magnetic circuit's differential permeability. Measurement of the hysteresis loops was performed by a magnetizing yoke, which is placed on the flat surface of the Charpy specimens. A C-shaped laminated Fe–Si transformer core was used. The yoke had a cross-section of 10 × 5 mm and a total external length of 18 mm. This size was chosen to fit the size of the investigated specimens. The specimen holder, designed for hot-cell measurements, is shown in Figure 6 [29]. The driving coil wound on the yoke produces triangular variations in the applied magnetic field, with stepwise increasing amplitudes and a fixed slope magnitude in all the triangles (Figure 7a).

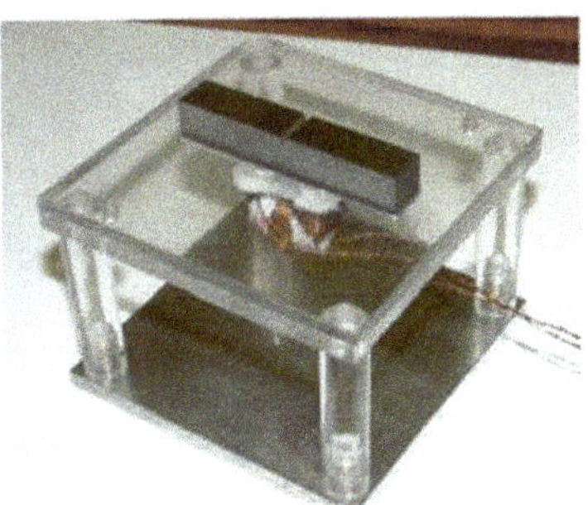

Figure 6. Photo of the specimen holder. A Charpy specimen is placed on the top (V-notch is opposite the magnetizing yoke) [29].

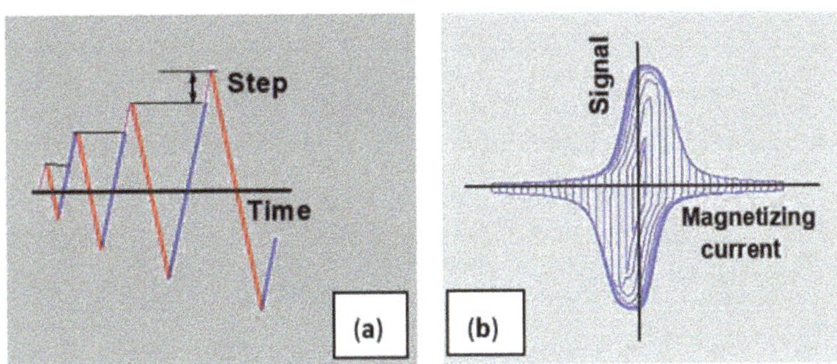

Figure 7. The time variation of the magnetizing current (**a**) and the measured permeability loops (**b**) [30].

The signal coil picks up the induced voltage, proportional to the differential permeability of the sample. This triangular variation of the magnetizing field, with time t and a voltage signal U, is induced in the pick-up coil for each kth sample:

$$U(\mathrm{d}F/\mathrm{d}t, F, A_j, \varepsilon_k) = K \times \partial B(\mathrm{d}F/\mathrm{d}t, F, A_j, \varepsilon_k)/\partial t = K \times \mu(\mathrm{d}F/\mathrm{d}t, F, A_j, \varepsilon_k) \times \mathrm{d}F/\mathrm{d}t, \quad (1)$$

where K is a constant determined by the geometry of the sample and by the experimental arrangement; ε is the independent degradation variable, in our case, Vickers hardness, yield strength, or transition temperature. As long as $F = F(t)$ sweeps linearly with time, i.e., $|\mathrm{d}F/\mathrm{d}t|$ is (the same) constant for measurement at each of the samples, Equation (1) states that the measured signal is simply proportional to the differential permeability μ of the measured magnetic circuit as it varies with the applied field F within each minor loop amplitude A_j for each kth measured sample. If we wish to obtain correct results that are not influenced by any previous remanence, each sample has to be thoroughly demagnetized before it is measured.

The permeameter works under the control of a notebook, which sends the steering information to the function generator and collects the measured data. An input/output data acquisition card accomplishes the measurement. The computer registers two data files for each measured family of minor loops. The first one contains detailed information about all the preselected parameters of the demagnetization and the measurement. The other file holds the course of the voltage signal U induced in the pick-up coil as a function of time t and of the magnetizing current I_F and/or field F. As an illustration, Figure 7b presents the families of permeability loops. Large amounts of data were generated, and our task was to compare them and find the most suitable data for characterizing the changes between samples.

Instead of keeping the signal and the magnetizing field in shapes of continuous time-dependent functions, it is practical to interpolate the family of data for each ε_k sample into a discrete square (i, j) matrix, $U(F_i,A_j,\varepsilon_k)$, with a suitably chosen step, $\Delta A = \Delta F$. Because dF/dt is a constant, identical for all measurements within one experiment, it is not necessary to write it explicitly as a variable of U. MAT is a relative method (practically all the nondestructive methods are relative), and the most suitable information regarding the degradation of the investigated material can be contained in the variation of any element of such matrices as a function of ε, relative to the corresponding element of the reference matrix $U(F_i,A_j,\varepsilon_0)$. So, we divided all $U(F_i, A_j, \varepsilon_k)$ elements by the corresponding elements $U(F_i,A_j,\varepsilon_0)$ of the reference sample matrix and obtained the normalized elements of the matrices of relative differential permeability $\mu(F_i,A_j,\varepsilon_k) = U(F_i, A_j, \varepsilon_k)/U(F_i, A_j, \varepsilon_0)$, and their proper sequences

$$\mu(F_i,A_j,\varepsilon) = U(F_i, A_j, \varepsilon)/U(F_i, A_j, \varepsilon_0) \quad (2)$$

as normalized μ-degradation functions of the inspected material. These μ elements are referred to as MAT descriptors, and they characterize the degradation of the material.

Numerous MAT descriptors are calculated during evaluation; their number depends on the density and how $\mu(F_i,A_j,\varepsilon_k)$ points are calculated. Normally, about 1000 MAT descriptors are calculated. The majority of them do not reflect the correlation between magnetic parameters and actual material degradation. The purpose of our evaluation is to choose the elements that provide the best correlation between MAT parameters and the independent parameter from this big data pool. These parameters are called optimally chosen MAT descriptors, and they are provided below in the Results section. The choice is correct if these parameters have the best sensitivity and yield good reproducibility simultaneously. The MAT evaluation process is described in detail in [23].

3MA-X8 and MAT measurements were carried out similarly: to adequately perform measurements in the hot cell, the specimens were placed on the top of the 3MA-X8 and MAT probes, respectively, to allow the easy replacement of specimens by a manipulator. The 3MA-X8 and MAT devices were placed outside the hot cell for remote control of the measurement.

Both 3MA-X8 and MAT are comparative-type measurement methods. This means that the measured magnetic parameters describe the material behavior from the magnetic point of view. For the quantitative characterization of specimens with unknown mechanical properties, an initial calibration/training process is needed on a well-defined calibration set of specimens (with known target properties, e.g., DBTT or hardness). A polynomial function describing the relation between all collected micromagnetic parameters, the magnetic fingerprint (MFP), and the target quantity (e.g., hardness, DBTT) of each material can be determined via regression analysis based on a database of calibration specimen measurements. Assuming a regression analysis is used, a simple calibration using a polynomial function can be written as $Y = a_0 + a_{1\times 1} + a_2X_2 + \ldots + a_NX_N$, where Y is the target quantity (e.g., DBTT or hardness), a_i are the coefficients, X_i are the measured parameters (components of the MFP), and N is the number of selected measured parameters [18]. The parameters are selected based on the least-squares algorithm. This polynomial function is generated empirically since there is no physical equation that describes the dependency between individual magnetic parameters and mechanical properties. The detailed calibration procedure is described in [18]. The trained system can be used to determine target quantities of specimens of the same material after a prior recording of the micromagnetic parameters.

3. Results

3.1. Mechanical Properties

The BR2 material test reactor at SCK CEN, with a nominal power of 125 MW and unique, adaptable core configuration, is the most powerful material test reactor currently operating in Europe. The neutron spectrum is typical of a research reactor, with thermal fluxes between 7×10^{13} and to 10^{15} n/(cm^{2}s) and fast fluxes (E > 0.1 MeV) between 1×10^{13} and 6×10^{14} n/(cm^{2}s). Like all nuclear reactors, there is no single neutron energy. The actual fast fluence values for all Charpy specimens were measured (Tables 3 and 4) by Fe dosimeters. The ^{235}U equivalent fission neutron flux was calculated from the ^{54}Mn activity formed by the ^{54}Fe(n,p)^{54}Mn reaction. The flux (fluence) was calculated using the ^{235}U fission spectrum averaged cross-section $<\sigma>$ = 81.7 mb, adopted from [31]. The equivalent fission fluence was converted to a fast fluence (E > 1 MeV) in Fe to provide the material damage of the reactor pressure vessel specimens.

First, the actual fluence values for all Charpy specimens were measured (Tables 3 and 4). The results of the Charpy impact tests are shown in Figure 8 for A508 Cl.2 and 15Kh2NMFA. An irradiation-induced embrittlement shift, ranging between 100 and 200 °C, was achieved.

A correlation usually exists between the yield strength increase, hardening, and embrittlement. To obtain some estimations on the correlation between yield strength increase and the embrittlement (ΔT_{41J}), some additional tests were performed. The first was Vickers hardness at HV5 (Figure 9). Correlations between hardness increase and neutron fluence were found for both steels. In Figure 10, the increase in yield strength $\Delta\sigma_Y$ is plotted for the two grades at the three neutron fluence levels. The increase in yield strength was similar for both grades at the tested fluence levels, even though the initial yield strength was different between the two grades. All test results are summarized in Tables 3 and 4 for A508 Cl.2 and 15kHNMFA steel, respectively.

Table 3. Fast fluence (E > 1 MeV), Vickers hardness (HV5), yield strength (hardening $\Delta\sigma_y$), and ductile to brittle transition temperature (DBTT; embrittlement ΔT_{41J}) for A508 Cl.2. RT, room temperature.

A508 Cl.2.	Fast Fluence (E > 1 MeV) ($\times 10^{19}$ n/cm^2)	Vickers Hardness HV5 (ΔHV)	Yield Strength @RT (MPa) ($\Delta\sigma_y$)	DBTT T_{41J} (°C) (ΔT_{41J})
Baseline	0	181.2 ± 2.8	448	−33 ± 9
Low fluence	1.55	237.9 ± 2.8 (56.7)	710 (262)	76 ± 15 (109)
Medium fluence	4.38	254.4 ± 1.5 (73.2)	793 (345)	125 ± 15 (158)
High fluence	7.04	258.5 ± 3.1 (77.3)	823 (375)	126 ± 15 (159)

Table 4. Fast fluence (E > 1 MeV), Vickers hardness (HV5), yield strength (hardening $\Delta\sigma_y$), and DBTT (embrittlement ΔT_{41J}) for 15Kh2NMFA.

15Kh2NMFA	Fast Fluence (E > 1 MeV) ($\times 10^{19}$ n/cm^2)	Vickers Hardness HV5 (ΔHV)	Yield Strength @RT (MPa) ($\Delta\sigma_y$)	DBTT T_{41J} (°C) (ΔT_{41J})
Baseline	0	226.1 ± 5.5	600	−51 ± 12
Low fluence	2.78	278.5 ± 3.4 (52.4)	875 (275)	88 ± 15 (139)
Medium fluence	6.83	292.8 ± 3.1 (66.7)	973 (373)	136 ± 15 (187)
High fluence	7.9	294.1 ± 4.5 (68.0)	987 (387)	124 ± 15 (175)

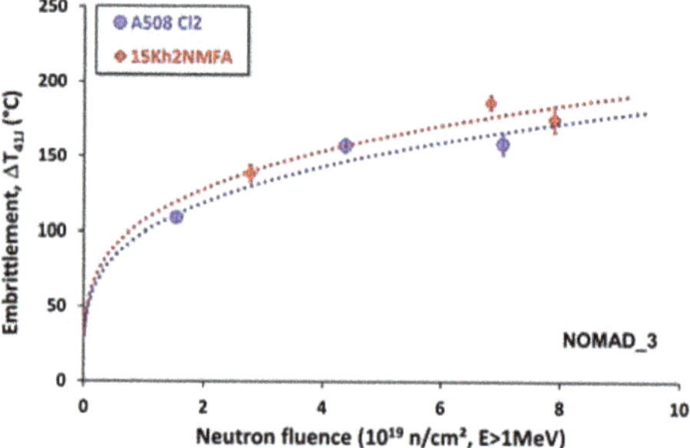

Figure 8. Embrittlement of 15Kh2NMFA and A508 Cl.2 after NOMAD_3 irradiation as a function of neutron fluence. The dashed lines are added to guide the eye (power law fit).

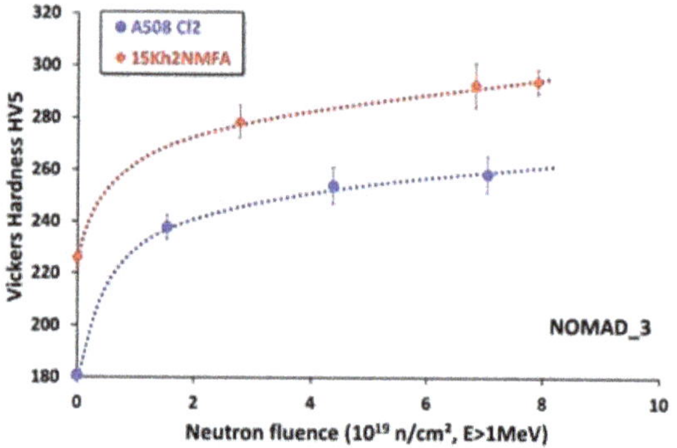

Figure 9. Vickers hardness (5kgf) for A508 Cl.2 and 15Kh2NMFA. The dashed lines are added to guide the eye.

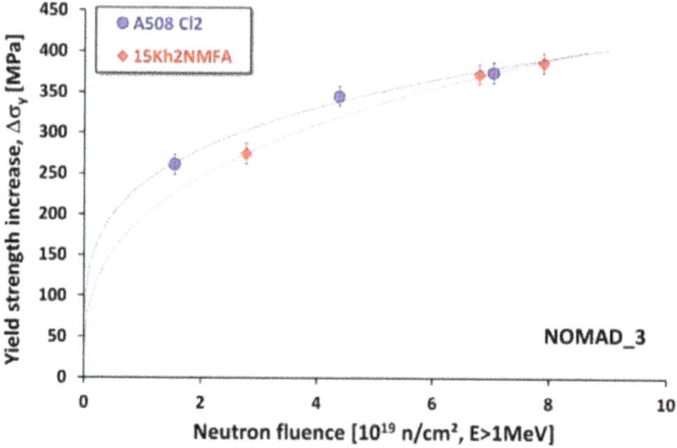

Figure 10. Tensile strength increase (hardening), tested at room temperature for A508 Cl.2 and 15Kh2NMFA steels, as a function of the fast neutron fluence. The dashed lines are added to guide the eye (power law fit).

All mechanical properties, such as hardness, yield strength, and DBTT, increased with increasing irradiation fluence. The increase in yield strength and DBTT caused by neutron irradiation for both materials was almost the same; the increase in the hardness of the eastern RPV material was slightly higher than that of the western RPV material. Additionally, we observed that hardness, yield strength, and DBTT slowly increased at middle and high fluence for the western RPV material. The same behavior was observed in terms of hardness and yield strength for the eastern RPV material. For this material, the DBTT at high fluence was slightly lower than at medium fluence. However, no conclusions can be drawn here due to the uncertainties in these values.

3.2. Micromagnetics Results

Both nondestructive techniques, 3MA-X8 and MAT, were applied to the specimens described in Section 2.1 using predefined measuring parameters (magnetic field amplitude and magnetization frequency) on the opposite side of the Charpy notch to avoid possible side effects that can alter the outcome of the measurements. All measurements were carried out at the Laboratory of Medium and High Activity (LMHA) at SCK CEN due to the high activation of the specimens. The exact same specimens were tested before and after neutron irradiation under the same environmental conditions: in a hot cell. Therefore, the disturbing influences of material inhomogeneity and scatter between different specimens of the same material were eliminated.

Both micromagnetic measurements recorded from Charpy specimens of RPV materials A508 Cl.2 (Figure 11) and 15Kh2NMFA (Figure 12) before and after neutron irradiation showed clear differences between the four irradiation conditions (nonirradiated; low, medium, and high fluence). We observed that the magnetic material properties of different specimens measured before irradiation were not exactly identical. After neutron irradiation, these differences were observed in the magnetic properties of specimens irradiated at the same fluence as well. Nevertheless, a trend was clearly identified: neutron irradiation caused easily measurable differences in magnetic parameters, although different procedures, 3MA-X8 or MAT, were used for measurements.

In the case of MAT measurements, the parameters are normalized by the corresponding parameter of the reference (nonirradiated) specimens. As shown in Figures 11 and 12, a clear increase in magnetic parameters was found due to neutron irradiation. This parameter depends on the material's condition. For A508 Cl.2 material, this descriptor was characterized by F_i = 780 mA and A_j = 1200 mA values, while 15Kh2NMFA material was characterized by F_i = 30 mA and A_j = 1080 mA.

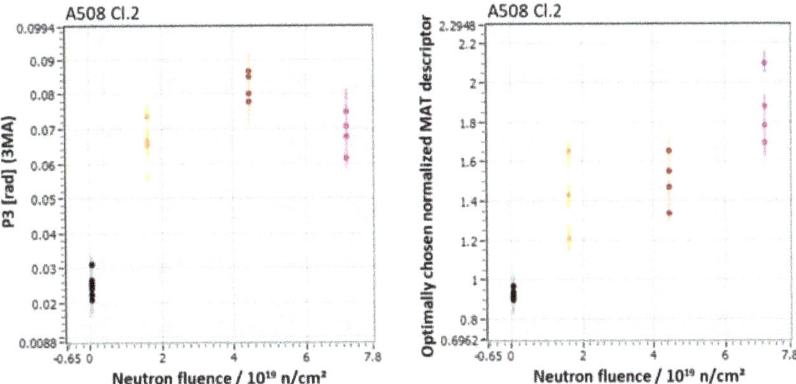

Figure 11. Dependency of the amplitude of third harmonics P3 (**left**) and of the magnetic adaptive testing (MAT) descriptor (**right**) on the fluence level for western reactor pressure vessel (RPV) material A508 Cl.2.

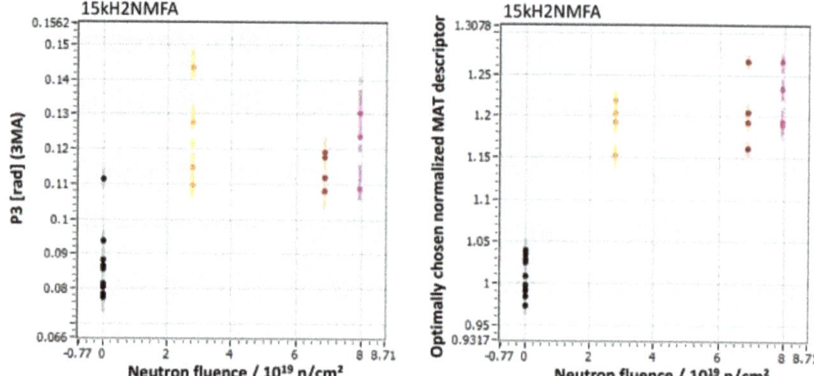

Figure 12. Dependency of the amplitude of third harmonics P3 (**left**) and of the MAT descriptor (**right**) on the fluence level for eastern RPV material 15Kh2NMFA.

In the case of 3MA-X8 measurements, several magnetic parameters (e.g., P3, A3, K, Ucdz) showed a clear trend with increasing neutron fluence. Figures 11 and 12 show the amplitude of the third harmonics P3-derived from upper harmonics analysis in the time domain signal of the magnetization current as a function of neutron fluence.

The results of the micromagnetic measurements were evaluated in terms of DBTT determined by Charpy tests, hardness, and yield strength. Trends in several quantities extracted from upper harmonics analysis and magnetic adaptive testing were observed in relation to different mechanical properties (DBTT, mechanical hardness, and yield strength).

Differences between the irradiation conditions can hardly be identified, especially between specimens irradiated at medium and high fluence. In this context, it has to be emphasized that for both materials, the mechanical properties of the specimens under the middle- and high-irradiation conditions were almost the same as well. In Figures 13–16, the same parameters as in Figures 11 and 12 are presented in terms of embrittlement (DBTT) and hardening (hardness and yield strength) for both materials and both testing methods.

We observed that the 3MA-X8 method is more sensitive to changes in the material properties of western RPV steel A508 Cl.2, whereas the MAT method is more sensitive to the changes in the material properties of eastern RPV steel 15Kh2NMFA.

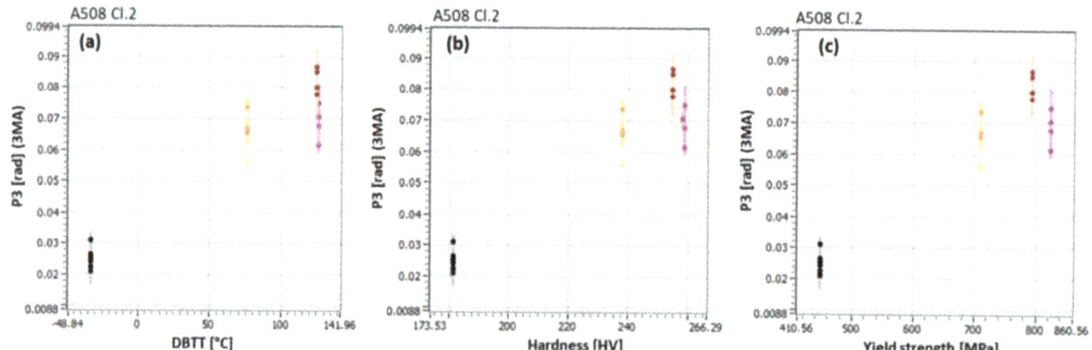

Figure 13. Dependency of the amplitude of third harmonics P3 on the ductile to brittle transition temperature (DBTT)(**a**), hardness (**b**), and yield strength (**c**) of western RPV material A508 Cl.2.

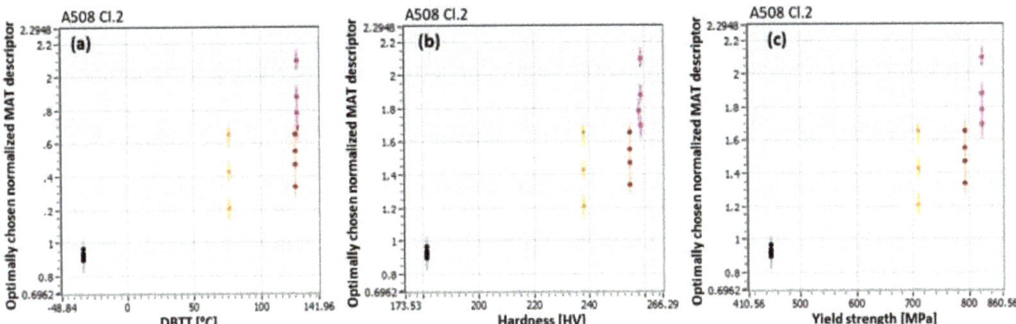

Figure 14. Dependency of the MAT descriptor on the DBTT (**a**), hardness (**b**), and yield strength (**c**) of western RPV material A508 Cl.2.

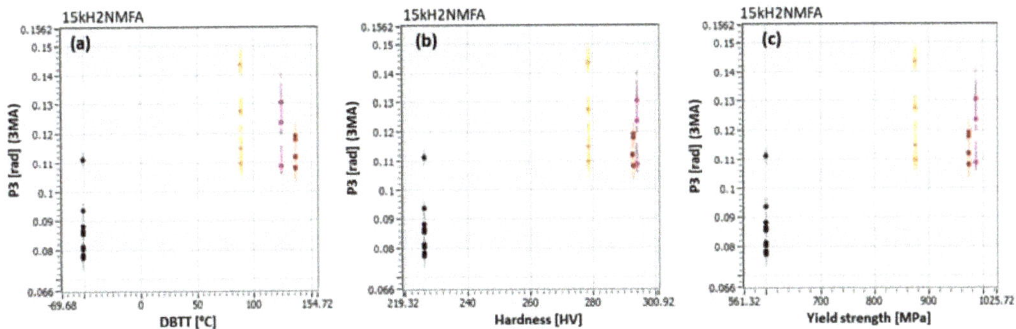

Figure 15. Dependency of the amplitude of third harmonics P3 on DBTT (**a**), hardness (**b**), and yield strength (**c**) of eastern RPV material 15Kh2NMFA.

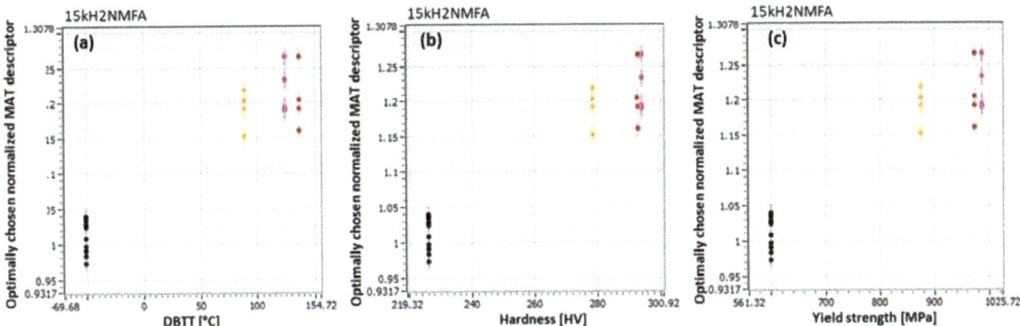

Figure 16. Dependency of the MAT descriptor on the DBTT (**a**), hardness (**b**), and yield strength (**c**) of eastern RPV material 15Kh2NMFA.

3.3. *Nondestructive Prediction of Embrittlement and Hardening*

All measured micromagnetic parameters were collected and classified in terms of irradiation condition, embrittlement, and hardening. The reason for using more than one measuring parameter for material characterization is the increased robustness against disturbing influences such as material variations and surface conditions (similar to the benefits of having and combining different human senses). Having collected all measured parameters obtained from 3MA-X8 and MAT measurements, parameters were separated into two data sets: one set for the calibration/training procedure and another set for testing. In the data set for calibration/training, all parameters obtained by means of MAT and 3MA-X8 methods on the predefined calibration set of specimens were merged with the corresponding mechanical properties. In the next step, the calibration/training procedure was conducted: polynomial functions empirically describing the relation between measured micromagnetic parameters and the target quantities (DBTT, mechanical hardness, and yield strength) were determined via regression analysis [18]. These polynomial functions quantitatively and empirically describe the correlation between mechanical properties and measured micromagnetic parameters. Finally, these polynomial functions, determined on the training set of specimens, were tested using the micromagnetic data obtained on the testing set of specimens; thus, the target quantities were estimated. Figures 17 and 18 show the results of the training (dark dots) and testing (light crosses) procedures for A508 Cl.2 and 15Kh2NMFA, respectively. The combination of both 3MA-X8 and MAT methods allows for the prediction of mechanical properties, independent of the difference between individual specimens under the same irradiation condition.

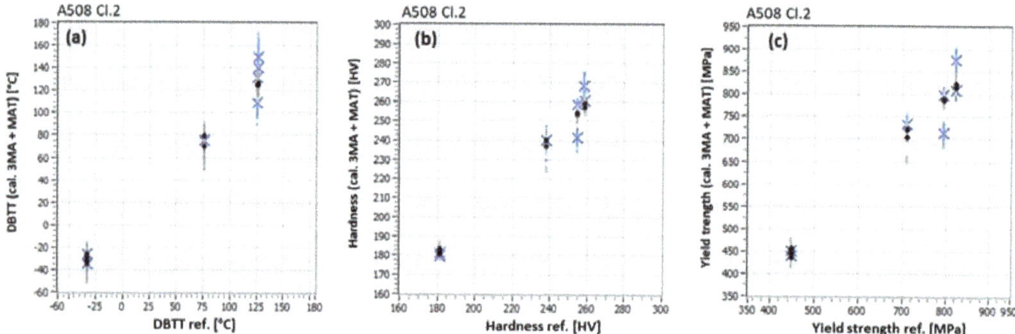

Figure 17. Prediction of DBTT (**a**), mechanical hardness (**b**), and yield strength (**c**) for A508 Cl.2 material using combined data from 3MA-X8 and MAT methods.

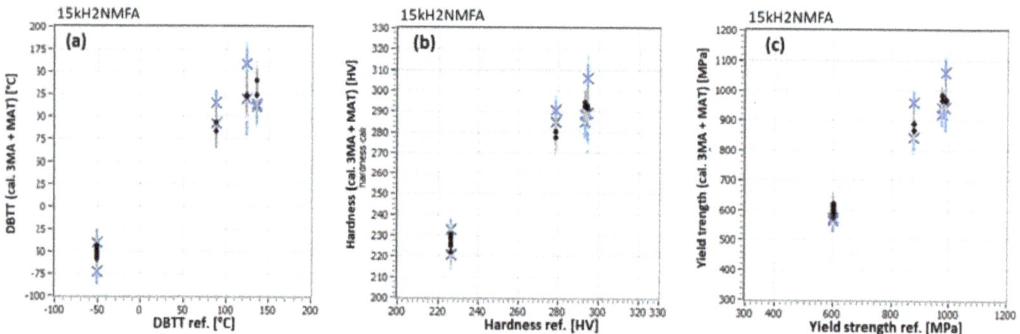

Figure 18. Results of the calibration procedure to predict DBTT (**a**), mechanical hardness (**b**), and yield strength (**c**) for 15Kh2NMFA.

Tables 5 and 6 show the correlation coefficients (R^2) and root mean square errors (RMSEs) for the estimation of all targeted quantities: DBTT, hardness, and yield strength. All targeted quantities were estimated with an accuracy higher than 91%, demonstrating the suitability of this procedure to estimate target quantities in a nondestructive manner.

Table 5. Correlation coefficients (R^2) and root mean square errors (RMSEs) for the estimation of DBTT, mechanical hardness, and yield strength of A508 Cl.2 material.

Parameter	R^2		RMSE	
	Training	Test	Training	Test
DBTT	99.8%	97.5%	2.9 °C	12.5 °C
Yield strength	99.9%	94.5%	6.2 MPa	38.0 MPa
Hardness	99.9%	96.4%	1.1 HV	6.6 HV

Table 6. Correlation coefficients and root mean square errors for the estimation of DBTT, mechanical hardness, and yield strength of 15Kh2NMFA material.

Parameter	R^2		RMSE	
	Training	Test	Training	Test
DBTT	99.6%	92.5%	5.2 °C	21.4 °C
Yield strength	99.4%	91.5%	13.2 MPa	51.6 MPa
Hardness	99.5%	92.8%	2.1 HV	7.9 HV

4. Discussion

In this study, several material properties were determined through mechanical tests and micromagnetic examinations. Changes in mechanical properties are caused by neutron-irradiation-induced microstructure defects, which, in turn, influence the magnetic properties measured using micromagnetic measurements.

It is well-known that an increase in neutron fluence yields an increase in the density and diameter of microstructure defects. The nature of the microstructure defects and of the degradation mechanism caused by neutron irradiation in RPV steels is dependent on, among other parameters (irradiation fluence, irradiation temperature), the chemical composition of the RPV steel, especially the Cu, Ni, and P contents. There are three important neutron-irradiation-induced damage mechanisms caused by matrix features, Cu-rich precipitates and Mn–Ni–Si-precipitates. The first one is responsible for hardening in both RPV steels, having Cu content below and above 0.04 wt %, generally at low fluences. The second one occurs at high fluences in RPV steels with Cu content higher than 0.04 wt %. The third one is typical for low-Cu RPV steels at high fluences [31]. In this study, the Cu contents of both RPV steels were higher than 0.04 wt % (Tables 1 and 2). Thus, for both RPV steels, the neutron-irradiation-induced embrittlement could be caused by matrix features at low fluence and Cu-rich precipitates at high fluences. However, microstructural examinations (e.g., small neutron angle scattering and high-resolution transmission electron microscopy) were not performed in this study.

For both materials, we observed that neutron irradiation at low fluences caused a significant modification in material properties: hardness, embrittlement (DBTT), and yield strength. A further increase in neutron fluence caused a much smaller further degradation of the mechanical properties, whereas a further increase in the neutron fluence did not cause any further change in the mechanical properties, indicating saturation that was observed (Tables 3 and 4). For 15Kh2NMFA steel, a small decrease in DBTT was observed at high neutron fluence, but it remained within the uncertainties of such measurements. All these material properties and their progress, induced by neutron irradiation, are strongly influenced by neutron-irradiation-induced microstructure changes.

Microstructure defects impede dislocation movement and Bloch-wall movement and cause changes in mechanical and magnetic properties, respectively. Micromagnetic parameters depend on Bloch-wall movement during magnetization and characterize their interaction with microstructure defects such as precipitates, voids, dislocations, or grain boundaries. Coercivity proportionally depends on the interaction intensity between Bloch walls and microstructure defects. Kersten's and Dijkstra and Wert's theories describe the correlation between coercivity, volume, and diameter of small microstructure defects [32]. All these theories describe an increase in magnetic hardness with the rising volume and diameter of microstructure defects. Since all other micromagnetic parameters strongly correlate with coercivity, they similarly depend on the microstructure defects' diameter and volume.

The clear difference in the results of magnetic measurements between the nonirradiated and irradiated conditions correlates well with the results of the mechanical tests and can be explained by the occurrence of microstructure defects due to neutron irradiation (matrix features at low fluence and Cu-rich precipitates at high fluence). Small microstructure defects impede Bloch-wall movement due to the foreign body effect and, therefore, cause an increase in magnetic hardness, as illustrated by the magnetic parameters in Figures 13–16. The foreign body effects describe the interaction between Bloch walls and small microstructure defects [31]. The higher the volume of small microstructure defects (matrix features), the higher the mechanical properties and the stronger the foreign body effect. The smaller change in mechanical properties at increased neutron fluence can be explained by an increase in the microstructure defects' (Cu-rich precipitates) diameter and a smaller increase of the volume of defects, which, in turn, similarly influence the magnetic properties. The foreign body effect and the interaction between Bloch walls and microstructure defects both weaken; thus, magnetic properties increase more slowly.

For the irradiated eastern RPV steel 15Kh2NMFA, we observed that parameters derived from both magnetic methods reached saturation. For western RPV material A508 Cl.2, the MAT descriptor continuously increased, but most 3MA-X8 parameters reached saturation.

Another conclusion is that by applying different nondestructive magnetic techniques, a similar correlation was found between mechanical properties and the modification of magnetic parameters. This finding is encouraging for future practical applications of magnetic NDE.

The innovative part of this study is the combination of the results obtained using both magnetic methods, followed by a training procedure to determine targeted material properties (hardness, DBTT, and yield strength). Excellent correlation coefficients and impressively low RMSEs were achieved for the prediction and quantification of hardness, ductile to brittle transition temperature, and yield strength of the tested specimens (Figures 17 and 18). Correlation coefficients larger than 91% were obtained between nondestructively predicted and destructively determined mechanical properties (Tables 5 and 6). Therefore, the two magnetic methods resulted in a strong correlation between their outcomes, and their combination can be used to precisely predict different mechanical properties.

5. Conclusions

Changes in mechanical properties due to irradiation of standard surveillance Charpy specimens have often been published according to standard surveillance programs [1].

As experimentally observed, the relationship between neutron fluence and the micromagnetic parameters is nonlinear since the dependence of the mechanical properties on neutron fluence is nonlinear as well. The results of the individual micromagnetic measurements, performed following MAT and 3MA-X8 methods, prove their suitability for characterizing the progressive degradation of RPV steels caused by simulated operation conditions in terms of low-temperature neutron irradiation. Differences between individual specimens at the same irradiation condition or the same damage stage (same DBTT, yield strength, or hardness) affect the outcome of both kinds of measurements.

The combination of both magnetic methods allows for the prediction of mechanical properties with high accuracy, independent of the microstructural differences between individual specimens at the same irradiation condition.

Author Contributions: Conceptualization, M.R. and A.G.; methodology, A.G., G.V., K.S., and I.S.; investigation, M.K., A.G., G.V., I.U., and I.S.; project administration, M.R. and A.G.; funding acquisition, M.R. and A.G. All authors have read and agreed to the published version of the manuscript.

Funding: This research was carried out within the frame of the "NOMAD" project. This project (Nondestructive Evaluation System for the Inspection of Operation-Induced Material Degradation in Nuclear Power Plants) has received funding from the Euratom research and training program 2014–2018 under grant agreement No. 755330.

Institutional Review Board Statement: Not applicable.

Informed Consent Statement: Not applicable.

Data Availability Statement: The data are contained within the article.

Conflicts of Interest: The authors declare no conflict of interest.

References

1. Ferreño, D.; Gorrochategui, I.; Gutiérrez-Solana, F. Degradation due to neutron embrittlement of nuclear vessel steels: A critical review about the current experimental and analytical techniques to characterise the material, with particular emphasis on alternative methodologies. In *Nuclear Power—Control, Reliability and Human Factors*; Tsvetkov, P., Ed.; IntechOpen Limited: London, UK, 2011; ISBN 978-9-53307-599-0. Available online: http://www.intechopen.com/articles/show/title/non-destructive-testing-for-ageing-management-of-nuclear-power-components (accessed on 14 May 2020).
2. Niffenegger, M.; Leber, H.J. Monitoring the embrittlement of reactor pressure vessel steels by using the Seebeck coefficient. *J. Nucl. Mater.* **2009**, *389*, 62. [CrossRef]
3. Niffenegger, M.; Reichlin, K.; Kalkhof, D. Application of the Seebeck effect for monitoring of neutron embrittlement and low-cycle fatigue in nuclear reactor steel. *Nucl. Eng. Des.* **2005**, *235*, 1777–1788. [CrossRef]
4. Seiler, G. Early detection of fatigue at elevated temperature in austenitic steel using electromagnetic ultrasound transducers. In Proceedings of the Seventh International Conference on Low Cycle Fatigue: LCF7, Deutscher Verband für Materialforschung und-Prüfung e.V. (DVM), Aachen, Germany, 9–11 September 2013; pp. S359–S364.
5. Smith, R.L.; Rusbridge, K.L.; Reynolds, W.N.; Hudson, B. Ultrasonic attenuation, microstructure and ductile to brittle transition temperature in Fe-C alloys. *Mater. Eval.* **1993**, *41*, 219–222.
6. Dobmann, G.; Kröning, M.; Theiner, W.; Willems, H.; Fiedler, U. Nondestructive characterization of materials (ultrasonic and magnetic techniques) for strength and toughness prediction and the detection early creep damage. *Nucl. Eng. Des.* **1995**, *157*, 137–158. [CrossRef]
7. Jiles, D.C. Magnetic methods in nondestructive testing. In *Encyclopedia of Materials Science and Technology*; Buschow, K.H.J., Ed.; Elsevier Press: Oxford, UK, 2001; p. 6021.
8. Blitz, J. *Electrical and Magnetic Methods of Nondestructive Testing*; Adam Hilger IOP Publishing, Ltd.: Bristol, UK, 1991.
9. Lo, C.C.H.; Jakubovics, J.P.; Scrub, C.B. Non-destructive evaluation of spheroidized steel using magnetoacoustic and Barkhausen emission. *IEEE Trans. Magn.* **1997**, *33*, 4035–4037. [CrossRef]
10. Kikuchi, H.; Ara, K.; Kamada, Y.; Kobayashi, S. Effect of microstructure changes on Barkhausen noise properties and hysteresis loop in cold rolled low carbon steel. *IEEE Trans. Magn.* **2009**, *45*, 2744–2747. [CrossRef]
11. Hartmann, K.; Moses, A.J.; Meydan, T. A system for measurement of AC Barkhausen noise in electrical steels. *J. Magn. Magn. Mater.* **2003**, *254–255*, 318–320. [CrossRef]
12. Pirfo Barroso, S.; Horváth, M.; Horváth, Á. Magnetic measurements for evaluation of radiation damage on nuclear reactor materials. *Nucl. Eng. Des.* **2010**, *240*, 722–725. [CrossRef]
13. Augustyniak, B.; Chmielewski, M.; Piotrowski, L.; Kowalewski, Z. Comparison of properties of magnetoacoustic emission and mechanical barkhausen effects for P91 steel after plastic flow and creep. *IEEE Trans. Magn.* **2008**, *44*, 3273–3276. [CrossRef]
14. Devine, M.K. Magnetic detection of material properties. *J. Min. Met. Mater. JOM* **1992**, *44*, 24–30. [CrossRef]
15. Kronmüller, H.; Fähnle, M. *Micromagnetism and the Microstructure of Ferromagnetic Solids*; Cambridge University Press: Cambridge, UK, 2003.
16. Szielasko, K.; Tschuncky, R.; Rabung, M.; Seiler, G.; Altpeter, I.; Dobmann, G.; Herrmann, H.G.; Boller, C. Early Detection of Critical Material Degradation by Means of Electromagnetic Multi-Parametric NDE. *AIP Conf. Proc.* **2014**, *1581*. [CrossRef]
17. Dobmann, G.; Altpeter, I.; Kopp, M.; Rabung, M.; Hubschen, G. ND-materials characterization of neutron induced embrittlement in German nuclear reactor pressure vessel material by micromagnetic NDT techniques. In *Electromagnetic Nondestructive Evaluation (XI)*; IOS Press: Amsterdam, The Netherlands, 2008; p. 54. ISBN 978-1-58603-896-0.
18. Altpeter, I.; Becker, R.G.; Dobmann, R.; Kern, W.A.; Theiner, A. Yashan: Robust Solutions of Inverse Problems in Eletromagnetic Non-Destructive Evaluation. *Inverse Probl.* **2002**, *18*, 1907–1921. [CrossRef]
19. Szielasko, K.; Wolter, B.; Tschuncky, R.; Youssef, S. Micromagnetic materials characterization using maschin learning—Progress in Nondestructive Prediction of Mechanical Properties of Steel and Iron. *Tech. Mess.* **2020**, *87*, 428–437. [CrossRef]

20. Takahashi, S.; Kobayashi, S.; Kikuchi, H.; Kamada, Y. Relationship between mechanical and magnetic properties in cold rolled low carbon steel. *J. Appl. Phys.* **2006**, *100*, 113908. [CrossRef]
21. Kobayashi, S.; Gillemot, F.; Horváth, Á.; Székely, R. Magnetic properties of a highly neutron-irradiated nuclear reactor pressure vessel steel. *J. Nucl. Mater.* **2012**, *421*, 112–116. [CrossRef]
22. Tomáš, I. Non-destructive Magnetic Adaptive Testing of ferromagnetic materials. *J. Mag. Mag. Mat.* **2004**, *268*, 178–185. [CrossRef]
23. Tomáš, I.; Vértesy, G. Magnetic Adaptive Testing. In *Nondestructive Testing Methods and New Applications*; Omar, M., Ed.; IntechOpen Limited: London, UK, 2012; Available online: http://www.intechopen.com/articles/show/title/magnetic-adaptive-testing (accessed on 8 November 2020)ISBN 978-953-51-0108-6.
24. Tomáš, I.; Vértesy, G.; Pirfo Barroso, S.; Kobayashi, S. Comparison of four NDT methods for indication of reactor steel degradation by high fluences of neutron irradiation. *Nucl. Eng. Des.* **2013**, *265*, 201–209. [CrossRef]
25. Uytdenhouwen, I.; Chaouadi, R.; Other NOMAD Consortium Members. NOMAD: Non-destructive Evaluation (NDE) system for the inspection of operation-induced material degradation in nuclear power plants—Overview of the neutron irradiation campaigns. In *ASME 2020 Pressure Vessels & Piping Conference*; Non-Destructive Examination, V007T07A003, PVP2020-21512; ASME: New York, NY, USA, 2020; Volume 7.
26. Uytdenhouwen, I.; Chaouadi, R. Effect of neutron irradiation on the mechanical properties of an A508 Cl.2 forging irradiated in a BAMI capsule. In *Pressure Vessels and Piping Conference*; Codes and Standards, V001T01A060, PVP2020-21513; ASME: New York, NY, USA, 2020; Volume 1.
27. Uytdenhouwen, I.; Chaouadi, R. Accepted to be Published in the ASME 2021 Proceedings PVP2021-61969 Paper: Effect of Neutron Irradiation on the Mechanical Properties of an A508 Cl.2 and 15Kh2NMFA Irradiated in the NOMAD_3 rig in the BR2 Cooling Water. Available online: https://proceedings.asmedigitalcollection.asme.org/PVP/ (accessed on 2 March 2021).
28. Tomáš, I.; Perevertov, O. Permeameter for Preisach approach to materials testing. In *JSAEM Studies in Applied Electromagnetics and Mechanics*, 9th ed.; Takagi, T., Ueasaka, M., Eds.; IOS Press: Amsterdam, The Netherlands, 2001; pp. 5–15.
29. Vértesy, G.; Gasparics, A.; Uytdenhouwen, I.; Szenthe, I.; Gillemot, F.; Chaouadi, R. Nondestructive investigation of neutron irradiation generated structural changes of reactor steel material by magnetic hysteresis method. *Metals* **2009**, *10*, 642. [CrossRef]
30. Vértesy, G.; Gasparics, A.; Griffin, J.M.; Mathew, J.; Fitzpatrick, M.E.; Uytdenhouwen, I. Analysis of surface roughness influence in non-destructive magnetic measurements applied to reactor pressure vessel steels. *Appl. Sci.* **2020**, *10*, 8938. [CrossRef]
31. *Integrity of Reactor Pressure Vessels in Nuclear Power Plants: Assessment of Irradiation Embrittlement Effects in Reactor Pressure Vessel Steels*; IAEA Nuclear Energy Series No.NP-T-3.11; Atomic Energy Agency: Vienna, Austria, 2009.
32. Seeger, A. *Moderne Probleme der Metallphysik*; Springer: Berlin/Heidelberg, Germany, 1966; p. 232.

 applied sciences

Article

Three-Dimensional Imaging of Metallic Grain by Stacking the Microscopic Images

Jinyi Lee [1,2,3], Azouaou Berkache [1,*], Dabin Wang [1] and Young-Ha Hwang [4]

1 IT-Based Real-Time NDT Center, Chosun University, Gwangju 61452, Korea; jinyilee@chosun.ac.kr (J.L.); anydabin@gmail.com (D.W.)
2 Department of Electronics Engineering, Chosun University, Gwangju 61452, Korea
3 Interdisciplinary Program in IT-Bio Convergence System, Chosun University, Gwangju 61452, Korea
4 Daegyeong Regional Division, Korea Institute of Industrial Technology, Daegu 42990, Korea; h69231@kitech.re.kr
* Correspondence: azouaoubrk@yahoo.fr

Abstract: Three-dimensional observation of metal grains (MG) has a wide potential application serving the interdisciplinary community. It can be used for industrial applications and basic research to overcome the limitations of non-destructive testing methods, such as ultrasonic testing, magnetic particle testing, and eddy current testing. This study proposes a method and its implementation algorithm to observe (MG) metal grains in three dimensions in a general laboratory environment equipped with a polishing machine and a metal microscope. An image was taken by a metal microscope while polishing the mounted object to be measured. Then, the metal grains (MGs) were reconstructed into three dimensions through local positioning, binarization, boundary extraction, (MG) selection, and stacking. The goal is to reconstruct the 3D MG in a virtual form that reflects the real shape of the MG. The usefulness of the proposed method was verified using the carbon steel (SA106) specimen.

Keywords: 3D imaging of metal grains; non-destructive testing methods; stacking images; SA106 carbon steel

Citation: Lee, J.; Berkache, A.; Wang, D.; Hwang, Y.-H. Three-Dimensional Imaging of Metallic Grain by Stacking the Microscopic Images. *Appl. Sci.* **2021**, *11*, 7787. https://doi.org/10.3390/app11177787

Academic Editors: Carosena Meola and Adel Razek

Received: 20 July 2021
Accepted: 23 August 2021
Published: 24 August 2021

1. Introduction

Predicting, comprehending, and monitoring the microstructure and mechanical properties, such as heterogeneous microstructure, ductility decreases, the concentration of residual stress, brittleness, and deterioration of toughness, of critical structures, such as heat exchangers, power plants, the oil and gas industry, chemical engineering, and especially nuclear power plants, is important to ensure safety and avoid failures that can lead to environmental disasters, causalities, and financial losses. To prevent and avert such risks, and to ensure the continued safety and integrity of these structures, periodic inspections are required [1–4].

Numerous non-destructive methods can be used for microstructure characterization. In most cases, non-destructive testing cannot directly determine material characteristics, which are established by destructive testing. To quantify material properties non-destructively, non-destructive methods must be correlated with destructively measured standard data. The limitation and disadvantage of destructive methods are that they do not allow the characterization of the evolution of material properties of the same specimen when successively damaged [3], and they are not easily applicable to characterize the mechanical behavior of materials, if possible, with restricted, difficult and expensive access, even using a numerical simulation, such as the finite element method.

The finite element method is used to analyze the relationship between the load, stress-strain, and resistance to failure of a structure [5–8]. Meanwhile, according to conventional studies, the size and shape of the microstructure affect the strength, toughness, and creep resistance of the structure [9–11]. However, it is not easy to reflect and demonstrate this with the finite element method [12–15]. This is because the factor used in the finite element method is the shape of the structure and does not include the microstructure of the structure itself. Therefore, we have not yet seen an example in which the size of the microstructure is inserted as an analysis condition in the finite element method (or two-dimensional analysis). Therefore, in order to analyze the shape and size of the microstructure, it is necessary to express the shape of the grain in three dimensions. On the other hand, the size and shape of the grains affect the analysis results of non-destructive tests [16–19], such as ultrasonic testing, magnetic particle testing, and eddy current testing. For instance, in the case of ultrasonic examination, research results have been reported that the size and shape of the microstructure influence the scattering of the sound [20]. However, analysis or qualitative matters considering only the average size of the microstructure are reported in [21], and it is difficult to find previous studies on ultrasound wave scattering targeting the size and shape of the actual microstructure. In the magnetic particle inspection method, the sample is magnetized, and the adsorption of magnetic particles by the magnetic flux leakage generated around the defect is observed. At this point, if a strong external magnetic field is applied and then removed, residual magnetization occurs. A similar fault signal is generated, in which magnetic particles are adsorbed even in non-defective areas due to residual magnetization [22–24]. Even though the deformation caused by an external force causes the magnetic moment inside the magnetic domain to be anisotropic in a specific direction and, therefore, residual magnetization occurs even if an external magnetic field is not applied [25–27]. In addition, residual magnetization is frequently generated in welds where heat and residual stress are generated by an external force. In the eddy current method of inspection, an induced current is applied to an object to be measured, and the distortion of the induced current occurring around a fault, that is, a change in impedance and a different phase due to eddy currents is measured [28,29]. However, it is difficult to determine the presence or size of a defect due to the edge effect that occurs at the end of the object to be measured [30]. This means that even when the grain is needle-shaped or the shape changes rapidly, the eddy current signal can be affected. It is important to provide a means of quantitatively measuring and expressing the three-dimensional shape of a grain in order to grasp the effects of such effects, i.e., the formation of magnetic domains, residual magnetization, and the effects of edge according to the shape and size of the grain. According to nano-radiography and tomography, the metal grain can be imaged in three dimensions [31,32]. Nevertheless, this requires very high accuracy and expensive equipment, including X-ray shielding. Furthermore, it is difficult to measure the distribution of magnetic domains within the grain in three dimensions. In addition, under the heat treatment influence, the application of external field and mechanical stress can rearrange the martensite variant and can change the magnetic domains. It is important to understand their effect not only at the macroscopic level but also at the microscopic level [33–36]. In this context, the use of an alternative method to overcome the limitations of non-destructive testing (NDT) techniques can make a valuable contribution to the characterization of materials and the evaluation of microstructure performance through rapid and efficient experiments and simulations. In this paper, we present a method to reproduce the real shape and size of the grains of the tested sample SA106 by pilling up two-dimensional images obtained by a metal microscope. To achieve this target, we introduce fixtures, procedures, and software algorithms based on mathematical formulas.

Several studies have previously developed 2D and 3D techniques to construct a way to study microstructure using different experimental and simulation techniques. The results of these studies lead to a good outcome that provides access to crucial structural information of the investigated materials, which facilitates and enhances the understanding of physical and chemical concepts at the microscopic level [37–43].

However, we have not seen any work similar to our work, which allows the reconstruction, extraction, and selection of one or more grains with the exact shape and size of the inspected material, without using extensive software and hardware, but just simple equipment and some algorithms. It should be emphasized that the method presented here permits: (1) 3D grain shape obtaining; (2) using conventional research laboratory environment; (3) easy preparing specimens; (4) multi-parallel data acquisition; (5) easy 3D image processing. This is particularly important for the industry and research laboratories.

The finding is encouraging for the near future application of the method for the analysis of 3D stress, strain deformation at grain level, and the analytical study of non-destructive evaluations using electromagnetic and ultrasound testing.

2. Specimen and Layer Capture

In this study, carbon steel SA106 was targeted. The sample was cut and cropped to a small volume of 25 × 9 × 25 mm (width × length × height), respectively, and mounted on a cylindrical resin with a diameter of 35 mm and a height of 27 mm. Then, the observation side and the backside of the mounted specimen were polished to be exposed outside the resin. This allows the "polished" height to be measured more accurately when viewed under a microscope. In other words, when measuring the height of the sample with a micrometer after polishing, errors likely to be caused by compression of the resin can be avoided when the resin is used as a measurement reference and not the surface of the sample. As shown in Figure 1a, the side surface was polished to deflect by 1.84 mm. This deflected side surface (hereafter "key-hole") was used as a reference surface for microscopic observation. Figure 1b is a jig for uniform grinding of the key-hole, and Figure 1c is a template (state) in which the key-hole is polished in the sample mounted in the jig. Figure 1d illustrates these mounted samples; they are inserted into the X-Y stage holder under the microscope and are fixed by a key (expressed in black in the figure). Due to this structure, the area observed by the microscope can fix the measurement surface with a repeatability of the order of 10 to 20 μm.

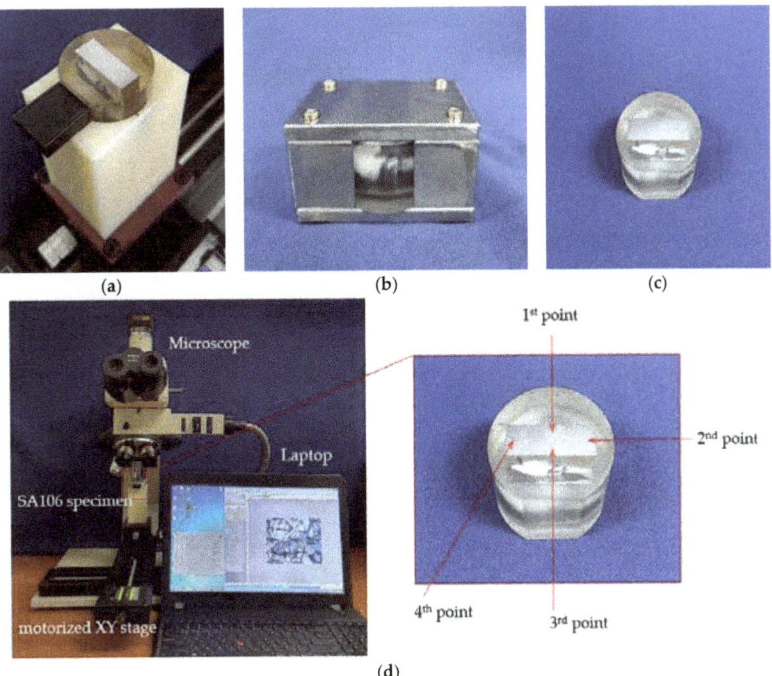

Figure 1. Specimen adjustment to reduce positioning error. (**a**) Mounted specimen; (**b**) Jig for key-hole; (**c**) After grinding Mounted specimen; and (**d**) Set specimen on the XY-stage using the key and key-hole.

The observation surface was polished using SiC polishing pads FEPA P200, P400, P800, P1200, and P2000. The last stage, P2000, has a grid size of 10.3 μm. Finally, fine polishing was carried out for 20 s using polishing cloths and polycrystalline diamond paste having a grid size of 1 μm. Once the micro-polishing was completed, an etchant solution obtained by diluting with nitric acid 3 g in 65 g ethanol was exposed onto the observation surface for 5 s, followed by ultrasonic cleaning with distilled water for 15 s and then cleaning with ethanol for 10 s. This chemical etching was performed a total of three times. The height of the four sides of the test sample was measured with a micrometer having an accuracy of ±0.01 μm, visualized through the microscope (500×), and recorded on the laptop. The observation by the microscope was repeated while repeating fine polishing, chemical etching, as well as the height measurement and can be directly linked to the 3D image processing. A total of 60 fine polishes were performed; Table 1 shows the heights measured in each layer.

Table 1. Height of the polished sample at each layer.

Layer	Height of Layer (mm)				
	1st Point	2nd Point	3rd Point	4th Point	Average
1	22.49370	22.47870	22.50430	22.51340	22.49753
2	22.49280	22.47680	22.50210	22.51190	22.49590
3	22.49160	22.47510	22.50000	22.50910	22.49395
4	22.48980	22.47290	22.49850	22.50800	22.49230
5	22.48820	22.47130	22.49740	22.50620	22.49078
6	22.48680	22.47090	22.49660	22.50440	22.48968
7	22.48510	22.46850	22.49450	22.50240	22.48763
8	22.48300	22.46730	22.49240	22.50030	22.48575
9	22.48160	22.46550	22.49120	22.49910	22.48435
10	22.48020	22.46410	22.48900	22.49700	22.48258
11	22.47910	22.46330	22.48750	22.49540	22.48133
12	22.47680	22.46140	22.48580	22.49320	22.47933
13	22.47570	22.46040	22.48330	22.49090	22.47758
14	22.47400	22.45810	22.48170	22.48880	22.47565
15	22.47220	22.45700	22.47990	22.48750	22.47415
16	22.47120	22.45550	22.47790	22.48630	22.47273
17	22.46920	22.45340	22.47640	22.48510	22.47103
18	22.46730	22.45250	22.47560	22.48420	22.46990
19	22.46586	22.45088	22.43980	22.48258	22.46828
20	22.46448	22.44918	22.47228	22.44048	22.46660
21	22.46338	22.44788	22.47038	22.47948	22.46528
22	22.46188	22.44578	22.46938	22.47798	22.46375
23	22.46088	22.44368	22.46688	22.47718	22.46215
24	22.45918	22.44238	22.46528	22.47568	22.46063
25	22.45788	22.44028	22.46358	22.47368	22.45885
26	22.45588	22.43848	22.46218	22.47188	22.45710
27	22.45398	22.43748	22.46038	22.47048	22.45558
28	22.45198	22.43548	22.45778	22.46968	22.45373
29	22.45038	22.43368	22.45578	22.46858	22.45210
30	22.44878	22.43258	22.45408	22.46608	22.45038
31	22.44728	22.43068	22.45258	22.46518	22.44893
32	22.44608	22.42948	22.45168	22.46338	22.44765
33	22.44498	22.42728	22.45008	22.46098	22.44583
34	22.44388	22.42518	22.44868	22.45968	22.44435
35	22.44278	22.42288	22.44748	22.45868	22.44295
36	22.44158	22.42188	22.44638	22.45618	22.44150
37	22.44008	22.42038	22.44368	22.45528	22.43985
38	22.44385	22.41880	22.44210	22.45370	22.43827
39	22.44364	22.41560	22.44090	22.45230	22.43630
40	22.43470	22.41440	22.43880	22.45090	22.43470

Table 1. *Cont.*

Layer	Height of Layer (mm)				
	1st Point	2nd Point	3rd Point	4th Point	Average
41	22.43230	22.41230	22.4376	22.44950	22.43292
42	22.43120	22.41070	22.43620	22.44720	22.43132
43	22.43000	22.40960	22.43550	22.44580	22.43022
44	22.42920	22.40870	22.43350	22.44460	22.42900
45	22.42830	22.40780	22.43150	22.44270	22.42757
46	22.42650	22.40620	22.42990	22.44030	22.42572
47	22.42570	22.40530	22.42900	22.43870	22.42467
48	22.42390	22.40380	22.42770	22.43740	22.42320
49	22.42210	22.40170	22.42670	22.43560	22.42152
50	22.42070	22.40070	22.42500	22.43430	22.42017
51	22.41910	22.39980	22.42350	22.43340	22.41895
52	22.41860	22.39840	22.42180	22.43210	22.41772
53	22.41700	22.39700	22.42000	22.43030	22.41607
54	22.41580	22.39510	22.41860	22.42880	22.41457
55	22.41420	22.39300	22.41670	22.42680	22.41267
56	22.41300	22.39160	22.41520	22.42550	22.41132
57	22.41230	22.39080	22.41340	22.42340	22.40997
58	22.41070	22.38880	22.41170	22.42150	22.40817
59	22.40860	22.38680	22.41050	22.42000	22.40650
60	22.40660	22.38590	22.40890	22.41850	22.40497

3. Algorithms for 3D Imaging of Grains and Results

3.1. Local Positioning

As mentioned above, the main goal of this study is to image MG in 3D. To reach our purpose, micrographs are taken while repeatedly polishing. At this time, the height information measured by the micrometer is taken as the z-axis. Then the location of the boundary of a specific MG selected in the photo is stored as the x-y coordinates. Therefore, it is possible to obtain three-dimensional coordinates for the node representing the boundary of the MG.

However, as the experimental methodology seems very simple, a challenge associated with this technique is a positional error within ±7 μm that occurred when the specimen was polished, and the same area was observed, even though we used very accurate equipment to avoid this (backlash play), because of the high special-resolution used (2 μm). As an example, Figure 2 is a photograph (hereinafter, nth layer photo) taken after polishing an area of 130 × 98 μm three times, four times, and fifth times, respectively, using a microscope at a magnification of 500. For mutual comparison, a white line with a + mark is indicated in the center. Based on the 3rd layer, the position of the 4th layer is shifted to the upper left. Further, based on the 4th layer, the 5th layer has moved to the right. Meaning the photos of each layer are not in the same position.

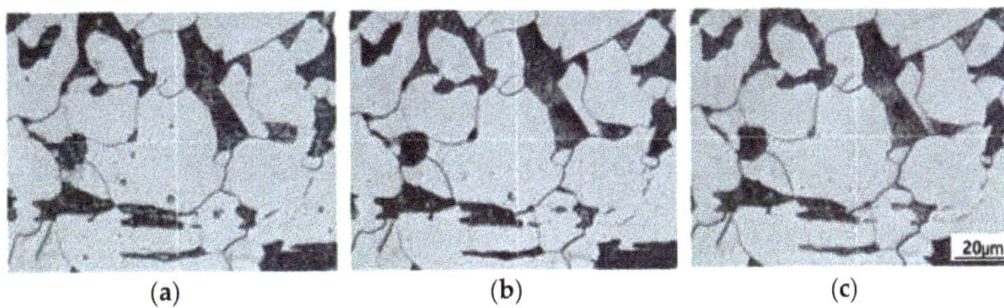

Figure 2. Microscopic photograph of the polished layers. (**a**) 3rd layer; (**b**) 4th layer; and (**c**) 5th layer.

To correctly stake the images, improve the accuracy, and achieve high precision of the final desired result, we have developed and used a set of algorithms that automatically corrects these positional errors. Figure 3a is an nth layer and is used as a reference. Figure 3b is a target as the $(n + 1)$th layer. Each microscope image is composed of 1536 and 2048 pixels in the row and column directions, respectively. The horizontal column is the x-direction, and the vertical row is the y-direction. The origin of the coordinates is (1,1), which is left-top based on the pixel, and the last coordinate is (2048,1536), which is the right-bottom.

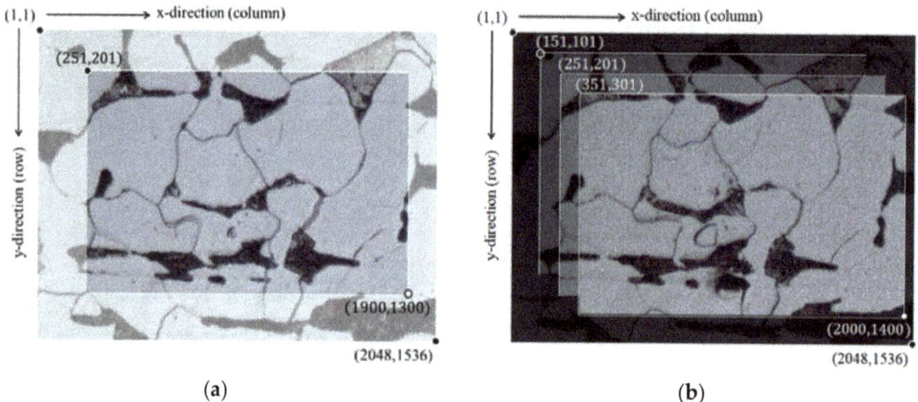

Figure 3. Area selection for local positioning. (**a**) Reference picture nth layer; (**b**) Target picture $(n + 1)$th layer.

The (x_1, y_1) and (x_2, y_2) shown in Equation (1) are the vertices in the diagonal direction of the area selected to apply the local positioning algorithm. Δx and Δy are the lengths of the horizontal and vertical sides of the analysis area. In the case of Figure 3a, (251,201) and (1900,1300) are the vertices, and the region of $\Delta x = 1650$ and $\Delta y = 1100$ is selected.

$$\left.\begin{array}{l}\Delta x = x_2 - x_1 \\ \Delta y = y_2 - y_1\end{array}\right\} \tag{1}$$

In Equation (2), the region of Figure 3a was selected from the first layer photo, which means that it was used as a reference photo. In other words, when stacking the photos, they should be positioned with the first one defined as the reference. In addition, Equations (3) and (4) involve selecting a region to be analyzed with respect to the reference photo in the second layer photo, which is a target for local positioning. Select a total of cases from the domain of $(x_1 - h, y_1 - h)$ and $(x_2 - h, y_2 - h)$ to the domain of $(x_1 + h, y_1 + h)$ and $(x_2 + h, y_2 + h)$. Where h represents the maximum position error of each layer. Figure 3b is the case of $h = 100$, which is a schematic diagram of the selection of a total of 40.000 cases from the domains of (151,101) and (1800,1200) to the domains of (351,301) and (2000,1400).

$$\left[P^1_{\Delta x \Delta y}\right] = \left[P^1_{ref}\,(x_1 : x_2, y_1 : y_2)\right] \tag{2}$$

$$\left[P^2_{\Delta x \Delta y}(i,j)\right] = \left[P^1_{target}(x_1 + i : x_2 + i, y_1 + j : y_2 + j)\right] \tag{3}$$

$$-h \le (i,j) \le +h \tag{4}$$

Equation (5) denotes $S_1(i,j)$, the sum of the absolute values of the differences between each element of the matrix of the selected region in the reference photo and the target photo, respectively. Equation (6) denotes the row ΔX_2 and column ΔY_2 having the minimum value in $S_1(i,j)$. In this instance, $(\Delta X_1, \Delta Y_1) = (0, 0)$. Furthermore, the minimum value of $S_1(i,j)$ is generated at the position where the reference photo and the target photo almost coincide.

$$S_1(i,j) = \sum_{(x,y)=(1,1)}^{(\Delta x, \Delta y)} \left[abs\left(\left[P^1_{\Delta x \Delta y}\right] - \left[P^2_{\Delta x \Delta y}(i,j)\right]\right)\right] \tag{5}$$

$$[\Delta X_2, \Delta Y_2] = min(S_1(i,j), \{(i,j) \in \pm h\}) \tag{6}$$

Thereafter, as shown in Equations (7)–(9), the reflecting region in the nth layer photo becomes the reference, and the local positioning is repeated by comparing it with the target $(n + 1)$th layer photo. Using this process along with ΔX_n and ΔY_n for each layer obtained using Equation (10), the region of interest of the first layer photo, to be the same region as the other target photos, each local positioning of the region having X_1, Y_1 and X_2, Y_2 as the diagonal vertices layer, can be observed, as shown in Equation (11).

$$\left[P^n_{\Delta x \Delta y}\right] = \left[P^n_{target}(x_1 + \Delta X_n : x_2 + \Delta X_n, y_1 + \Delta Y_n : y_2 + \Delta Y_n)\right] \tag{7}$$

$$\left[P^{n+1}_{\Delta x \Delta y}(i,j)\right] = \left[P^{n+1}_{target}(x_1 + i : x_2 + i, y_1 + j : y_2 + j)\right] \tag{8}$$

$$S_n(i,j) = \sum_{(x,y)=(1,1)}^{(\Delta x, \Delta y)} \left[abs\left(\left[P^n_{\Delta x \Delta y}\right] - \left[P^{n+1}_{\Delta x \Delta y}(i,j)\right]\right)\right] \tag{9}$$

$$[\Delta X_{n+1}, \Delta Y_{n+1}] = min(S_n(i,j), \{(i,j) \in \pm h\}) \tag{10}$$

$$\left[P^n_{align}\right] = \left[P^n_{target}(X_1 + \Delta X_n : X_2 + \Delta X_n, Y_1 + \Delta Y_n : Y_2 + \Delta Y_n)\right] \tag{11}$$

On the other hand, X_1, X_2, Y_1, Y_2 of Equation (11) to be selected for observing the change of a specific cross section of the MG are independent of x_1, x_2, y_1, y_2 of Equations (1), (2), (7) and (8), which were selected for local positioning. As an example, local positioning was performed at $(x_1, y_1) = (250, 200)$, $(x_2, y_2) = (2000, 1000)$, $\Delta X_{17} = 55$ and $\Delta Y_{17} = 167$ in the 3rd layer photo in Figure 4a,c. Further, values of $\Delta X_{18} = 35$ and $\Delta Y_{18} = 170$ were obtained in the 4th layer photo of Figure 4b,d. Then, by adjusting (X_1, X_2, Y_1, Y_2), an area of 1700 × 1200 pixels in Figure 4a,b, and also 420 × 420 pixels in Figure 4c,d were selected, respectively.

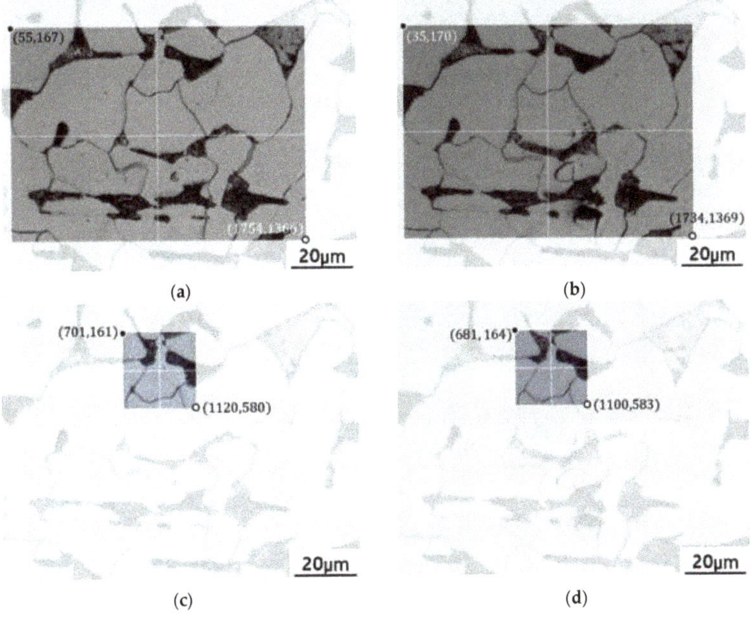

Figure 4. Microscopic photographs with local positioning in a certain area. (**a**) 17th layer, (55,167) ⦣ (1754,1368); (**b**) 18th layer, (35,170) ⦣ (1734,1369); (**c**) 17th layer, (701,161) ⦣ (1120,560); and (**d**) 18th layer, (681,164) ⦣ (1100,583).

3.2. Selection and Binarization

In order to select a specific MG and image it in 3D, it is necessary to select the boundary surface of the corresponding MG. First, a point inside a specific MGF is read, it is called point-1, and the brightness of the corresponding coordinate is read. Next, a point is read in the boundary area, it is called point-2, and the brightness of the corresponding coordinate is read. Half of the difference between the two coordinates is used as the threshold, and the resulting ones are separated into a double collection of pixels (two levels of color), namely bright and dark. If it is bright, it is replaced by "1", and if it is dark, it is replaced by "0".

After fixing the column at point-1, the row with the first 0 is read while shrinking the row. In Figure 5a, the horizontal direction is the column, and the vertical direction is the row. The column number increases as the column moves to the right. On the other hand, the row number increases as the row goes down. Hence, the upward arrow in Figure 5b means that the row becomes smaller. Therefore, the starting point of the boundary can be found by adding 1 to the row where the first 0 occurs. Furthermore, according to the micrograph, voids may occur in the layer or be expressed as black dots (points) due to decidualization. The presence of these dots may confuse the position of the first zero in the row. As a matter of fact, in this study, as can be seen in Figure 5c, all the small "0" surrounded by the wide distribution of "1" were changed to "1". By these measures, the black dots inside each layer in Figure 5b become collectively "1", as depicted in Figure 5c, such that the position of the boundary can be clearly found.

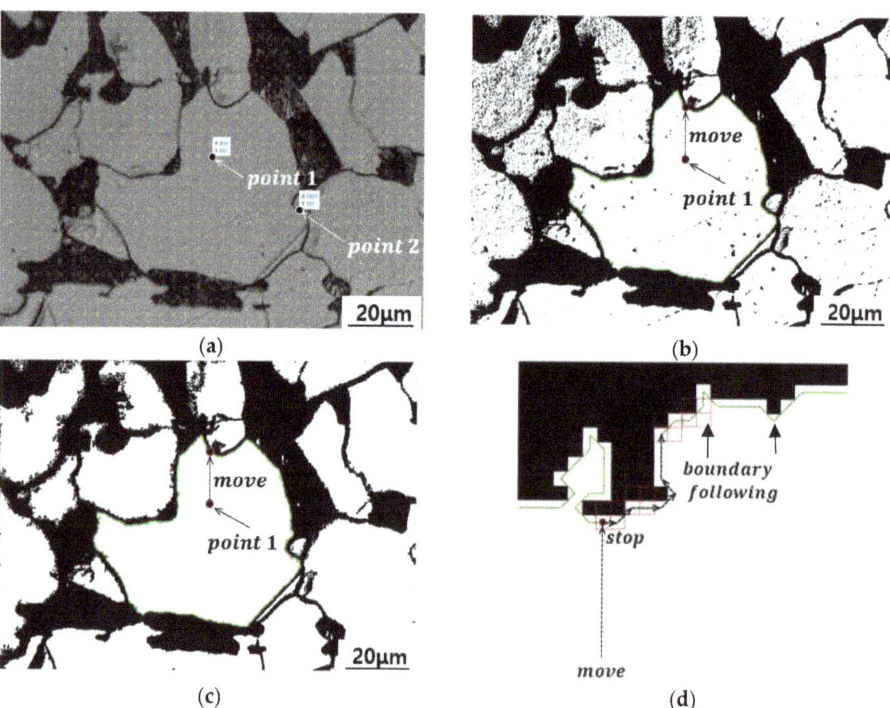

Figure 5. Algorithms for selection, binarization, and boundary extraction. (**a**) Select a structure and background; (**b**) Boundary extraction with voids; (**c**) Boundary extraction without voids; and (**d**) Boundary extraction algorithm.

Figure 5d is a diagram explaining the algorithm for finding the coordinates of the boundary after finding the starting point of the boundary by the algorithm of Figure 5c. We know the positions of the 'first 0' and 'last 1'. The presence or absence of "1" and "0" is checked in four adjacent pixels. There are four cases. Point-1 moves to the upper side and stops at the boundary of "1" and "0", so it is the fixed condition. We can assume four cases, including this fixed condition, right upper, and right side will be 1-1, 0-1, 0-0, 1-0. In the case of 0-1, the boundary is to the right, as shown in the following case of the breakpoint in Figure 5d. In the case of 1-1, the boundary will be right-side up, as shown in the following case of 0-1 in Figure 5d. In the case of 0-0, as shown in Figure 5d, the peak at the following arrows. The MATLAB command boundary is used.

3.3. Nodes and Stacking

When the metallographic microscope used in this paper is set to 500× magnification, the resolution per pixel is approximately 0.063 μm; the micrometer accuracy for measuring the height is 0.01 μm. Consequently, the boundary expressed by the (x,y) coordinates obtained from the MG picture is drawn with a relatively large number of nodes compared to the height z, and higher spatial resolution than necessary. This reduces the computational speed during three-dimensional imaging, as well as forming needle-like surfaces up and down. To overcome this problem, if the number of nodes forming the boundary is reduced, the shape of the layer may be distorted. Accordingly, in this section, we examined how effective it is to reduce the number of nodes forming the boundary of the organization.

Let the pixels of the camera (microscope) be as defined by Equation (12). Generally the ratio of vertical and horizontal pixels is such that $m/n = 4/3$. The spatial resolution of the image, which can be obtained using the microscope, depends on the optical lens system and can be expressed by Equation (13).

$$R_c = m \times n \quad \text{(pixels)} \tag{12}$$

$$R_p = w \times h \quad (\mu\text{m}) \tag{13}$$

where w and h are the width and height of each image. The quantitative distance between each pixel can be derived from Equation (14)

$$P = w/n \quad (\mu\text{m/pixels}) \tag{14}$$

Here we were using $m = 2048$ and obtaining $w = 130$ (μm). Therefore, P would be 130/2048 (μm/pixels), corresponding to 0.063 (μm/pixels). However, as indicated in Table 1, the height of each layer after polishing was 1–2 μm. If we use $P = 0.063$ (μm/pixels), the lateral surface of the crystal will be expressed as shallow needle-like squares, and the surface lacks uniformity, due to the small value of P. In order to reproduce the result well, we defined P' as expressed in Equation (15), with $P' = P \times r$ and $r = 30 \sim 60$.

$$P' = w'/n' \quad (\mu\text{m/pixels}) \tag{15}$$

Figure 6a is a polygonal figure drawn by the boundary coordinates obtained by the above-mentioned algorithm using the picture shown in Figure 4c. There are 3084 polygonal vertices, represented by solid blue lines. Then, the polygons expressed by 1028 vertices skipping three each were expressed by green dot-lines, and the polygons expressed by skipping nine by 343 dots were expressed by circular lines marked in red. A black dot-line is a polygon expressed by 114 dots that are skipped by 27. In the case of a bay with a complex shape located above the structure, it cannot be expressed by the number of vertices being skipped by 27, but the shape of the bay can be expressed by the marked red circular solid line that is expressed by skipping every nine. This means that detailed regions of the structure can be expressed with a spatial resolution of about 0.57 μm. Meanwhile, Figure 6b shows the area of the polygon as a function of the number of vertices of the polygon. The area of the polygon drawn with 114 vertices by skipping 27 is reduced by about 50%; however, when less than 13 are skipped, the area ratio hardly changes. Therefore, in this study, a boundary with a spatial resolution of 0.57 μm was applied by skipping nine each.

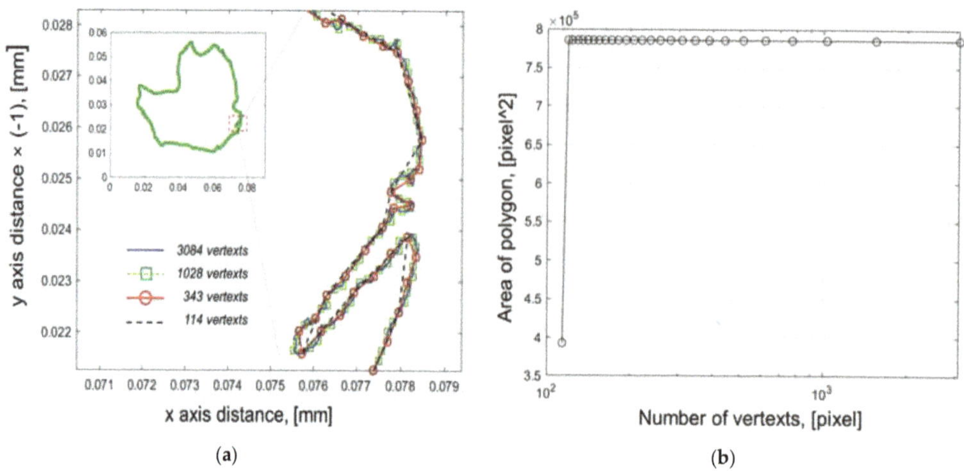

(a) (b)

Figure 6. Changes of shape and area due to reducing the number of nodes which consists of the boundary. (**a**) Shape changes; (**b**) Area changes.

Figure 7a–d shows the boundary with a spatial resolution of 0.57 μm of the microscope's xy section image for the shape of three layers (#1, #2, #3) along the z-axis height, respectively, which is the result of extraction. In the case of #1 and #2, we see an example of a layer that gradually narrows in width as it is polished. If the layer of #1 in Figure 7d becomes smaller, this means that it can be eventually be divided into several layers as in #2. Figure 8a–c is the results of stacking by extracting the coordinates of the layer's boundaries #1, #3, #5, and #6 by height. #1 indicates a shape in which the cross-sectional area increases gradually and then decreases again. #3 decreases gradually, and at the same time, #5 increases gradually, and #6 appears. Figure 8d is a top view, and #2 and #3 partially overlap to show that a part of the organization has been merged. On the plus side, the upper part of #1 represents a phenomenon in which the layer is separate as the polishing proceeds into a state in which the upper left layer is integrated.

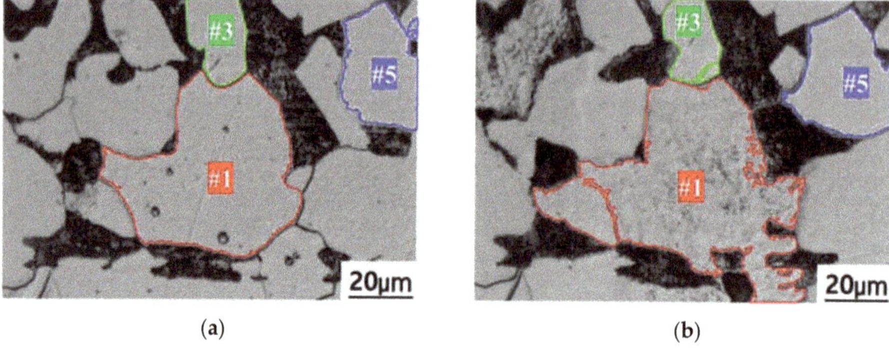

(a) (b)

Figure 7. *Cont.*

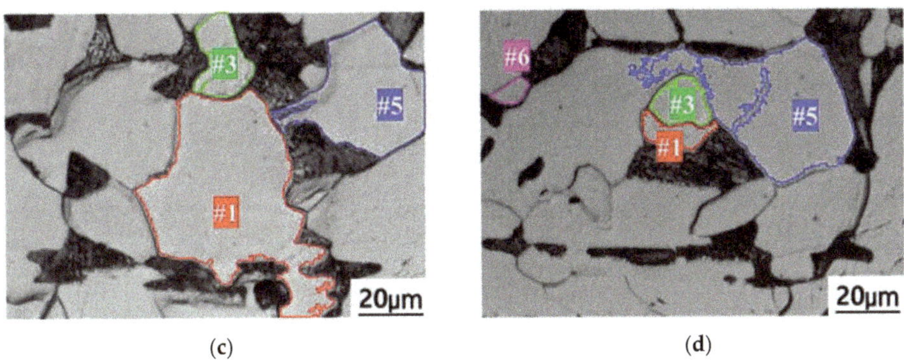

(c) (d)

Figure 7. Boundaries at several layers. (**a**) 3rd layer; (**b**) 7th layer; (**c**) 10th layer; and (**d**) 26th layer.

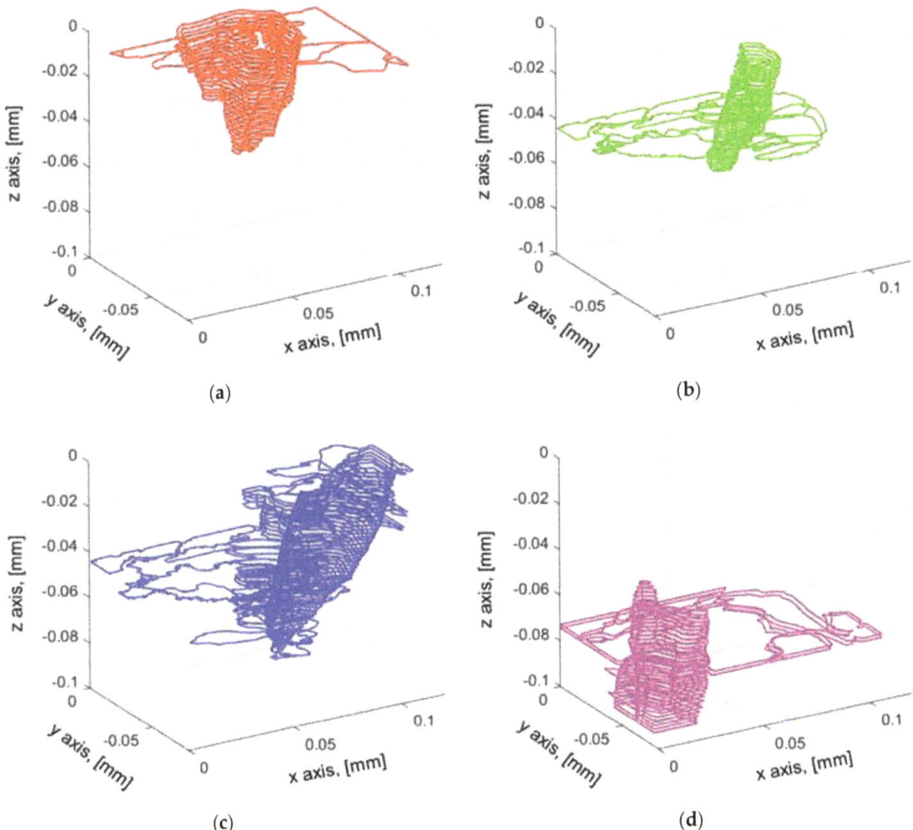

(a) (b)

(c) (d)

Figure 8. *Cont.*

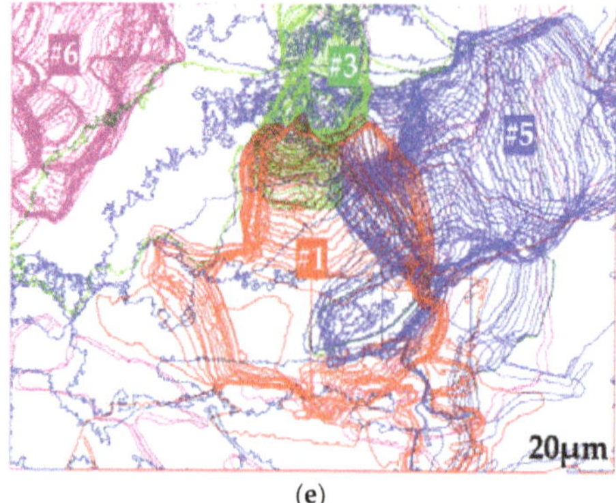

(e)

Figure 8. Stacking boundaries at each height. (**a**) Stacking of #1; (**b**) Stacking of #3; (**c**) Stacking of #5; (**d**) Stacking of #6; and (**e**) Top-view of stacking boundaries with four grains.

3.4. Post-Processing

The above result is the stacking of the boundary in each layer, and to understand it as a volume, it is necessary to form a surface connecting the coordinate point of the boundary and the position of the adjacent height. In this study, the results shown in Figures 9 and 10 was obtained using "alpha Shape", a function that creates a boundary area or volume surrounding a set 2D or 3D point in MATLAB. This 3D graphical function may help to understand the layer or MG by imaging the shape of the layer in 3D.

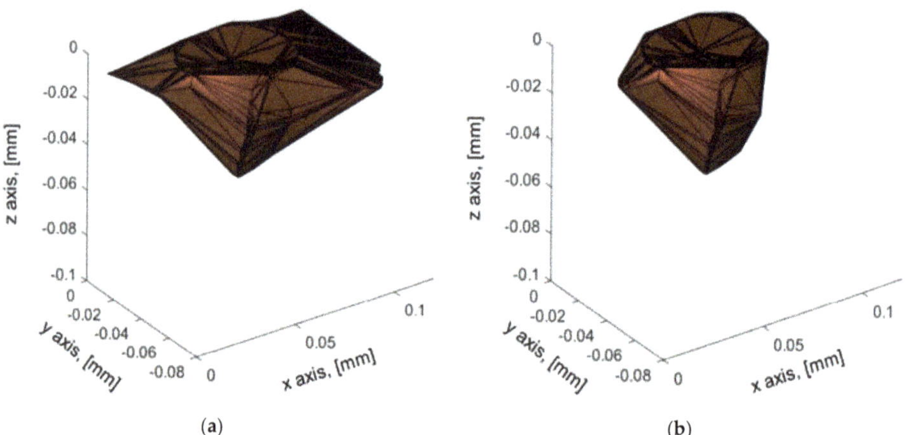

Figure 9. Three-dimensional imaging of microstructure. (**a**) 3D image of #1 including layer-8; (**b**) 3D image of #1 except layer-8.

Due to the large grain selected in layer 8, as shown in Figure 9, it is difficult to understand the original shape of the 3D image. Furthermore, we do not know which data is correct in layer 8, that is to say, whether it is connected to the neighboring grain or not. Therefore, we have to compare four figures, namely the original microscopic image, the extracted boundary image, and two types of stacked volumetric images. The reconstructed 3D grains volume is illustrated in Figure 10.

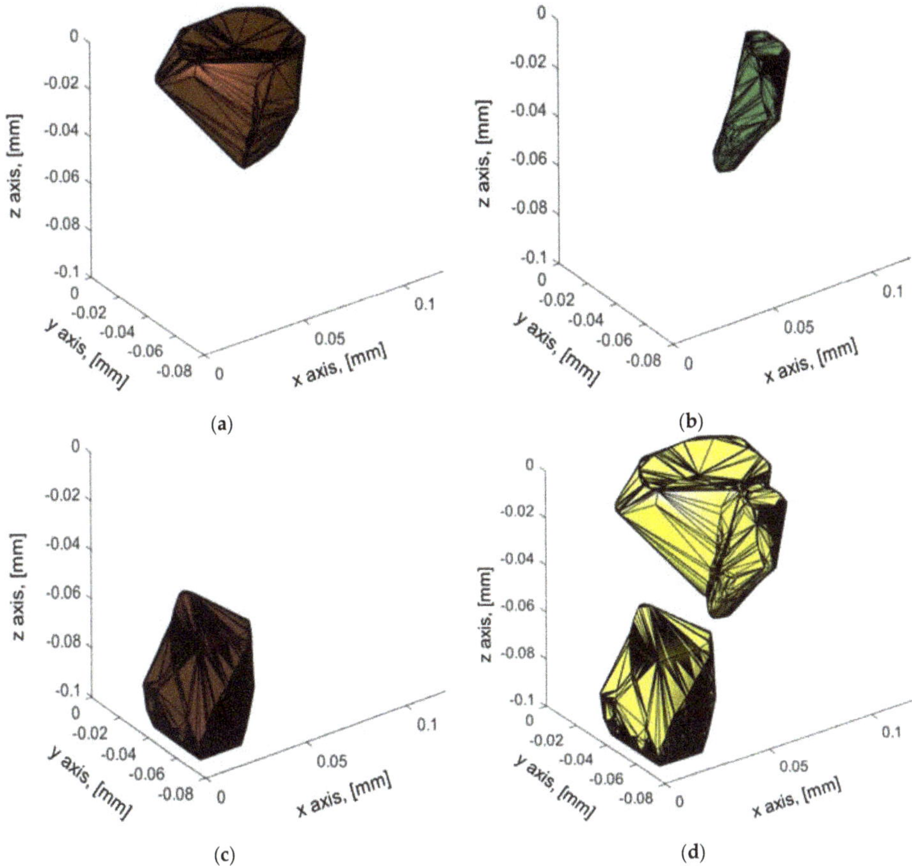

Figure 10. Three-dimensional volumetric imaging of grains with except discontinuous layers. (**a**) 3D image of #1; (**b**) 3D image of #3; (**c**) 3D image of #6; and (**d**) 3D image of three grains.

4. Discussion

In this study, a method and its implementation algorithm for the three-dimensional observation of the microstructure of carbon steel (SA106) was proposed while explaining and illustrating its principle.

The visualization of the shape of the 3D grains with simple equipment, namely a polishing machine, a metallurgical microscope, and some algorithms, is a great advantage. Moreover, it is inexpensive and easy to implement due to its simple technique and does not involve extensive hardware and software.

While the mounted object was being polished, a photograph of the layer was taken using a metallurgical microscope. Subsequently, the metallic MG was reconstructed in three dimensions by assembling and processing the collected 2D images through local positioning, MG selection, binarization, resizing, stacking (superimposing), and surfacing, which provides very accurate data. The assembly and processing techniques are based on mathematical formulae as described above; the program derived from these formulae is straightforward. It is easily accessible and provides a quick result, making it a simple resource to operate and can be implemented in general-purpose software, e.g., MATLAB, C, C++, Python.

As shown in the example in Figure 6, the MG image is designed by a large number of nodes and will require a lot of computational power. To overcome this, skipping is necessary. However, there is a limit, which should not be exceeded, to not compromise the accuracy of the final results. Skipping by nine, in this case, leads to both minimal resources simulation time and accurately obtained results. Organizational photos may contain personal opinions. In other words, the shredded pieces may, in fact, be the tip of a large layer. If this personal subjectivity is expressed in three dimensions, a somewhat objective judgment is possible.

Notwithstanding non-destructive testing techniques, three-dimensional imaging grains will be one of the most widely used techniques in the near future. Firstly, for its numerous advantages listed in the introduction. Another interest is to be able to reconstruct the shape of the microstructure, which may allow the evaluation and prediction of the material's behavior under stress, damage, heat treatment effects, etc., and a variety of research has been carried out in this area [19,44,45].

The reconstructed 3D grains volume, as illustrated in Figure 10, from the processed 2D images is of significant interest and may be useful for our future research in the field of non-destructive evaluation using mainly electromagnetic and ultrasonic testing to better understand structural properties at the microstructural level. In addition, the generation of a three-dimensional grains structure may also broaden the scope of applications, especially in the field of materials science, designing structural materials, and investigation of engineering materials.

In the last few years, thanks to numerical models and the increase in computing and processing capacity, it has become possible to obtain precise models and simulations of complex material properties, especially microstructures, using advanced software and modeling techniques [46–48]. The software algorithms developed under MATLAB employed are complementary to the experimental work, and the combination of the two leads to more accurate and realistic grain representations with a very acceptable simulation time. In addition, the method allows the creation of microstructures with the exact shape and size of the grains and offers the possibility of selecting one or more grains, as illustrated in Figures 7, 8 and 10.

5. Conclusions

Three-dimensional imaging of metallic grain by stacking the microscopic images was successfully applied to obtain 3D grain volume from processed 2D images. This method can be applied as a useful tool for the characterization of metal grains. It is ready to be used in industrial steel applications. It has been pointed out that the method presented here allows quick and easy use. Furthermore, it does not require many experimental resources.

An efficient technique to avoid positional errors in image piling has been established. Owing to the size of a wide number of nodes, an alternate method was adopted and presented without jeopardizing the precision of the results.

The research finding confirmed that the 3D imaging technique would be suitable for the mechanical properties and also very interesting and promising for studying the visualization of 3D magnetic domains, which is the next step of this research.

Algorithms and experimental techniques are constantly being refined to further improve and enhance the reliability and efficiency of the desired 3D microstructures.

Author Contributions: Conceptualization, J.L.; methodology, J.L., A.B. and Y.-H.H.; software, A.B., J.L. and D.W.; validation, A.B. and J.L.; formal analysis, A.B., J.L. and Y.-H.H.; investigation, J.L. and A.B.; resources, J.L. and D.W.; data curation, J.L., A.B. and D.W.; writing—original draft preparation, A.B. and J.L.; writing—review and editing, A.B. and J.L.; visualization, D.W. and J.L.; supervision, J.L.; project administration, J.L.; funding acquisition, J.L. This paper was prepared with the contributions of all authors. All authors have read and agreed to the published version of the manuscript.

Funding: This work was supported by the National Research Foundation of Korea (NRF) grant funded by the Korea government (MSIT) (No.NRF-2019R1A2C2006064).

Institutional Review Board Statement: Not applicable.

Informed Consent Statement: Not applicable.

Data Availability Statement: Not applicable.

Conflicts of Interest: The authors declare no conflict of interest.

References

1. Kumar, S.; Shahi, A. Effect of heat input on the microstructure and mechanical properties of gas tungsten arc welded AISI 304 stainless steel joints. *Mater. Des.* **2011**, *32*, 3617–3623. [CrossRef]
2. Jang, D.; Kim, K.; Kim, H.C.; Jeon, J.B.; Nam, D.G.; Sohn, K.Y.; Kim, B.J. Evaluation of Mechanical Property for Welded Austenitic Stainless Steel 304 by Following Post Weld Heat Treatment. *Korean J. Met. Mater.* **2017**, *55*, 664–670. [CrossRef]
3. Rabung, M.; Kopp, M.; Gasparics, A.; Vértesy, G.; Szenthe, I.; Uytdenhouwen, I.; Szielasko, K. Micromagnetic Characterization of Operation-Induced Damage in Charpy Specimens of RPV Steels. *Appl. Sci.* **2021**, *11*, 2917. [CrossRef]
4. Berkache, A.; Lee, J.; Choe, E. Evaluation of Cracks on the Welding of Austenitic Stainless Steel Using Experimental and Numerical Techniques. *Appl. Sci.* **2021**, *11*, 2182. [CrossRef]
5. Cook, R.D. *Finite Element Modeling for Stress Analysis*, 1st ed.; John Wiley & Sons, Inc.: Hoboken, NJ, USA, 1995.
6. Thomas, D.J. Using Finite Element Analysis to Assess and Prevent the Failure of Safety Critical Structures. *J. Fail. Anal. Prev.* **2016**, *17*, 1–3. [CrossRef]
7. Müzel, S.D.; Bonhin, E.P.; Guimarães, N.M.; Guidi, E.S. Application of the Finite Element Method in the Analysis of Composite Materials: A Review. *Polymers* **2020**, *12*, 818. [CrossRef] [PubMed]
8. Xing, J.; Du, C.; He, X.; Zhao, Z.; Zhang, C.; Li, Y. Finite Element Study on the Impact Resistance of Laminated and Textile Composites. *Polymers* **2019**, *11*, 1798. [CrossRef]
9. Raabe, D.; Sun, B.; Da Silva, A.K.; Gault, B.; Yen, H.-W.; Sedighiani, K.; Sukumar, P.T.; Filho, I.R.S.; Katnagallu, S.; Jägle, E.; et al. Current Challenges and Opportunities in Microstructure-Related Properties of Advanced High-Strength Steels. *Met. Mater. Trans. A* **2020**, *51*, 5517–5586. [CrossRef]
10. Wu, Q.; Miao, W.-S.; Zhang, Y.-D.; Gao, H.-J.; Hui, D. Mechanical properties of nanomaterials: A review. *Nanotechnol. Rev.* **2020**, *9*, 259–273. [CrossRef]
11. Estrada, N.; Oquendo, W.F. Microstructure as a function of the grain size distribution for packings of frictionless disks: Effects of the size span and the shape of the distribution. *Phys. Rev. E* **2017**, *96*, 042907. [CrossRef] [PubMed]
12. Ashcroft, I.A.; Mubashar, A. Numerical Approach: Finite Element Analysis. In *Handbook of Adhesion Technology*; Springer: Berlin/Heidelberg, Germany, 2011; pp. 629–660. [CrossRef]
13. Paknia, A.; Pramanik, A.; Dixit, A.R.; Chattopadhyaya, S. Effect of Size, Content and Shape of Reinforcements on the Behavior of Metal Matrix Composites (MMCs) Under Tension. *J. Mater. Eng. Perform.* **2016**, *25*, 4444–4459. [CrossRef]
14. Takeo, K.; Aoki, Y.; Osada, T.; Nakao, W.; Ozaki, S. Finite Element Analysis of the Size Effect on Ceramic Strength. *Materials* **2019**, *12*, 2885. [CrossRef] [PubMed]
15. Gad, S.; Attia, M.; Hassan, M.; El-Shafei, A. Predictive Computational Model for Damage Behavior of Metal-Matrix Composites Emphasizing the Effect of Particle Size and Volume Fraction. *Materials* **2021**, *14*, 2143. [CrossRef]
16. Ebuchi, T.; Kitasaka, J.; Nagai, T. Non-Destructive Evaluation of Weld Structure Using Ultrasonic Imaging Technique. *Mater. Trans.* **2012**, *53*, 604–609. [CrossRef]
17. Savin, A.; Craus, M.L.; Bruma, A.; Novy, F.; Malo, S.; Chlada, M.; Steigmann, R.; Vizureanu, P.; Harnois, C.; Turchenko, V.; et al. Microstructural Analysis and Mechanical Properties of $TiMo_{20}Zr_7Ta_{15}Si_x$ Alloys as Biomaterials. *Materials* **2020**, *13*, 4808. [CrossRef]
18. International Atomic Energy Agency. *Non-Destructive Testing: A Guidebook for Industrial Management and Quality Control Personnel*; International Atomic Energy Agency (IAEA): Vienna, Austria, 1999; p. 296.
19. Ono, K. A Comprehensive Report on Ultrasonic Attenuation of Engineering Materials, Including Metals, Ceramics, Polymers, Fiber-Reinforced Composites, Wood, and Rocks. *Appl. Sci.* **2020**, *10*, 2230. [CrossRef]
20. Masumura, R.; Hazzledine, P.; Pande, C. Yield stress of fine grained materials. *Acta Mater.* **1998**, *46*, 4527–4534. [CrossRef]
21. Kosoń-Schab, A.; Szpytko, J. Magnetic Metal Memory in the Assessment of the Technical Condition of Crane Girders for the Needs of Safety. *J. Konbin* **2019**, *49*, 49–73. [CrossRef]
22. Kosoń-Schab, A.; Smoczek, J.; Szpytko, J. Magnetic Memory Inspection of an Overhead Crane Girder—Experimental Verification. *J. KONES* **2019**, *26*, 69–76. [CrossRef]
23. Murakami, K.; Ishimoto, K.; Senoo, T.; Ishikawa, M. Human Robot Hand Interaction with Plastic Deformation Control. *Robotics* **2020**, *9*, 73. [CrossRef]
24. Rogovoy, A.A.; Stolbov, O.V.; Stolbova, O.S. The Microstructural Model of the Ferromagnetic Material Behavior in an External Magnetic Field. *Magnetochemistry* **2021**, *7*, 7. [CrossRef]
25. Feinberg, B.; Gould, H. Placed in a steady magnetic field, the flux density inside a permalloy-shielded volume decreases over hours and days. *AIP Adv.* **2018**, *8*, 035303. [CrossRef]
26. Cullity, B.D.; Graham, C.D. *Introduction to Magnetic Materials*, 2nd ed.; John Wiley & Sons: Hoboken, NJ, USA, 2011; p. 568.

27. Nestleroth, J.; Davis, R.J. Application of eddy currents induced by permanent magnets for pipeline inspection. *NDT E Int.* **2007**, *40*, 77–84. [CrossRef]
28. Xie, Y.; Li, J.; Tao, Y.; Wang, S.; Yin, W.; Xu, L. Edge Effect Analysis and Edge Defect Detection of Titanium Alloy Based on Eddy Current Testing. *Appl. Sci.* **2020**, *10*, 8796. [CrossRef]
29. Grishina, D.A.; Harteveld, C.A.M.; Pacureanu, A.; Devashish, D.; Lagendijk, A.; Cloetens, P.; Vos, W.L. X-ray Imaging of Functional Three-Dimensional Nanostructures on Massive Substrates. *ACS Nano* **2019**, *13*, 13932–13939. [CrossRef] [PubMed]
30. Yan, H.; Voorhees, P.W.; Xin, H.L. Nanoscale x-ray and electron tomography. *MRS Bull.* **2020**, *45*, 264–271. [CrossRef]
31. Siodlak, D.; Lotter, U.; Kawalla, R.; Schwich, V. Modelling of the Mechanical Properties of Low Alloyed Multiphase Steels with Retained Austenite Taking into Account Strain-Induced Transformation. *Steel Res. Int.* **2008**, *79*, 776–783. [CrossRef]
32. Kim, J.; Le, M.; Park, J.; Seo, H.; Jung, G.; Lee, J. Measurement of residual stress using linearly integrated GMR sensor arrays. *J. Mech. Sci. Technol.* **2018**, *32*, 623–630. [CrossRef]
33. Suzuki, M.; Kim, K.-J.; Kim, S.; Yoshikawa, H.; Tono, T.; Yamada, K.T.; Taniguchi, T.; Mizuno, H.; Oda, K.; Ishibashi, M.; et al. Three-dimensional visualization of magnetic domain structure with strong uniaxial anisotropy via scanning hard X-ray microtomography. *Appl. Phys. Express* **2018**, *11*, 36601. [CrossRef]
34. Kent, A.D.; Yu, J.; Rüdiger, U.; Parkin, S. Domain wall resistivity in epitaxial thin film microstructures. *J. Phys. Condens. Matter* **2001**, *13*, R461–R488. [CrossRef]
35. D'Silva, G.J.; Feigenbaum, H.P.; Ciocanel, C. Visualization of Magnetic Domains and Magnetization Vectors in Magnetic Shape Memory Alloys Under Magneto-Mechanical Loading. *Shape Mem. Superelasticity* **2020**, *6*, 67–88. [CrossRef]
36. Lebensohn, R.A.; Kanjarla, A.K.; Eisenlohr, P. An elasto-viscoplastic formulation based on fast Fourier transforms for the prediction of micromechanical fields in polycrystalline materials. *Int. J. Plast.* **2012**, *32–33*, 59–69. [CrossRef]
37. Knezevic, M.; Drach, B.; Ardeljan, M.; Beyerlein, I.J. Three dimensional predictions of grain scale plasticity and grain boundaries using crystal plasticity finite element models. *Comput. Methods Appl. Mech. Eng.* **2014**, *277*, 239–259. [CrossRef]
38. Tasan, C.; Hoefnagels, J.; Diehl, M.; Yan, D.; Roters, F.; Raabe, D. Strain localization and damage in dual phase steels investigated by coupled in-situ deformation experiments and crystal plasticity simulations. *Int. J. Plast.* **2014**, *63*, 198–210. [CrossRef]
39. Ardeljan, M.; Knezevic, M.; Nizolek, T.; Beyerlein, I.J.; Mara, N.; Pollock, T.M. A study of microstructure-driven strain localizations in two-phase polycrystalline HCP/BCC composites using a multi-scale model. *Int. J. Plast.* **2015**, *74*, 35–57. [CrossRef]
40. Zhang, D.; Prasad, A.; Bermingham, M.J.; Todaro, C.J.; Benoit, M.J.; Patel, M.N.; Qiu, D.; StJohn, D.H.; Qian, M.; Easton, M.A. Grain Refinement of Alloys in Fusion-Based Additive Manufacturing Processes. *Met. Mater. Trans. A* **2020**, *51*, 4341–4359. [CrossRef]
41. Todaro, C.J.; Easton, M.A.; Qiu, D.; Zhang, D.; Bermingham, M.J.; Lui, E.W.; Brandt, M.; StJohn, D.H.; Qian, M. Grain structure control during metal 3D printing by high-intensity ultrasound. *Nat. Commun.* **2020**, *11*, 1–9. [CrossRef] [PubMed]
42. Qayyum, F.; Guk, S.; Kawalla, R.; Prahl, U. On Attempting to Create a Virtual Laboratory for Application-Oriented Microstructural Optimization of Multi-Phase Materials. *Appl. Sci.* **2021**, *11*, 1506. [CrossRef]
43. Darbandi, P.; Bieler, T.R.; Pourboghrat, F.; Lee, T.-K. The Effect of Cooling Rate on Grain Orientation and Misorientation Microstructure of SAC105 Solder Joints Before and After Impact Drop Tests. *J. Electron. Mater.* **2014**, *43*, 2521–2529. [CrossRef]
44. Liu, W.; Lianbc, J.; Aravasde, N.; Münstermanna, S. A strategy for synthetic microstructure generation and crystal plasticity parameter calibration of fine-grain-structured dual-phase steel. *Int. J. Plast.* **2020**, *126*, 102614. [CrossRef]
45. Rolchigo, M.R.; Mendoza, M.Y.; Samimi, P.; Brice, D.A.; Martin, B.; Collins, P.C.; Lesar, R. Modeling of Ti-W Solidification Microstructures Under Additive Manufacturing Conditions. *Met. Mater. Trans. A* **2017**, *48*, 3606–3622. [CrossRef]
46. Mendoza, M.Y.; Samimi, P.; Brice, D.A.; Martin, B.W.; Rolchigo, M.R.; Lesar, R.; Collins, P.C. Microstructures and Grain Refinement of Additive-Manufactured Ti-xW Alloys. *Met. Mater. Trans. A* **2017**, *48*, 3594–3605. [CrossRef]
47. Wong, S.L.; Madivala, M.; Prahl, U.; Roters, F.; Raabe, D. A crystal plasticity model for twinning- and transformation-induced plasticity. *Acta Mater.* **2016**, *118*, 140–151. [CrossRef]
48. Qayyum, F.; Guk, S.; Prüger, S.; Schmidtchen, M.; Saenko, I.; Kiefer, B.; Kawalla, R.; Prahl, U. Investigating the local deformation and transformation behavior of sintered X3CrMnNi16-7-6 TRIP steel using a calibrated crystal plasticity-based numerical simulation model. *Int. J. Mater. Res.* **2020**, *111*, 392–404. [CrossRef]

Article

THz-TDS Techniques of Thickness Measurements in Thin Shim Stock Films and Composite Materials

Kwang-Hee Im [1,*], Sun-Kyu Kim [2], Young-Tae Cho [3], Yong-Deuck Woo [1] and Chien-Ping Chiou [4]

1 Department of Automotive Engineering, Woosuk University, Wanju-kun, Jeonbuk 55338, Korea; wooyongd@woosuk.ac.kr
2 Division of Mechanical System Engineering, Jeonbuk National University, Jeonju, Jeonbuk 54896, Korea; sunkkim@jbnu.ac.kr
3 Department of Basic Science, Jeonju University, Jeonju, Jeonbuk 55069, Korea; choyt@jj.ac.kr
4 Aerospace Engineering and Center for Nondestructive Evaluation, Iowa State University, Ames, IA 50011, USA; cchiou@iastate.edu
* Correspondence: khim@woosuk.ac.kr; Tel.: +82-63-290-1473

Abstract: Terahertz wave (T-ray) scanning applications are one of the most promising tools for nondestructive evaluation. T-ray scanning applications use a T-ray technique to measure the thickness of both thin Shim stock films and GFRP (glass fiber-reinforced plastics) composites, of which the samples were selected because the T-ray method could penetrate the non-conducting samples. Notably, this method is nondestructive, making it useful for analyzing the characteristics of the materials. Thus, the T-ray thickness measurement can be found for both non-conducting Shim stock films and GFRP composites. In this work, a characterization procedure was conducted to analyze electromagnetic properties, such as the refractive index. The obtained estimates of the properties are in good agreement with the known data for poly methyl methacrylate (PMMA) for acquiring the refractive index. The T-ray technique was developed to measure the thickness of the thin Shim stock films and the GFRP composites. Our tests obtained good results on the thickness of the standard film samples, with the different thicknesses ranging from around 120 μm to 500 μm. In this study, the T-ray method was based on the reflection mode measurement, and the time-of-flight (TOF) and resonance frequencies were utilized to acquire the thickness measurements of the films and GFRP composites. The results showed that the thickness of the samples of frequency matched those obtained directly by time-of-flight (TOF) methods.

Keywords: terahertz waves; refractive index; thickness measurement; Shim stock films; composite materials; reflection mode

Citation: Im, K.-H.; Kim, S.-K.; Cho, Y.-T.; Woo, Y.-D.; Chiou, C.-P. THz-TDS Techniques of Thickness Measurements in Thin Shim Stock Films and Composite Materials. *Appl. Sci.* **2021**, *11*, 8889. https://doi.org/10.3390/app11198889

Academic Editors: Jinyi Lee, Hoyong Lee and Azouaou Berkache

Received: 3 August 2021
Accepted: 1 September 2021
Published: 24 September 2021

1. Introduction

Terahertz waves (T-ray) have recently been utilized for technical applications [1]. Along with the recent progress of T-ray technology and monitoring instruments, defect inspection methods have emerged based on the electronic spectrum. Moreover, the T-ray has a relatively higher resolution. In addition, the T-ray has led to advanced progress for spectroscopic monitoring in security areas, food inspection, water, the mechanical field, and materials. Terahertz time domain spectroscopy (THz-TDS) has been utilized to inspect various delamination or foreign materials in advanced non-contact composites. THz-TDS is based on photoconductivity, and this depends on low-cycle formations with the utilization of a photoconductive antenna (Femtosecond (10–15 s) laser) [2].

It is possible to create THz waves in less than a pico-second. Therefore, detection techniques using a high signal-to-noise (S/N) ratio are available, which affects the broad bandwidth. A temporary change in the T-ray emitter occurs due to the resistance of the photoconductive switch on the T-ray timescale [3,4]. In addition, another method, known as optical anisotropic conversion or optical mixing, can be utilized along with two continuous wave (CW) lasers [5]. When these two lasers are mixed, a beating is generated, and this beating can modulate the conductance of the photoconductive switch using the terahertz differential frequency [6]. Continuous-wave terahertz (CW-THz) can be obtained using this method. In some cases, a T-ray image can also show the chemical components of a target material [7]. Owing to these characteristics, the T-ray image has attracted significant attention. The T-ray image has commercial applications in various fields, including humidity analysis, quality management of plastic products, and packing inspection (monitoring) [8–10].

Owing to its broad utilization and far-ranging applications, the THz-TDS techniques could have the possibility to become a portable THz image. This approach is composed of two sections, which both involve the use of the T-ray. First, the importance of fiber-reinforced plastics (FRP) in the space and civil aviation fields is generally well known, and the FRP-laminated plate is widely used. In addition, the waveforms of terahertz pulses in the TDS mode have a strong resemblance to those of ultrasonic tastings. Regarding wave propagation concepts such as time of flight (TOF), transmission and reflection coefficients, refraction and diffraction are common to both waves. However, there are also fundamental differences when materials are probed with terahertz radiation, an electromagnetic wave, and with ultrasound, a mechanical wave [11]. In order to measure the thickness of a specimen using conventional ultrasonic waves, a couplant medium is always required, which makes the ultrasonic waves easily propagated. In the case of using air as a couplant medium, selecting ultrasonic frequency is narrowly ranged; thus, there is a limitation to measuring thinner samples. Therefore, due to the couplant medium, the factors affecting the accuracy of the measurements should be considered such as attenuation, diffraction, and dispersion of the samples [12]. By the way, the terahertz wave used in this study requires no couplant medium and is utilized under the mode of noncontact. Thus, the terahertz wave could make better reproducibility of data produced and also a higher frequency could be selected, which could bring the stronger measurement of thickness in case of thin samples.

The other is composed of the refractive index (n), the electrical conductivity of fiber-embedded epoxy matrix composite material, and the measurement of T-ray thickness for both glass fiber-reinforced plastics (GFRP) and thin Shim stock films, which are produced as a standard sample with an arbitrary thickness (ranging from tens of μm to hundreds of μm) [13]. Thus, the thicknesses for both GFRP and thin Shim stock films are measured using T-ray technology. Carbon fiber-reinforced plastics (CFRP) are conductive, but epoxy matrix is non-conductive [14,15]. However, the carbon fiber of the CFRP-laminated plate has conductivity, enabling the T-ray characteristics evaluation of glass fiber and carbon fibers [16].

In this study, the results of the experiment on the T-ray were obtained based on the non-destructive evaluation methods using FRP composite materials. In addition, the correlation was performed between the fiber direction and the E-field of the GFRP composites and the CFRP composite-laminated plate according to the refractive index measurement technique, which shows the properties of various materials and the existence of conductivity. A new numerical method of measurement of refractive indexes in reflection and transmission modes was proposed. In addition, we performed a fundamental experimentation and brought a simple testing procedure for acquiring the thickness of samples as an existing NDE method. Here, the measured thickness and the reference thickness of the Shim stock films, which had a standard reference thickness, were compared. In addition, the thicknesses of the GFRP composites with non-conductivity were measured.

Therefore, a difference in the time-of-flight (TOF) was utilized to measure the thickness of the GFRP composites using the T-ray. The effectiveness of a T-ray examination was successfully evaluated by comparing and reviewing the specimens using the resonance frequency.

2. Fundamental Theory

2.1. Measurement of Refractive Index

Using the refractive index measurement technique, the reflection mode was applied in the time domain of the T-ray, and the refractive index was induced by picking up a signal reflected through the specimen. The progress direction of the T-ray signal is shown in Figure 1. Here, T is the transmitter of T-ray and R is the receiver of T-ray.

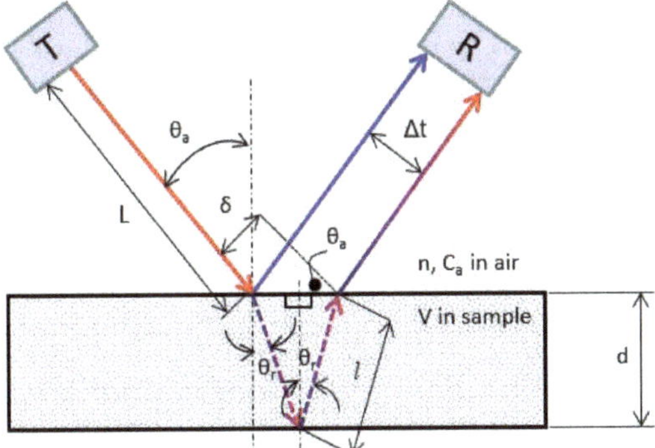

Figure 1. Diagram showing the geometry of the reflection mode [2].

The refractive index can be obtained by calculating the time of the T-ray, which is reflected from the terahertz pulsed emitter's arrival at the pulsed receiver and the time-of-flight (TOF) when the T-ray passes through a specimen of a certain thickness [1].

This reflection mode obtains the refractive index by calculating each length of optical fiber reflected on the top and bottom of the specimen in the T-ray time-of-flight (TOF). Figure 1 shows the shape and path of T-ray. At first, if it is assumed that T-ray is projected on the specimen vertically, a time difference (Δt) can be obtained as follows:

$$\Delta t = \frac{2d}{v} \tag{1}$$

In consideration of the path of the oblique T-ray and the shape delay time in the reflection mode, as shown in Figure 1, a time difference (Δt) between the surface-reflected wave and back-reflected wave on the specimen can be obtained as follows:

$$\Delta t = \frac{2l}{v} - \frac{\delta}{C_a} \tag{2}$$

Here, $l = \frac{d}{cos\theta_r}$, $\delta = 2l sin^2\theta_a = 2\frac{d}{cos\theta_r} sin^2\theta_a$, C_a is the velocity in air, d is the sample thickness, v is the sample velocity, θ_a is the angle of inclination in the reflection mode, θ_r is the refractive angle in the sample, and n is the refractive index. When the shape delay time and the path of oblique T-ray are traced, both the time difference (Δt) and resonance frequency (Δf) can be expressed as follows [13]:

$$\Delta t = \left(\frac{2d}{v\ cos\theta_r} - \frac{\delta}{C_a}\right) = \frac{2d}{cos\theta_r}\left(\frac{1}{v} - \frac{sin^2\theta_a}{C_a}\right) \tag{3}$$

$$\Delta f = \frac{1}{\left(\frac{2d}{v\ cos\theta_r} - \frac{\delta}{C_a}\right)} = \frac{1}{\frac{2d}{cos\theta_r}\left(\frac{1}{v} - \frac{sin^2\theta_a}{C_a}\right)} \tag{4}$$

Here, l is the refracted length in the sample, d is the thickness of specimen, v is the velocity in the specimen, C_a is the velocity in the air, l is the refracted length in the sample, δ is the skip length of refractive waves in the sample, θ_r is the refraction angle in the specimen, and θ_a is the refraction angle in the air. The refractive index, which is one of the electromagnetic properties, can be calculated by following the steps above.

The refractive index can be obtained with the approximate solution as follows:

$$n^4 - An^4 - Asin^2\theta_{p1} = 0 \quad (5)$$

where d is the sample thickness, V_{air} is the light speed in air, and V_s is the light speed in the sample. $\Delta t\ (T)$ is the difference time between with sample and without sample, and $A = \frac{T^2V_{air}^2}{4d_2^2}$. Here, assuming that the normal reflection mode is vertical on the sample, the refractive index (n) should be $v\Delta t/2d$. However, this reflection mode was composed with some angles. Therefore, a correction factor needs to be considered to obtain a better solution, as shown in Equation (5).

2.2. Measurement of Refractive Index

In through-transmission mode, the index of refraction (n) can be calculated using the following equation, according to [2]:

$$\therefore n = 1 + \frac{\Delta t\ v_{air}}{d} \quad (6)$$

where Δt is the time cap between with sample and without sample, d is the sample thickness, V_{air} is the light speed in air, and L is the distance between the pulsed emitter and pulsed receiver.

3. Experiment System and Measurement

3.1. Measurement System

Figure 2 shows a photo of the THz-TDS system, which is a non-destructive testing device. This system is used to collect the material characteristics and scan the image of the specimen. The T-ray system used in this study was produced by TeraView Ltd. Cambridge in the United Kingdom. This system was composed of the time domain spectroscopy (TDS) pulse device and the frequency domain continuous wave (CW) device. It was composed of TDS technologies for generating, controlling, and searching terahertz pulses. The THz-TDS system can obtain an image and improve data acquisition, and its unique structural characteristics for manipulating the T-ray have a direct influence on the image production experiment. This TDS system had a frequency range from 50 GHz to 4 THz, and the delay time reached up to 300 ps. The T-ray beam was concentrated on the focal distances of 50 mm and 150 mm, and the full width at half-maximum (FWHM) values were 0.8 mm and 2.5 mm, respectively. This TDS device can be set for measuring the penetration or reflection (small-angle pitch-catch). The frequency range of the CW device was between 50 GHz and 1.5 THz. The focal distances of the CW device were also 50 mm and 150 mm. The TDS and CW devices were connected to each other through the optical fiber. Figure 3 shows the schematic diagram of the T-ray system.

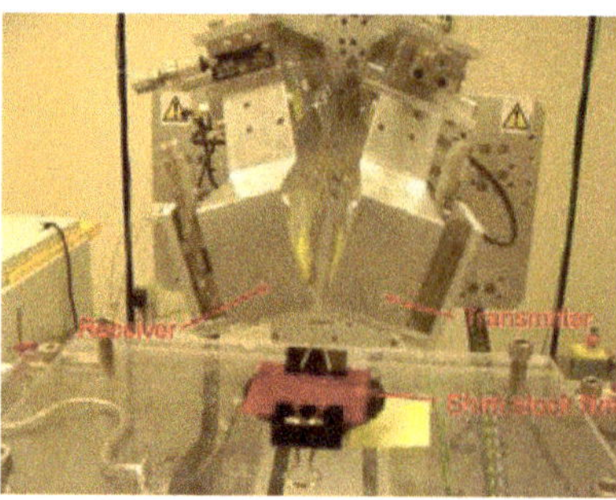

Figure 2. A photo of the THz-TDS system for imaging and measuring material parameters.

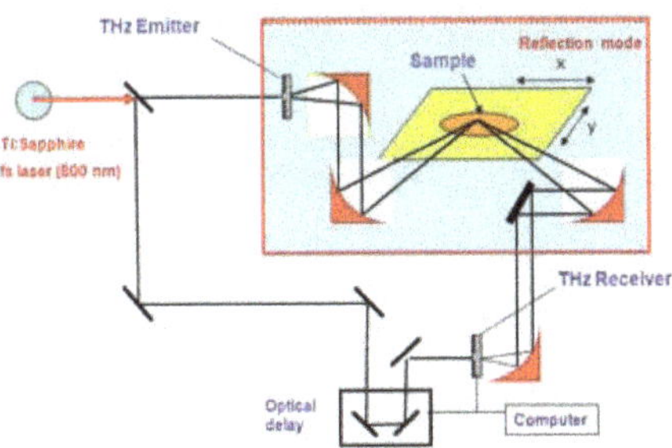

Figure 3. Overview of the THz measurement method [2].

3.2. Measurement Method

Figure 3 exhibits the T-ray measurement system, which demonstrates the reflection mode. When the test was carried out in this system, the T-ray was created from the emitter and sent to the receiver. At this time, the test was carried out by matching the focal point of the emitter and the receiver with the desired specimen. Then, the angle of inclination of the T-ray lens was determined as 16.6°. Figure 4 shows the Shim stock films and GFRP composites. The thicknesses of the Shim stock films were 0.127 mm, 0.254 mm, 0.381 mm, 0.508 mm, and 0.762 mm, and the thicknesses of GFRP composites were 2.02 mm, 3.08 mm, 5.74 mm, and 5.92 mm, respectively. Figure 5 shows typical A-scan data, which is the reflection mode of the GFRP composites of the T-ray. The thickness of the specimen was 3.0 mm.

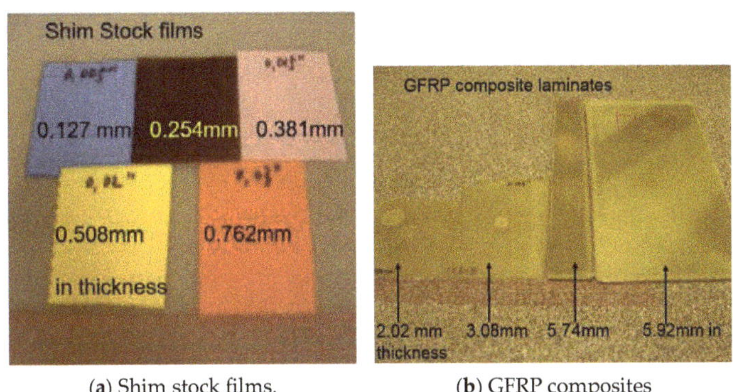

(**a**) Shim stock films. (**b**) GFRP composites

Figure 4. Samples of (**a**) Shim Stock films and (**b**) GFRP composites with various thicknesses.

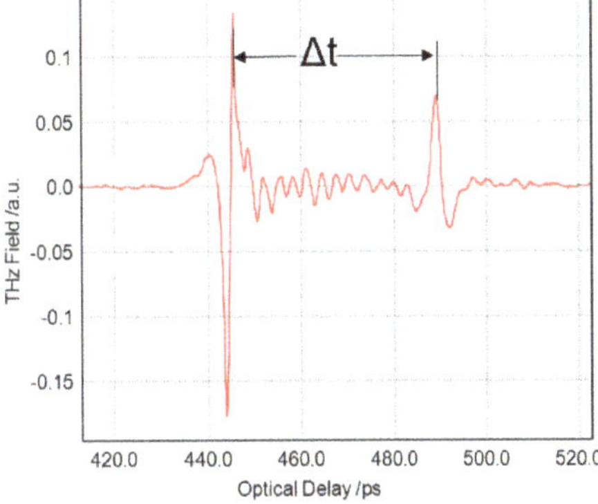

Figure 5. T−ray pulses from the transmitted GFRP composites (n = 2.13 Δt = 42.7 ps, t = 3.0 mm).

4. Results and Discussion

4.1. Measurement of Terahertz Refractive Index

To measure the T-ray parameter which indicated the material properties, the THz pulse was obtained from the Shim stock films and GFRP composites in the reflection mode. Figure 5 clearly shows the time difference (Δt) between the surface and the back of the GFRP composites in the reflection mode. GFRP was used for the specimen, and the thickness of specimen was approximately 3.0 mm. The time difference (Δt), which was obtained according to the thickness of the specimen, was 42 ps. Therefore, the optical time difference was calculated using the reflection mode, which is a measurement technique used to obtain the refractive index. The optimal time difference was calculated using Equation (4). In addition, the Shim stock films, GFRP composites, PMMA, and fused quartz specimens were measured under the reflection mode method, as shown in Table 1. When the results were compared with those from the previous references, only a difference within ±1% was found [1,6].

Table 1. Averaged THz refractive indices of the individually studied materials.

Materials	Refractive Index (n) *	Refractive Index (n)
		Reflection Mode
PMMA	1.60 ± 0.08	1.59 ± 0.07
Shim Stock films	-	1.52 ± 0.03
Fused quartz	1.95 ± 0.05	1.94 ± 0.09
GFRP	-	2.17 ± 0.05

* Data in References [6,11].

Here, the reflection mode measurement techniques of the terahertz were performed in one direction, and experiments were carried out considering various aspects. In addition, since the measurement methods and the characteristics of the GFRP composites and Shim stock films were different, it was difficult to compare them with the previous data.

4.2. Electric Field Evaluation of Carbon Fiber

Unlike non-conductive materials, the T-ray has limited penetrating power against conductive materials [17]. At first, the test was carried out by applying the T-ray GFRP composites composed of non-conductive materials and the CFRP composites composed of conductive materials partially. The CFRP composites are composed of carbon fiber with conductive and non-conductive resin. When the CFRP-laminated composite plate is observed with a microscope, it is composed of various fibers and resins that could affect conductivity significantly, so the quantitative characteristic evaluation of carbon fiber composite material of T-ray is necessary. According to the previous reference, the radial conductivity of carbon fiber is approximately three times larger in the case of the electrical conductivity on the carbon fiber axis. The CFRP composites are composed of unidirectional composites, and the conductivity of the CFRP laminated plate composed of various lamination layers is affected. A transverse (vertical to the fiber axis) conductive generator depends on the fiber contact that occurs between adjacent fibers. Studies regarding the electrical conductivity of carbon fiber composite material are scarce. In some references, researchers have reported that the value of longitudinal conductivity (σl) ranges between 1×10^4 s/m and 6×10^4 s/m, and the value of transverse conductivity (σt) ranges between approximately 2 s/m and 600 s/m, which is much wider [18].

The transverse conductivity value of the laminated plate using the unidirectional Prepreg sheet varies significantly according to the production process and the quality of the laminated plate. The plane conductivity on the flowing current, while forming the θ angle with the fiber axis in the unidirectional CFRP composites, is given as follows [19]:

$$\sigma = \sigma l cos^2 cos\theta + \sigma t sin^2\theta \quad (7)$$

Since it is significantly higher than the longitudinal conductivity of the fiber (σl >> σt), the T-ray which penetrates the unidirectional CFRP composites significantly varies according to the angle between the electric field vector and the axis of the carbon fiber. When the electric field of the T-ray is parallel to the axial direction of carbon fiber, the conductivity becomes the largest and the penetrating power becomes the smallest. On the contrary, when the electric field vector is perpendicular to the axis of fiber, the conductivity becomes the smallest and the penetrating power becomes the largest. The surface depth of the unidirectional oriented CFRP composites on the T-ray using the value of 10 s/m is 0.2 mm in 1 THz and 0.5 mm in 0.1 THz when the direction of electric field is vertical to the fiber axis. The effect of the penetrating power on the angle in the 24-ply unidirectional CFRP composite-laminated plate was experimentally evaluated using the CW terahertz device.

Figure 6 exhibits the amplitude profile of the penetrating power of both the GFRP and CFRP composites by the function of angle under the frequency of 0.1 THz. The amplitude profile of power was obtained, with values ranging from 0° to 90° for both the GFRP composites and CFRP composites. Notably, in the case of the GFRP composites, there was no change in the amplitude profile. However, the CFRP composites showed a higher amplitude of penetrating power at 90°, although they showed almost no amplitude of penetrating power at 0°. When the measurement was made, the GFRP composites were not dependent on any angle, but the CFRP composites were dependent on the angle of the carbon fibers.

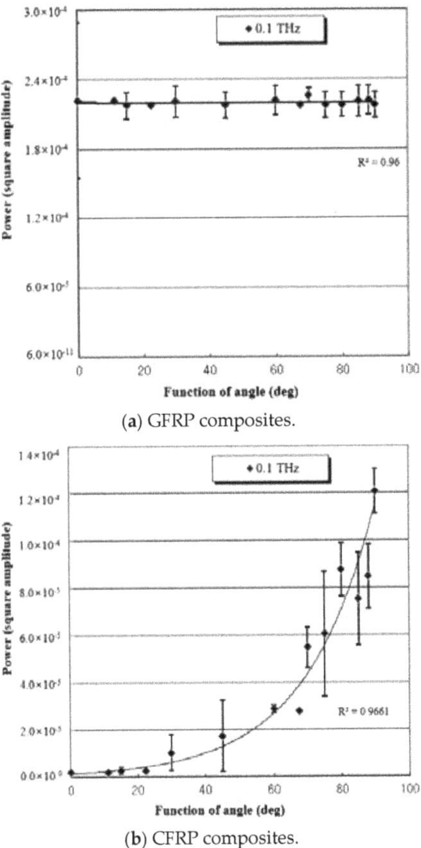

(a) GFRP composites.

(b) CFRP composites.

Figure 6. Amplitude profile of the penetrating power of both the GFRP and CFRP composites by the function of angle (a 24−ply glass and carbon composites).

4.3. Measurement of Thickness Using the Reflection Mode

The THz-TDS reflection mode was applied to measure the thickness for both the Shim stock films and GFRP composites with the one-side direction. Figure 7 exhibits the T-ray scan images of the thin Shim stock films. The thicknesses of the thin Shim stock films were 0.127 mm, 0.254 mm, 0.381 mm, 0.508 mm, and 0.762 mm. The values of the thicknesses were utilized as the standard samples of the films. Figure 7a exhibits the difference (Δt) in the time-of-flight (TOF), which indicates the difference between the surface and the back of the Shim stock films. Figure 7b represents Figure 7a as the FFT domain, and Δf refers to the resonance frequency, which is correlated with the thickness of the thin Shim stock films. Here, Δt is the difference time in the time-of-flight (TOF). Namely, $1/\Delta t$ should be Δf. Here, the example thickness in the thin Shim stock film was 0.381 mm.

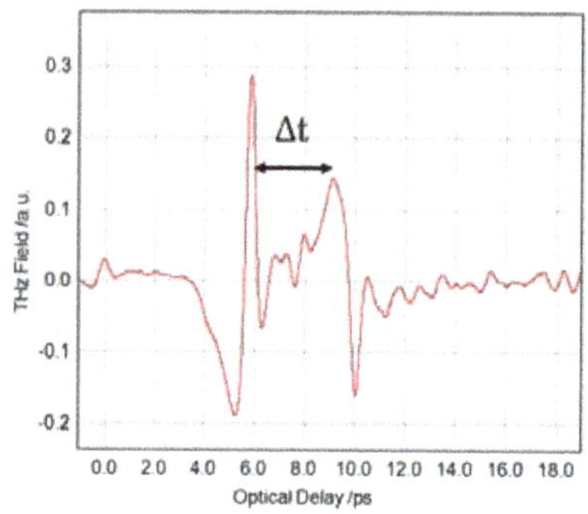

(**a**) A-scan image.

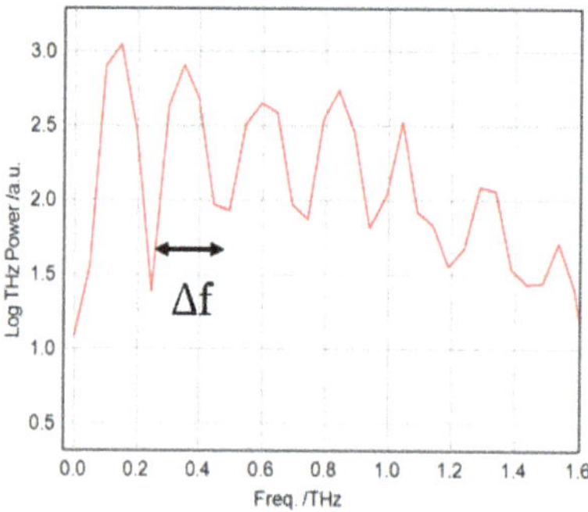

(**b**) Frequency-domain signal.

Figure 7. A TOF and FFT image of the thin Shim stock films under the reflection mode (0.381 mm in thickness).

Table 2 shows the comparison of the TOF difference (Δt) of the thin Shim stock films, resonance frequency (Δf), and T-ray measurement and reference thickness. Figure 8 shows the data after the T-ray scanning GFRP composites. The thicknesses of the GFRP composites were 2.02 mm, 3.08 mm, 5.74 mm, and 5.92 mm. The values of the thicknesses were used as the standard thickness of the samples.

Figure 8a shows the difference (Δt) in the time-of-flight (TOF), which indicates the difference between the surface and the back of the GFRP composites. Figure 8b represents Figure 8a as the FFT domain, and Δf exhibits the resonance frequency, which is related to the thickness of the GFRP composites. Here, Δt is the TOF difference. Namely, $1/\Delta t$ should be Δf. The thickness of the GFRP composites was 2.02 mm.

Table 3 shows the comparison of the TOF difference (Δt) of the GFRP composites, resonance frequency (Δf), T-ray measurement, and reference thickness.

Table 2. Measurements of the Shim stock films with various thicknesses using the THz techniques.

Sample No.	Delay Time (Δt, ps)	Resonance Frequency (Δf)	T-ray Measure ment (mm)	Reference Thickness (mm)	Others
1	1.322	0.750	0.137	0.127	Shim stock Co., Ltd. (Edenvale, South Africa)
2	2.551	0.392	0.250	0.254	
3	3.846	0.260	0.396	0.381	
4	5.881	0.170	0.539	0.508	
5	8.600	0.115	0.789	0.762	

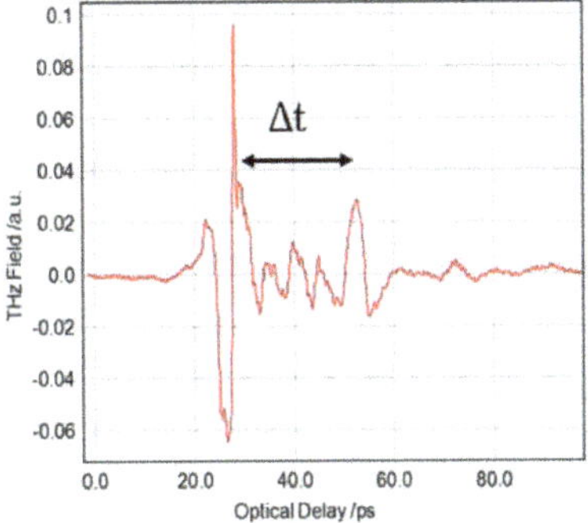

(**a**) A-scan image.

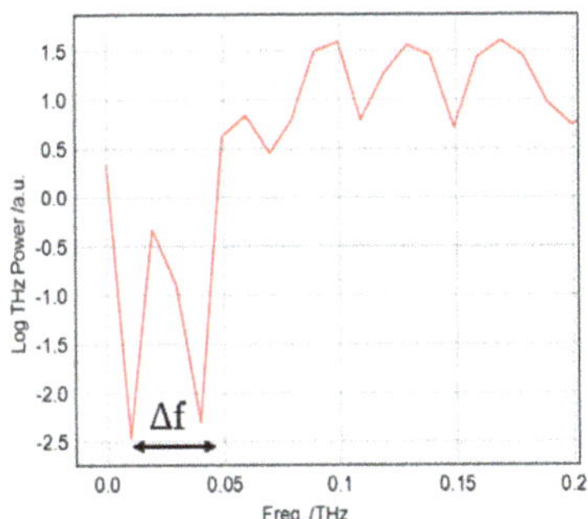

(**b**) Frequency-domain signal.

Figure 8. A TOF and FFT image of the GFRP composites under the reflection mode (2.02 mm in thickness).

Table 3. Measurements of the GFRP composites with various thicknesses using the THz techniques.

Sample No.	Delay Time (Δt, ps)	Resonance Frequency (Δf)	T-ray Measurement (mm)	Reference Thickness (mm)	Others
1	24.480	0.041	2.000	2.020	Shim Stock Co., Ltd.
2	50.000	0.020	3.180	3.080	
3	75.130	0.013	5.600	5.740	
4	83.30	0.012	5.920	5.920	

4.4. Relation between Nominal Thickness and Thickness Measured from T-ray Techniques

The Shim stock films and GFRP plates with non-conductivity were not dependent on the direction of the T-ray, so the measurement was possible. In addition, the T-ray reflection mode which could enable the measurement in one direction was adopted. Figure 9 exhibits the comparison between the nominal thickness in the thin Shim stock films and the thickness measured using the T-ray. As shown in Figure 9, the thicknesses of the thin Shim stock films were 0.127 mm, 0.254 mm, 0.381 mm, 0.508 mm, and 0.762 mm. The thickness of the thin Shim stock films was shown in a straight, solid line. This line shows the proportional relation with the standard thickness. Here, — represents the nominal thickness; □ represents the measured data in the case of the measurement, assuming that the T-ray was vertical to the specimen; and △ represents the measured data in the case of the inclined T-ray. Here, to effectively obtain the Refractive index (n), a suitable sample is the case with a thickness of several ones of mm. In case of the films, we did not prepare such a thicker sample. In this testing, this value is the average value of all the samples.

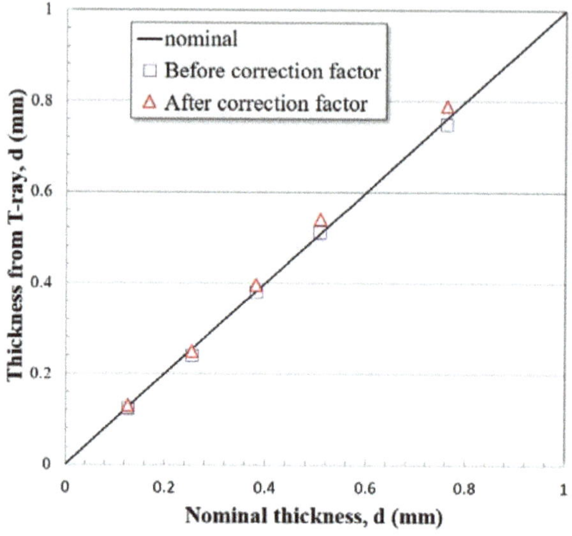

Figure 9. The relation between the nominal thickness and the thickness measured from the T-ray techniques in the Shim stock films.

Thus, the difference of data could be considered to be due to the average value of the refractive index to some degree. Even though there was a difference of ±2%, the results tended to be in agreement in the linear aspect.

Figure 10 shows the comparison between the nominal thickness of the GFRP composites and the thickness measured using the T-ray. The thicknesses of the GFRP composites were 2.02 mm, 3.08 mm, 5.74 mm, and 5.92 mm. In Figure 9, — represents the nominal thickness; □ represents the case of the measurement, assuming that the T-ray was vertical to the specimen; and △ represents the case of the inclined T-ray. Unlike the thickness at the microgram scale, the case of the inclined T-ray matched with the standard thickness at the millimeter scale. This can be attributed to the thickness of the specimen, the relatively small effect from the error, the strong received signal in electric field according to the fiber orientation of the GFRP composites, and the high penetration ratio of T-ray, enabling us to optimize a reception strong signal. Therefore, we found that it had potential reproducibility.

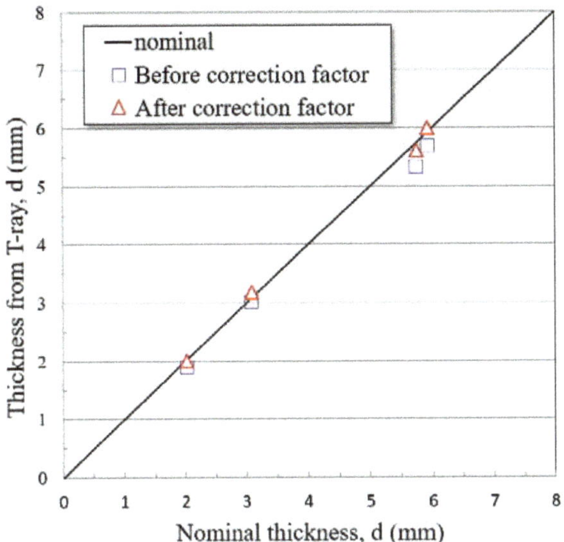

Figure 10. The relation between the nominal thickness and the thickness measured from the T-ray techniques in the GFRP composites.

5. Conclusions

In this approach, the refractive index measurement technique was established to calculate the material properties regarding the utilization of the T-ray for the non-destructive examination of Shim stock films and GFRP composites. In addition, the T-ray limitation in the energy penetrating power was discussed with respect to the conduction characteristics of the GFRP composites and the fiber lamination angle of CFRP plates. Possible THz-TDS techniques are summarized for measuring the thickness of the thin Shim stock films and GFRP composites as follows:

(1) It was possible to solve the refractive index of the thin Shim stock films and GFRP composites utilizing T-ray techniques under the reflection mode.

(2) The T-ray showed a constant level of penetrating power in the glass fiber class composites, which led to a very high penetration ratio and enabled the optimization of a strong reception signal. Therefore, it was found that it had potential reproducibility.

(3) The values of the measured thicknesses for both Shim stock films and GFRP composites were in agreement with those of the nominal thicknesses. The values were successfully measured through the correlation between the TOF cap and the resonance frequency under the reflection mode.

(4) We expect that the manufactured thickness measurement device using T-ray techniques may be very useful for non-destructive examinations in future applications in the advanced aerospace field.

Author Contributions: K.-H.I. suggested and designed the experiments; S.-K.K., Y.-T.C., and Y.-D.W. performed the experiments; C.-P.C. helped in the accomplishment of ideas and the administration of the experiments. The data were discussed and analyzed, and the manuscript was written and revised by all members. All authors have read and agreed to the published version of the manuscript.

Funding: This research was supported by the Basic Science Research Program through the National Research Foundation of Korea (NRF) (No. 2021R1I1A3042195) and also experimentally helped by the CNDE at Iowa State University, Ames, IA, USA.

Institutional Review Board Statement: Not applicable.

Informed Consent Statement: Not applicable.

Data Availability Statement: The data are contained within the article.

Conflicts of Interest: The authors declare no conflict of interest.

References

1. Chiou, C.P.; Blackshire, J.L.; Thompson, R.B.; Hu, B.B. Terahertz Ray System Calibration and Material Characterzations. *Rev. QNDE* **2009**, *28*, 410–417.
2. Im, K.H.; Lee, K.S.; Yang, I.Y.; Yang, Y.J.; Seo, Y.H.; Hsu, D.K. Advanced T-ray Nondestructive Evaluation of Defects in FRP Solid Composites. *Int. J. Precis. Eng. Manuf.* **2013**, *14*, 1093–1098. [CrossRef]
3. Li, M.W.; Liang, C.P.; Zhang, Y.B.; Yi, Z.; Chen, X.F.; Zhou, Z.G.; Yang, H.; Tang, Y.J.; Yi, Y.G. Terahertz Wideband Perfect Absorber Based on Open Loop with Cross Nested Structure. *Results Phys.* **2019**, *15*, 102603. [CrossRef]
4. Chunlian, C.; Zhang, Y.; Chen, X.; Yang, H.; Yi, Z.; Yao, W.; Tang, Y.; Yi, Y.; Wang, J.; Wu, P. A Dual-Band Metamaterial Absorber for Graphene Surface Plasmon Resonance at Terahertz Frequency. *Phys. E Low-Dimens. Syst. Nanostruct.* **2020**, *117*, 113840. [CrossRef]
5. Zhang, Y.; Wu, P.; Zhou, Z.; Chen, X.; Yi, Z.; Zhu, J.; Zhang, T.; Jile, H. Study on Temperature Adjustable Terahertz Metamaterial Absorber Based on Vanadium Dioxide. *IEEE Access* **2020**, *8*, 85154–85161. [CrossRef]
6. Im, K.-H.; Kim, S.-K.; Jung, J.-A.; Cho, Y.-T.; Woo, Y.-D.; Chiou, C.-P. NDE Detection Techniques and Characterization of Aluminum Wires Embedded in Honeycomb Sandwich Composite Panels Using Terahertz Waves. *Materials* **2019**, *12*, 1264. [CrossRef] [PubMed]
7. Huber, R.; Brodschelm, A.; Tauser, A.; Leitenstorfer, A. Generation and Field-Resolved Detection of Femtosecond Electromagnetic Pulses Tunable up to 41 THz. *Appl. Phys. Lett.* **2000**, *76*, 3191–3199. [CrossRef]
8. Rudd, J.V.; Mittleman, D.M. Influence of Substrate-Lens Design in Terahertz Time-Domain Spectroscopy. *J. Opt. Soc. Am. B* **2000**, *19*, 319–329. [CrossRef]
9. Gregory, I.S.; Baker, C.; Tribe, W.; Bradley, I.V.; Evans, M.J.; Linfield, E.H. Optimization of Photomixers and Antennas for Continuous-Wave Terahertz Emission. *IEEE J. Quantum Electron.* **2000**, *41*, 717–728. [CrossRef]
10. Brown, E.R.; Smith, F.W.; McIntosh, K.A. Coherent Millimeterwave Generation by Heterodyne Conversion in Low-Temperature-Grown GaAs Photoconductors. *J. Appl. Phys.* **1993**, *73*, 1480–1484. [CrossRef]
11. Hsu, D.K.; Lee, K.S.; Park, J.W.; Woo, Y.D.; Im, K.H. NDE Inspection of Terahertz Waves in Wind Turbine Composites. *Int. J. Precis. Eng. Manuf.* **2012**, *13*, 1183–1189. [CrossRef]
12. Hsu, D.K.; Hughes, M.S. Simultaneous Ultrasonic Velocity and Sample Thickness Measurement and Applications in Composites. *J. Acoust. Soc. Am.* **1992**, *92*, 669–675. [CrossRef]
13. Brown, E.R.; McIntosh, K.A.; Nichols, K.B.; Dennis, C.L. Photomixing up to 3.8 THz in Low-Temperature-Grown GaAs. *Appl. Phys. Lett.* **1995**, *66*, 285–287. [CrossRef]
14. Zhang, J.; Li, W.; Cui, H.L.; Shi, C.; Han, X.; Ma, Y.; Chen, J.; Chang, T.; Wei, D.; Zhang, Y.; et al. Nondestructive Evaluation of Carbon Fiber Reinforced Polymer Composites Using Reflective Terahertz Images. *Sensors* **2016**, *16*, 875. [CrossRef] [PubMed]
15. Li, L.; Yang, W.; Shui, T.; Wang, X. Ultrasensitive Sizing Sensor for a Single Nanoparticle in a Hybrid Nonlinear Microcavity. *IEEE Photonics J.* **2020**, *12*, 1–8. [CrossRef]
16. Wu, P.; Chen, Z.; Jile, H.; Zhang, C.; Xu, D.; Lv, L. An Infrared Perfect Absorber Based on Metal-Dielectric-Metal Multi-Layer Films with Nanocircle Holes Arrays. *Results Phys.* **2020**, *16*, 102952. [CrossRef]
17. Schueler, R.; Joshi, S.P.; Schulte, K. Damage Detection in CFRP by Electrical Conductivity Mapping. *Compos. Sci. Technol.* **2001**, *61*, 921–930. [CrossRef]
18. Hsu, D.K. Characterization of a Graphite/Epoxy Laminate by Electrical Resistivity Measurements. *Rev. QNDE* **1985**, *4*, 1219–1228.
19. Tse, K.W.; Moyer, C.A.; Arajs, S. Electrical Conductivity of Graphite Fiber epoxy Resin Composites. *Mater. Sci. Eng.* **1981**, *49*, 41–46. [CrossRef]

Article

Time-Resolved Neutron Bragg-Edge Imaging: A Case Study by Observing Martensitic Phase Formation in Low Temperature Transformation (LTT) Steel during GTAW

Axel Griesche [1,*], Beate Pfretzschner [1], Ugur Alp Taparli [1] and Nikolay Kardjilov [2]

[1] Bundesanstalt für Materialforschung und Prüfung (BAM), Unter den Eichen 87, 12205 Berlin, Germany; bepfr@gmx.de (B.P.); ugur-alp.taparli@bam.de (U.A.T.)
[2] Helmholtz-Zentrum-Berlin für Materialien und Energie (HZB), Hahn-Meitner-Platz 1, 14109 Berlin, Germany; kardjilov@helmholtz-berlin.de
* Correspondence: axel.griesche@bam.de

Abstract: Polychromatic and wavelength-selective neutron transmission radiography were applied during bead-on-plate welding on 5 mm thick sheets on the face side of martensitic low transformation temperature (LTT) steel plates using gas tungsten arc welding (GTAW). The in situ visualization of austenitization upon welding and subsequent α'-martensite formation during cooling could be achieved with a temporal resolution of 2 s for monochromatic imaging using a single neutron wavelength and of 0.5 s for polychromatic imaging using the full spectrum of the beam (white beam). The spatial resolution achieved in the experiments was approximately 200 μm. The transmitted monochromatic neutron beam intensity at a wavelength of λ = 0.395 nm was significantly reduced during cooling below the martensitic start temperature M_s since the emerging martensitic phase has a ~10% higher attenuation coefficient than the austenitic phase. Neutron imaging was significantly influenced by coherent neutron scattering caused by the thermal motion of the crystal lattice (Debye–Waller factor), resulting in a reduction in the neutron transmission by approx. 15% for monochromatic and by approx. 4% for polychromatic imaging.

Keywords: neutron radiography; Bragg-edge imaging; gas tungsten arc welding (GTAW); low transformation temperature (LTT) steel; austenite-to-martensite transformation; Debye–Waller factor

Citation: Griesche, A.; Pfretzschner, B.; Taparli, U.A.; Kardjilov, N. Time-Resolved Neutron Bragg-Edge Imaging: A Case Study by Observing Martensitic Phase Formation in Low Temperature Transformation (LTT) Steel during GTAW. *Appl. Sci.* **2021**, *11*, 10886. https://doi.org/10.3390/app112210886

Academic Editors: Jinyi Lee and Ana M. Camacho

Received: 11 August 2021
Accepted: 15 November 2021
Published: 18 November 2021

1. Introduction

Welding residual stresses can have a crucial influence on the crack resistance of a steel component under service load. It was found that phase transformations during the cooling of the weld seam can have a significant influence on the residual stresses around the weld seam. Therefore, it is advantageous to control such phase transformations during cooling to minimize residual stresses and consequently improve the crack resistance of the welds [1]. Compressive residual stresses can thereby have a positive influence on crack prevention. A unique possibility of generating compressive residual stresses already during the welding procedure is offered by so-called low transformation temperature (LTT) filler wires [2–4]. Compared to conventional filler wires, these materials show lower phase transformation temperatures, which can work against the cooling-specific contraction. In consequence, distinct compressive residual stresses can be observed within the weld and adjacent areas.

The use of neutrons allows for examining samples with thicknesses in the cm-range, e.g., for measuring residual stresses around weld seams by neutron diffraction [5,6], or for visualizing grain orientations by tomographic neutron diffraction imaging [7,8]. Another imaging technique based on diffraction contrast is neutron Bragg-edge imaging (NBEI). This technique allows the acquisition of monochromatic images at distinct wavelengths. Depending on the crystalline structure of the measured material, sudden transmission intensity changes, so-called Bragg edges, occur at wavelength positions λ equal to twice the lattice plane distances d_{hkl} in polycrystalline samples, as stated by Bragg's law (Equation (1)), considering the transmission geometry with $\theta = 90°$. Hence, taking radiographic images before and after a Bragg edge achieves a contrast, which relates to the crystallographic properties:

$$n\lambda = 2d_{hkl}\sin\theta \tag{1}$$

NBEI was already used by our team for visualizing the γ-austenite to α'-martensite phase transformation, where a super martensitic stainless steel sample was heated until complete austenitization and was subsequently cooled down to room temperature. The phase transformation process was visualized by acquiring the transmission images of a specific wavelength with a temporal resolution of 30 s and a spatial resolution of 100 μm [9].

In the present study, the aim was to film for the first time in situ the phase transformation of an LTT steel induced by bead-on-plate GTAW. Therefore, we used neutron imaging to visualize the sample remelting during the welding process using specific neutron wavelengths to two-dimensionally visualize the phase transformations in the samples. Additionally, some weldments were filmed using a polychromatic neutron beam.

2. Materials and Methods

The experiments were carried out at the Cold Neutron RADiography 2 (CONRAD 2) instrument located at the research reactor BER II of the Helmholtz-Zentrum Berlin für Materialien und Energie [10]. The imaging set-up consists of the sample holder system and the welding torch located approx. 10 cm in front of a 200 μm thick ^{6}LiF:ZnS scintillator. The scintillator converts the transmitted neutrons into visible light and a mirror reflects the visible light image of the sample out of the neutron beam and towards a high-sensitive CCD camera with magnifying optics. A double-crystal monochromator allows the adjustment of the neutron wavelength in the range between 0.2 nm and 0.6 nm with a resolution of $\Delta\lambda/\lambda \approx 3\%$. More technical information regarding the CONRAD 2 instrument can be found elsewhere [11].

The material used was LTT martensitic steel. The sample sheets were made by build-up welding using LTT steel wire and subsequent machining to achieve plates of the required dimensions. The sample dimensions in length, width, height are 100 mm, 5 mm, 13 mm, where the width of 5 mm was the penetration length for the neutrons.

The microstructure consists mainly of cellular martensite in rows because of the build-up welding process. The primary solidification state of the LTT steel was austenitic. Micrographs showed that some retained austenite can be found as a fine network in the microstructure due to segregation effects of Ni and Cr. The martensite transformation temperature for this LTT steel upon cooling is $M_s \approx 250$ °C and the transformation is finalized at around $M_f \approx 80$ °C. The chemical composition of the LTT steel is given in Table 1. The composition was measured by spark emission spectroscopy with a relative measurement error of $\Delta c/c \approx 10\%$.

Table 1. Chemical composition of the LTT steel in wt.-%.

C	Ni	Cr	Si	Mn	Mo	Fe
0.054	10.2	10.3	0.55	0.90	0.28	Bal.

A type K thermocouple covered by Inconel™ sheathing was positioned in the sample side 2 mm below the surface and 20 mm behind the starting point of welding (see Figure 1). The temperature was monitored during the complete welding process.

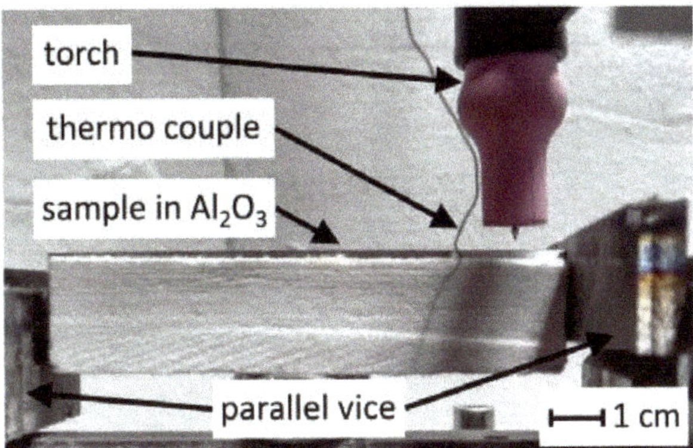

Figure 1. Image of the welding set-up. Direction of view is in neutron flight direction. The sample is fixed edgewise in a parallel vice and is thermally isolated with alumina felt (white). Welding takes place on the long narrow side of the specimen.

The scintillator in the background is also protected against weld spatter by alumina felt, which is almost transparent for neutrons. Welding direction is from right to left.

Gas tungsten arc welding (GTAW) was used to perform remote-controlled bead-on-plate welds with argon as a shielding gas. The arc welding power source used was a Castolin CastoTIG 1611 DC. The welding length for each parameter set was 45 mm, which allowed for four weldments per sample (two on each side). The set-up is shown in Figure 1. The used welding parameters together with the respective imaging parameters are listed in Table 2. The energy input could be controlled by varying both the welding current and the welding velocity.

Table 2. Welding parameters and neutron imaging parameters.

Welding Current/A	Welding Velocity/cm min^{-1}	Neutron Wavelength/nm	Exposure Time/s
40	5	0.395/0.44	2
60	5	0.395/0.44	2
80	10	0.395/0.44 polychromatic	2 0.5

The welding experiments were performed twice, each time with a different neutron wavelength as shown in Table 2. To obtain good contrast between the austenitic and martensitic crystallographic phases, a neutron wavelength of $\lambda = 0.395$ nm was used. This wavelength has been defined by performing a wavelength scan using the original set-up.

Although the monochromatic imaging allowed the recording of the phase transformation step by step, the needed exposure time of $t = 2$ s to gain a reasonable signal-to-noise ratio allowed only for a few frames per weldment. To prolong the cooling state and thus gain more images during the phase transformation of austenite to martensite, the cooling rate was reduced by thermal isolation of the samples. For comparison, some weldments were also filmed using a polychromatic beam that allowed the recording of images continuously with an exposure time of $t = 0.5$ s.

In the preprocessing step of the image analysis, the acquired images were normalized by background and flat-field corrections, which is standard procedure in quantitative X-ray and neutron radiography [12]. In order to eliminate the attenuation of the ceramic insulation and to further enhance the contrast inside the sample, each image series was normalized by dividing a complete series by its first image. Thus, a transmission value of "1" means a fully transparent sample, whereas "0" means no transmission of neutrons through the sample. The very short exposure times of 0.5 s for the polychromatic images led to the very low signal-to-noise ratio of the data. This ratio could be further improved by applying a sliding median filter using a set of three images for each welding experiment image series.

3. Results

3.1. Ex Situ Monochromatic Imaging

The result of NBEI wavelength scans prior to and after welding with a current of 60A at a welding speed of 5 cm/min is shown in Figure 2.

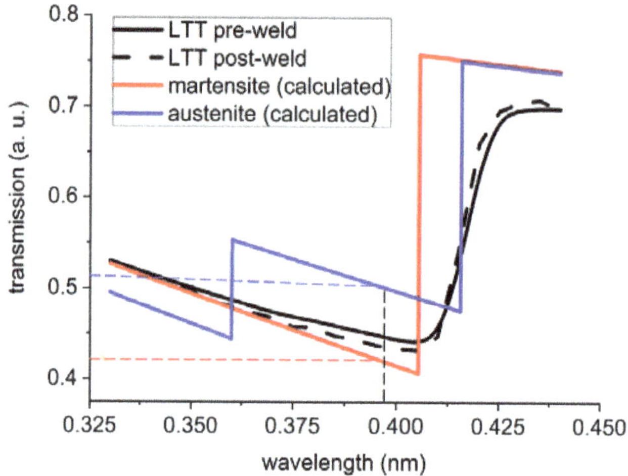

Figure 2. Calculated (colored) and measured (black) Bragg-edge curves for martensite and austenite at room temperature for sample thickness of 5 mm as used in the experiments. The transmission levels at 0.395 nm for the two phases are shown as colored dashed lines. The theoretical data were calculated with the help of a program library [13].

The almost identical black curves measured before and after the welding clearly show that the sample transformed back after welding into martensite. This finding is further supported by the missing austenite Bragg edge at around 0.36 nm in both cases. Additionally, retained austenite could not be detected at grain boundaries by light microscopy in micrographs of the welded samples. This clearly indicates that only a very small volume portion of retained austenite, if any, remains in the sample.

The shift of both measured curves towards higher wavelengths with respect to the calculated martensite Bragg edge at $\lambda \approx 0.405$ nm by about $\Delta\lambda \approx 0.012$ nm can be related to a slight miscalibration of the setup because there is experimental evidence from experiments of other groups using other materials but the same imaging set-up for such a systematic wavelength shift [14].

3.2. In Situ Monochromatic Imaging

As shown in Figure 2, martensite and austenite have different attenuation coefficients in the wavelength interval between approx. 0.36 nm and 0.405 nm just below the Bragg-edge wavelengths of both phases. Although the difference of both attenuation coefficients is not as large as for the coefficients between both Bragg edges, a wavelength in this interval is used for monochromatic imaging due to the abovementioned limits of the experimental wavelength resolution. Thus, we present here the result of monochromatic imaging during GTAW at 0.395 nm for the visualization of the phase transformations. For comparison, we additionally measured after the Bragg edge at 0.44 nm using the same welding and imaging parameter.

Figure 3 shows the mean value of the transmitted intensity approx. 1 mm underneath the sample top surface as a function of time detected in the rectangular yellow Region-of-Interest (ROI). The welding current was 80A, the welding velocity was 10 cm/min, and images were acquired every 2 s with an exposure time of 2 s. The transmission intensity increases drastically as soon as the torch passes by the ROI from approx. 0.42 a.u. to more than 0.56 a.u., i.e., by approx. 60%, and decreases slightly slower back to 0.42 a.u. at t ≈ 150 s. This behavior can be even quantified using the theoretical data presented in Figure 2, where for a neutron wavelength of 0.395 nm the transmission for the martensitic phase is 0.42 a.u. and for the martensitic phase is 0.51 a.u. Thus, this transmission change can be attributed to the transformation from martensite to austenite, where the transmission decrease is due to the almost complete re-transformation to martensite upon cooling. The asymmetry of the transmission curve is due to the different heat fluxes during heating and cooling. Whereas the heat input from the electric welding arc fosters a very high heating rate, the heat flow upon cooling is somewhat slower and limited by the thermal isolation of the sample material. The slope of the transmission curve implies that the martensite reformation is finished within 1–1.5 min.

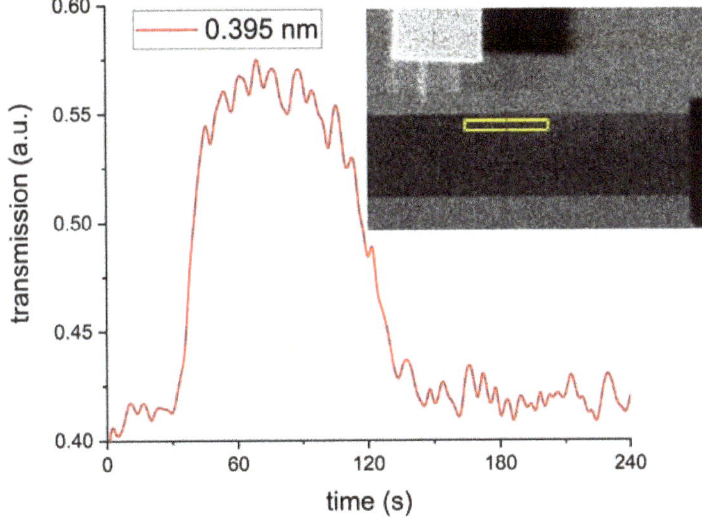

Figure 3. Plot of the transmitted intensity at λ = 0.395 nm measured as mean value from the ROI (yellow rectangle) as a function of time. The insert shows the monochromatic image taken at t ≈ 130 s. The very low neutron intensity results in a poor image quality, showing essentially no visual image contrast in the sample for all times.

For checking purposes, an additional in situ imaging experiment using the same welding and imaging parameters was performed at a wavelength of 0.44 nm. Since this wavelength is larger than the Bragg-edge wavelengths of both lattice types, we expected that the transmitted intensity remains constant. The result is shown in Figure 4. Indeed, we observed an intensity drop of approx. 15% during passing by of the ROI by the welding arc. Subsequently, the measured intensity increases exponentially within approx. 30 s and reaches the initial transmission of approx. 0.66 a.u., which correlates well with the transmission for the martensitic phase in Figure 2.

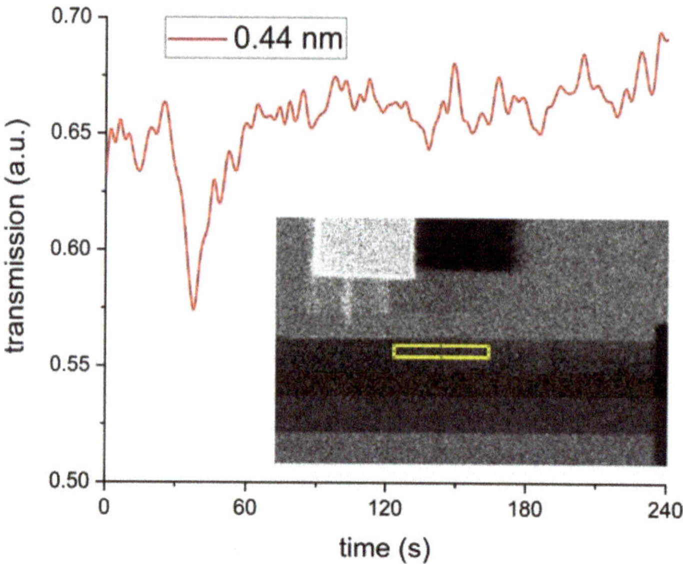

Figure 4. Plot of the transmitted intensity at $\lambda = 0.44$ nm taken as mean value from the ROI (yellow rectangle) as a function of time. The insert shows the monochromatic image taken at $t \approx 130$ s. The very low neutron intensity results in a poor image quality, showing essentially no visual image contrast in the sample.

A possible explanation of the observed drop in the transmitted neutron intensity is the so-called Debye–Waller factor [15,16]. The Debye–Waller factor describes the scattering of neutrons at an oscillating lattice. The immense heat input by the traveling welding arc melts the material at the surface and excites massive lattice oscillations in the subjacent crystalline material. Since the ROI is located 1 mm underneath the surface, most of the material measured should be solid during analysis. These massive lattice oscillations can influence the scattering cross section for neutrons. This effect of lattice oscillations is commonly considered when using neutron scattering techniques. Until now, this effect was not considered explicitly for neutron imaging. Whether neutrons are scattered with increasing temperature depends on the scattering mechanism. Whereas the elastic neutron cross section decreases with increasing temperature, the inelastic neutron cross section increases. Coherent inelastic scattering, however, has the most significant effect and deflects more neutrons at elevated temperatures [17]. Thus, the overall transmitted intensity starts decreasing at high temperatures.

3.3. Polychromatic Imaging

The CONRAD beamline provides a cold neutron spectrum, which has a maximum intensity peak at around 0.25 nm [18]. This resulting high neutron flux allows us to observe the martensite–austenite phase transformation with very short exposure times and provides a much better signal-to-noise ratio, i.e., a much better image contrast compared to monochromatic imaging.

The same welding parameters as for the abovementioned measurements with monochromatic neutrons were used for the polychromatic imaging experiment. The exposure time could be as low as 0.5 s, allowing a four times higher image acquisition rate. Figure 5 shows the transmitted intensity for an equal ROI as used for monochromatic imaging. A sheathed type K thermocouple, inserted in a bore hole close to the welded surface, allowed the measurement of the temperature.

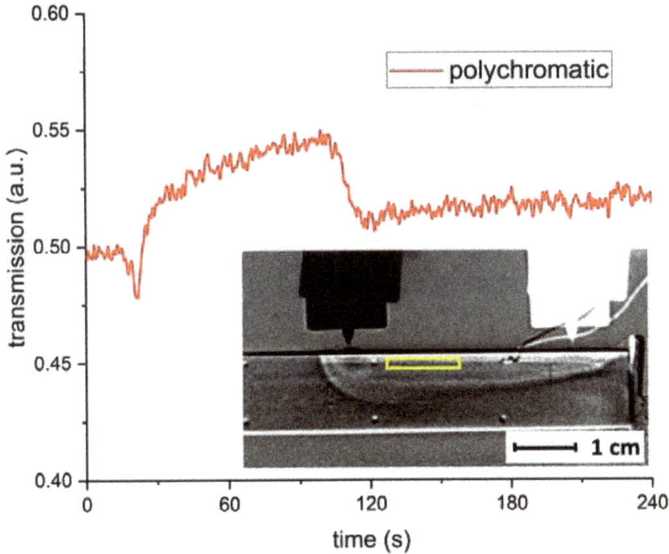

Figure 5. Plot of the transmitted intensity as a function of time taken as mean value from the ROI (yellow rectangle). The insert shows the normalized polychromatic image taken just before switching off the welding torch at the end position.

The intensity dip of approx. 4% relatively at t $\approx$ 25 s can be attributed again to the increasing sample temperature due to the traveling welding torch and the associated change of the Debye–Waller factor. The following intensity increase starting at t > 30 s is due to the change in transmission because of the evolving austenitic phase. The austenitic phase re-transforms to martensite between t $\approx$ 102 s and t $\approx$ 120 s, resulting in a drop in the transmitted neutron intensity by around 7%.

Again, the welding nozzle can be seen twice due to the normalization of all images by dividing all images by the first image. The starting position is to the right (white nozzle), whereas the final position is to the left (black nozzle). The thermocouple was inserted in the upper right hole and had also moved slightly during welding, showing its initial position (white wire) in all images. The bright and dark streaks at the bottom and the top of the sample were also caused by the image normalization. The thermal expansion bended the sample upwards, and dividing all images by the image of the cold sample at t = 0 results in such streaks at the sample edges.

The bright area in the upper part of the sample is the austenitic phase in the heat-affected zone (HAZ), which is still present, and which is in the process of re-transformation from austenite to martensite. The peak temperature reading, when the torch is passing by the inserted thermocouple, is larger than 1200 °C. Measurements with type K thermocouples above such temperatures are not possible. The temperature reading dropped down to around 600 °C by the end of the welding. Another 2.5 min later the M_s temperature is reached, and it takes another 7.5 min to reach M_f.

The increasing vertical dimension of the HAZ from the right to the left is due to the heat accumulation during welding. Additionally, after stopping the torch movement, the electric arc burns for another two seconds, introducing even more energy at this position. The heat protection of both sides of the metal plate restricts the heat dissipation mainly in the two dimensions perpendicular to the neutron flight direction, shaping the temperature profile in the sample.

4. Discussion

Neutron Bragg-edge imaging allows us to film phase changes during the welding of martensitic steel and to film the spatial spread of the heat-affected zone during gas tungsten arc welding. The use of a polychromatic beam results in a four times higher temporal resolution and in an order of magnitude higher signal-to-noise ratio. This leads to a significantly better image quality than for monochromatic imaging. Further, imaging with a polychromatic beam allows us to film the spreading of the heat-affected zone that corresponds to the moving welding torch. The use of monochromatic neutrons allows for phase-specific imaging, but neutron scattering by thermally induced lattice vibrations may influence the image quality.

In this investigation, the lattice vibrations at high temperatures disturbed the monochromatic imaging. From this, it follows that it should, in principle, be possible to measure the temperature of a sample by energy-selective neutron radiography [19] using the temperature dependence of the Debye–Waller factor. A prerequisite for such measurements is the knowledge of these temperature dependence data for all involved crystallographic phases. However, systematic studies of the temperature dependence of such data are not known to the authors up to now.

Generally, NBEI is a useful tool to study in situ phase changes in metals. This might help in clarifying material-related problems, e.g., in components that show strain-induced phase transformations [20] during service. For selected cases, NBEI is a useful non-destructive testing method that might help in damage prevention and damage handling.

Author Contributions: Conceptualization, A.G. and N.K.; methodology, N.K.; software, B.P.; validation, B.P., N.K. and A.G.; formal analysis, B.P.; investigation, B.P., U.A.T. and A.G.; resources, A.G.; data curation, B.P.; writing—original draft preparation, A.G.; writing—review and editing, N.K. and A.G.; visualization, B.P.; supervision, A.G.; project administration, A.G. All authors have read and agreed to the published version of the manuscript.

Funding: This research received no external funding.

Institutional Review Board Statement: Not applicable.

Informed Consent Statement: Not applicable.

Data Availability Statement: The data presented in this study are available on request from the corresponding author.

Acknowledgments: The BAM authors thank the Helmholtz-Zentrum Berlin für Materialien und Energie for the allocation of neutron radiation beamtime.

Conflicts of Interest: The authors declare no conflict of interest.

References

1. Kromm, A.; Kannengiesser, T.; Gibmeier, J. In Situ studies of phase transformation and residual stresses in LTT alloys during welding using synchrotron radiation. In *In-Situ Studies with Photons, Neutrons and Electrons Scattering*; Kannengiesser, T., Babu, S.S., Komizo, Y.I., Ramirez, A.J., Eds.; Springer: Berlin/Heidelberg, Germany, 2010; pp. 13–26.
2. Hosseini, S.A.; Gheisari, K.; Moshayedi, H.; Warchomicka, F.; Enzinger, N. Basic alloy development of low-transformation-temperature fillers for AISI 410 martensitic stainless steel. *Sci. Technol. Weld. J.* **2019**, *25*, 243–250. [CrossRef]
3. Gach, S.; Olschok, S.; Arntz, D.; Reisgen, U. Residual stress reduction of laser beam welds by use of low-transformation-temperature (LTT) filler materials in carbon manganese steels—In situ diagnostic: Image correlation. *J. Laser Appl.* **2018**, *30*, 032416. [CrossRef]
4. Ozdemir, O.; Cam, G.; Cimenoglu, H.; Kocak, M. Investigation into mechanical properties of high strength steel plates welded with low temperature transformation (LTT) electrodes. *Int. J. Surf. Sci. Eng.* **2012**, *6*, 157–173. [CrossRef]
5. Kromm, A. Evaluation of weld filler alloying concepts for residual stress engineering by means of Neutron and X-ray diffraction. *Adv. Mater. Res.* **2014**, *996*, 469–474. [CrossRef]
6. Yu, Z.Z.; Feng, Z.L.; Woo, W.C.; David, S. Application of In Situ Neutron Diffraction to Characterize Transient Material Behavior in Welding. *JOM* **2013**, *65*, 65–72. [CrossRef]
7. Peetermans, S.; Bopp, M.; Vontobel, P.; Lehmann, E.H. Energy-selective neutron imaging for three-dimensional non-destructive probing of crystalline structures. *Phys. Proc.* **2015**, *69*, 189–197. [CrossRef]

8. Woracek, R.; Penumadu, D.; Kardjilov, N.; Hilger, A.; Boin, M.; Banhart, J.; Manke, I. 3D Mapping of Crystallographic Phase Distribution using Energy-Selective Neutron Tomography. *Adv. Mater.* **2014**, *26*, 4069–4073. [CrossRef] [PubMed]
9. Dabah, E.; Pfretzschner, B.; Schaupp, T.; Kardjilov, N.; Manke, I.; Boin, M.; Woracek, R.; Griesche, A. Time-resolved Bragg-edge neutron radiography for observing martensitic phase transformation from austenitized super martensitic steel. *J. Mater. Sci.* **2017**, *52*, 3490–3496. [CrossRef]
10. Kardjilov, N.; Hilger, A.; Manke, I.; Woracek, R.; Banhart, J. CONRAD-2: The new neutron imaging instrument at the Helmholtz-Zentrum Berlin. *J. Appl. Crystallogr.* **2015**, *49*, 195–202. [CrossRef]
11. Kardjilov, N.; Hilger, A.; Manke, I.; Griesche, A.; Banhart, J. Imaging with Cold Neutrons at the CONRAD-2 Facility. *Phys. Proc.* **2015**, *69*, 60–66. [CrossRef]
12. Griesche, A.; Solorzano, E.; Beyer, K.; Kannengiesser, T. The advantage of using in situ methods for studying hydrogen mass transport: Neutron radiography vs. carrier gas hot extraction. *Int. J. Hydrogen Energ.* **2013**, *38*, 4725–4729. [CrossRef]
13. Boin, M. nxs: A program library for neutron cross section calculations. *J. Appl. Crystallogr.* **2012**, *45*, 603–607. [CrossRef]
14. Al-Falahat, A.M.; Kardjilov, N.; Khanh, T.V.; Markötter, H.; Boin, M.; Woracek, R.; Salvemini, F.; Grazzi, F.; Hilger, A.; Alrwashdeh, S.S.; et al. Energy-selective neutron imaging by exploiting wavelength gradients of double crystal monochromators -Simulations and experiments. *Nuc. Instr. Meth. Phys. Res. Sect. A-Accel. Spectr. Detect. Assoc. Equip.* **2019**, *943*, 162477. [CrossRef]
15. Paradezhenko, G.V.; Melnikov, N.B.; Reser, B.I. Debye-Waller Factor in Neutron Scattering by Ferromagnetic Metals. *Theor Math. Phys.* **2018**, *195*, 572–583. [CrossRef]
16. Ramadhan, R.S.; Kockelmann, W.; Minniti, T.; Chen, B.; Parfitt, D.; Fitzpatrick, M.E.; Tremsin, A.S. Characterization and application of Bragg-edge transmission imaging for strain measurement and crystallographic analysis on the IMAT beamline. *J. Appl. Crystallogr.* **2019**, *52*, 351–368. [CrossRef]
17. Al-Falahat, A.M. Wavelength-Selective Neutron Imaging for Materials Science. Ph.D. Thesis, Technische Universität Berlin, Berlin, Germany, 2019. [CrossRef]
18. Kardjilov, N.; Hilger, A.; Manke, I.; Strobl, M.; Dawson, M.; Williams, S.; Banhart, J. Neutron tomography instrument CONRAD at HZB. *Nuc. Instr. Meth. Phys. Res. Sec. A-Accel. Spectrom. Detect. Assoc. Equip.* **2011**, *651*, 47–52. [CrossRef]
19. Al-Falahat, A.M.; Kardjilov, N.; Woracek, R.; Boin, M.; Markter, H.; Theil Kuhn, L.; Makowska, M.; Strobl, M.; Pfretzschner, B.; Banhart, J.; et al. Temperature dependence on Bragg edge neutron transmission measurements. *J. Appl. Crystallogr.* **2021**. submitted.
20. Polatidis, E.; Morgano, M.; Malamud, F.; Bacak, M.; Panzner, T.; Van Swygenhoven, H.; Strobl, M. Neutron Diffraction and Diffraction Contrast Imaging for Mapping the TRIP Effect under Load Path Change. *Materials* **2020**, *13*, 1450. [CrossRef] [PubMed]

MDPI
St. Alban-Anlage 66
4052 Basel
Switzerland
Tel. +41 61 683 77 34
Fax +41 61 302 89 18
www.mdpi.com

Applied Sciences Editorial Office
E-mail: applsci@mdpi.com
www.mdpi.com/journal/applsci

www.ingramcontent.com/pod-product-compliance
Lightning Source LLC
LaVergne TN
LVHW071010170726
843515LV00006B/1120

* 9 7 8 3 0 3 6 5 3 5 9 9 9 *